READINGS ABOUT
The Social Animal

Fifth Edition

READINGS ABOUT
The Social Animal

Edited by
Elliot Aronson
University of California, Santa Cruz

W. H. Freeman and Company
New York

Cover drawing by Tom Durfee.

Library of Congress Cataloging-in-Publication Data

Readings about the social animal / edited by Elliot Aronson. —5th ed.
    p. cm.
    Includes bibliographies and indexes.
    ISBN 0-7167-1935-5.   ISBN 0-7167-1936-3 (pbk.)
    1. Social psychology.   I. Aronson, Elliot.
HM251.R36 1988   87-25270
302—dc 19   CIP

Printed in the United States of America

1 2 3 4 5 6 7 8 9 0   VB   6 5 4 3 2 1 0 8 9 8

To my bigger, smarter brother,
Jason Aronson (1929–1961)

Contents

III. MASS COMMUNICATION, PROPAGANDA, AND PERSUASION   87

IV. SELF-JUSTIFICATION   149

V. HUMAN AGGRESSION   233

VI. PREJUDICE AND ATTRIBUTION   303

VII. ATTRACTION: WHY PEOPLE LIKE EACH OTHER   323

VIII. INTERPERSONAL COMMUNICATION AND SENSITIVITY   427

*"The most incomprehensible thing about
the world is that it's comprehensible."*

—*Albert Einstein*

Preface

In my textbook *The Social Animal*, I attempt to paint a clear picture of the current state of our social-psychological knowledge and how such knowledge might be applied to alleviate some problems plaguing us in the world today. *The Social Animal* is intended to be concise, brisk, and lively. It is almost totally unencumbered by graphs, charts, tables, statistical analysis, or detailed methodological discussions. Although that kind of presentation provides an easy and even enjoyable introduction to the world of social psychology, many readers express a need to delve more deeply into the details of the research that forms the backbone of *The Social Animal*. To meet that need, I have edited this book, *Readings About the Social Animal*, Fifth Edition.

I have selected the readings that appear in this book in such a way that they both complement and supplement the material contained in *The Social Animal*. Not only are the sections organized so as to coincide with chapters in *The Social Animal*, but the specific readings also represent an attempt to amplify and elaborate on the major themes covered in that book. Moreover, I have been especially careful to choose readings that provide a mixture of classic and contemporary research. Some of the articles were already classics when I first read them as a student. At the other end of the continuum are articles I encountered in the form of prepublication reports. This combination will provide the reader with a historical sweep as well as with the most contemporary ideas in the field.

There is another way of classifying the articles in this collection. Most of the articles are reports of specific research as originally published in technical journals; others are more general pieces summarizing several studies on a given topic written by one of the major contributors to that area. A specific report, though not always easy to read, has the advantage of providing the detail necessary for enabling the reader to gain some understanding of exactly what goes into a piece of research. The summary article is usually less technical and, therefore, easier to read, offering a more panoramic overview of the area by the people who know it best. In effect, it enables the reader to look over an investigator's shoulder and see how he or she views an array of research on a given topic.

For this, the fifth edition, I have continued the policy of blending the classic with the contemporary — and the specific with the panoramic. As the years roll by, it is gratifying to note that some of the articles I first selected as contemporary in 1972 have now taken their place as genuine classics in 1988. My hope is that *today's* contemporaries will likewise take their place as classics in the decades to come.

I am indebted to Marti Hope Gonzales for her wise counsel in helping me select new articles for the fifth edition.

Elliot Aronson

An Open Letter to the Reader

Welcome backstage. As mentioned in the Preface, there are two kinds of articles contained in this volume. Some selections are descriptions of research programs. These make exciting reading in that they describe in some detail a series of experiments aimed at explicating or extending a single idea. Other selections are reports of individual pieces of empirical research. These are equally exciting but sometimes get rather technical. I'm sure some of you (teachers, graduate students, statisticians, and other dedicated types) will want to understand thoroughly every sentence of every article on the following pages, perhaps in the hope of planning some research of your own. *Bon voyage!* For your benefit, I have not abridged or changed a line of the original.

My guess is that most of you do not require that amount of detail. Chances are, what you would like to get out of these articles is an understanding of what the investigator was trying to find out, how he or she went about his or her task, and how successful the outcome was. There is no better way to understand the research process than to read original reports. The adventure of reading an original report lies in your ability as the reader to put yourself in the shoes of the investigator as he or she transforms an idea into a viable set of research operations and tries to make sense of the results, which occasionally do not conform precisely to the predictions. Each of the original research reports contains four principal sections: (1) First there is an *introduction,* in which the author states the idea, where it came from, and why it's important, and distills the idea down to a hypothesis or series of hypotheses. (2) This is followed by a *procedure* section, in which the author tests the hypothesis by translating the idea into a concrete set of operations. In social psychology this frequently becomes a full-blown scenario designed to provide the participant (or subject) with a reasonable justification for responding to events without being allowed to know the true purpose of the procedure. The procedure section of a good piece of research is often the most interesting part, because it requires a great deal of ingenuity for the investigator to achieve precision without sacrificing realism or impact. (3) In the *results* section the investigator states as clearly and succinctly as possible what the findings were. The investigator uses various statistical procedures to ascertain the extent to which the data are reliable. (4) Finally there is a *discussion* section, in which the researcher evaluates and interprets the data presented in the preceding section and tries to make sense of them in the context of previous research. The creative researcher can also use this section to speculate about the implications of the data and to point the way toward future research.

To those of you who do not yet have much experience in reading research reports, I offer a few suggestions about which parts of the study to read carefully and which parts to skim. If the article contains a summary (either at the beginning or at the end), I would read that first in order to familiarize myself with the general idea behind the piece of research and to learn quickly what the results were. Next, I would read the introduction carefully in order to learn the history of the idea and to understand the hypothesis thoroughly. I would then read the procedure section pretty carefully. I would skim the results section just to see the extent to which the findings agreed with the predictions. If the findings did not fit the predictions, I would look closely at the discussion section to see how the author had made sense of the results he or she did get and whether or not the explanation seemed plausible to me. Unless you are adept at and/or intrigued by statistical analyses, I would advise you to skim that section. For those of you who have little or no knowledge of statistical procedures, it would be terribly frustrating and would serve no useful purpose to bog yourself down in some of the details of the statistical analyses. These articles were selected because they were well done. It is probably safe for you to accept on faith that the analysis was performed competently.

The Authors

Paul R. Abramson
University of Connecticut

Dane Archer
University of California, Santa Cruz

Arthur P. Aron
University of British Columbia, Vancouver, Canada

Elliot Aronson
University of California, Santa Cruz

Solomon E. Asch
Rutgers University

Danny Axsom
Princeton University

Curtis Banks
Stanford University

Roberta L. Baral
University of Kansas

C. Daniel Batson
University of Kansas

Leonard Berkowitz
University of Wisconsin

Ellen Berscheid
University of Minnesota

Diana Bolen
Northwestern University

Philip Brickman
Northwestern University

Diane Bridgeman
University of California, Santa Cruz

Joel Cooper
Princeton University

Barbara Critchlow
University of Washington

James M. Dabbs, Jr.
Georgia State University

John M. Darley
Princeton University

Michael Jay Diamond
University of Hawaii

Donald G. Dutton
University of British Columbia, Vancouver, Canada

Scott C. Fraser
University of Southern California

Jonathan L. Freedman
Columbia University

Rosemary Gartner
University of Wisconsin, Madison

Russell G. Geen
University of Missouri, Columbia

Philip A. Goldberg
Connecticut College

Marc Gottesdiener
Connecticut College

Craig Haney
University of California, Santa Cruz

Sara B. Kiesler
National Academy of Sciences

John F. Layton
University of Utah

Howard Leventhal
University of Wisconsin

W. Charles Lobitz
University of Oregon

John Manzolati
Stanford University

David R. Mettee (late)
University of Denver

Stanley Milgram
City University of New York

Richard Miller
Northwestern University

Neal Osherow
University of California, Santa Cruz

Nancy Ostrove
University of Maryland

Thomas F. Pettigrew
University of California, Santa Cruz

Davis P. Phillips
University of California, San Diego

Jane Piliavin
University of Wisconsin

Steven Prentice-Dunn
University of Alabama

Ronald W. Rogers
University of Alabama

Irwin Rubin
Massachusetts Institute of Technology

Lynn Schmidt
University of Wisconsin

Jerrold Lee Shapiro
University of Hawaii

Muzafer Sherif
Pennsylvania State University

Gary L. Shope
University of Missouri, Columbia

Harold Sigall
University of Maryland

Lynn Stanley Simons
University of Utah

Mark Snyder
University of Minnesota

Lillian L. Southwick
University of Washington

Claude M. Steele
University of Washington

David Stonner
University of Missouri, Columbia

Elizabeth Decker Tanke
University of Santa Clara

Charles W. Turner
University of Utah

Elaine Walster (Hatfield)
University of Hawaii

G. William Walster
University of Wisconsin

Carl O. Word
Wright Institute

Mark P. Zanna
University of Waterloo

Philip Zimbardo
Stanford University

READINGS ABOUT
The Social Animal

I

INTRODUCTION: REFLECTIONS ON THE RESEARCH PROCESS

1

Research in Social Psychology as a Leap of Faith

Elliot Aronson

A year ago I was asked to deliver a talk to this group. The suggested topic was: "Whatever Became of Elliot Aronson?" I couldn't make it last year but was given a rain check—and here I am. But, while on the plane to Los Angeles, I realized that I didn't have any clear idea of what I was going to talk about. To follow the assigned topic would be presumptuous. I (or should I say "even I") lack the chutzpah to presume that very many of you are holding your breaths waiting to learn the answer to that question. So I thought that I might describe some of my current research—an action research project aimed at exploring the consequences of building cooperative learning groups in elementary schools. This is a project I'm very excited about: Among other things, we are finding that the loss of self-esteem among ethnic children following desegregation (discovered by Gerard and Miller, 1975) is largely a function of the competitiveness of the traditional class-room. When we placed kids in cooperative learning groups (of a special kind) we reversed this trend and produced sharp increases in self-esteem and test perfor-

This chapter was delivered as an invited address at the 1976 Meeting of the Society of Experimental Social Psychologists at the University of California at Los Angeles. It is reprinted from *Personality and Social Psychology Bulletin,* Vol. 3, No. 2, Spring 1977, pp. 190–195, by permission of Division 8 of the American Psychological Association and of the Publisher, Sage Publications, Inc.

mance among black and Chicano children (Aronson et al., 1975; Lucker et al., in press; Aronson et al., in press).

But after I arrived here, I realized that a heavy dose of data would be inappropriate for an after-dinner speech. My data might be too heavy a dessert. Besides, ever since I've been here, several of you have been urging me to keep it light and perhaps even to make it humorous. So, I decided to talk to you about philosophy of science. After all, graduate students consider my approach to the philosophy of science to be extraordinarily light and some even find it to be hilarious.

In 1968 Merrill Carlsmith and I wrote a chapter in *The Handbook of Social Psychology* on experimentation in social psychology. We spent most of that chapter talking about sticky problems: the trials and tribulations involved in designing and conducting an experiment in this area. We wrote about ethical issues, about experimenter bias, about random samples, experimental realism, debriefing, and all kinds of things to watch out for. After we finished it, we read the thing over and we realized that it made experimentation sound like a terrible drag. Further, we realized that this impression didn't reflect our own excitement and enthusiasm about doing research in social psychology. So we included another paragraph in which we said, in effect, "Hey, it *does* sound like doing research in social psychology is difficult, problematical, and occasionally a pain in the ass—and this is all true— but we will have really misled you if we don't convey our major feeling about doing research in social psychology—and that is that, all things considered, it is great fun.

It's now eight years since that chapter was published, and I think that it's about time that I explained at least some of what we meant by the statement that experimentation is fun. And I'd like to do it in terms of a metaphor.

The metaphor I'd like to make use of occurs in a novel by Camus called *The Plague*. The backdrop of the novel involves bubonic plague that is ravaging a town on the coast of Algeria. One of the major characters is Monsieur Grand—an amiable sort of man who is writing a book. He wants that book to be absolutely perfect. He wants every sentence to be flawless, every paragraph to be magnificent, every page to be fantastically beautiful. He wants it to be perfect to the extent that, when he sends it to a publisher, the publisher will read the first sentence and be so struck by it that he will stand up and say to his colleagues, "Ladies and Gentlemen, hats off."

Monsieur Grand spends a lot of time writing that first sentence—the sentence is about a woman riding a horse in the park, but not just any woman, horse, or park. "One fine morning in the month of May an elegant young horsewoman might have been seen riding a handsome sorrell mare along the flowery avenues of the Bois de Boulogne." But he's not satisfied. Does each noun have too many qualifiers? Too few? Does each qualifier convey precisely what he intends? Does each word properly convey the rhythm of a cantering horse? Would "flower-strewn" be preferable to "flowery"? Eight or nine months go by and he's still working on that first sentence—he has 50 manuscript pages all on that first sentence, because, as you know, he wants to be so good that the publisher will say to his colleagues, "Ladies and Gentlemen, hats off."

One day as he is working on his manuscript, he becomes ill; as his symptoms develop, it soon becomes clear that he has contracted bubonic plague. His physician—who is also his friend—examines him and says, "I'm really sorry to have to tell you this, but you're going to die; you don't have much longer to live." So Monsieur Grand orders his physician to destroy his manuscript. He issues this order with such assertiveness and such strength that the physician immediately takes the manuscript—this 50 pages of one sentence meticulously honed and sharpened—and throws it into the fire.

The next day Monsieur Grand recovers. And he says to his friend, "I think I acted too hastily." This is incredibly ironic, of course; the one action on his part that was spontaneous was in the interest of *destroying* the thing that he had created with anything but spontaneity.

So much for the metaphor. Basically, I believe that there are two very different ways to do science. One is in the slow, methodical, one-step-at-a-time manner exemplified by Monsieur Grand. This involves a great deal of meticulous honing, sharpening, and polishing of the design and operations. This may require several months. When the researcher is ready to leave the drawing board, he runs the study, and after he runs a few subjects he realizes that there are some things about it that are not perfect. So he stops and brings it back to the drawing board and works some more, sharpening and honing. He then runs a few more subjects and notices something else that can be improved. Several years later he may have the kind of experiment that, when he submits it to the journal, the editor will say to his colleagues, "Ladies and Gentlemen, hats off." [1]

There's another way to do science. This involves, in effect, sketching a study in quick, broad strokes, pilot testing it—seeing where you went wrong, recasting it, and then running it as best as you can at the moment. As you finish the study and begin to write it up, you may realize that if you had it all to do over again you would have done it better. Of course; learning always takes place as a function of experience—even in the researcher. But in this approach, instead of going back to the drawing board to design that mythical "perfect" study that will produce a "hats off" response, you finish the writeup and submit it for publication—relying on the notion that science is a self-corrective enterprise. Self-corrective in the following sense: I know that if I do a study that *isn't* perfect, it will soon be improved upon by others. Thus, my goal is to get it into the literature to give my colleagues a chance to look at it, be stimulated by it, be provoked by it, annoyed by it, and then go ahead and do it better—even if their intent is to prove me wrong, and even if they succeed in proving me wrong. That's the exciting thing about science; it progresses by people taking off on one another's work. This is what

---

[1] Even under these circumstances the "hats off" response is unlikely; chances are the editor will find some flaws that the author overlooked.

William James (1956) called the "leap of faith." I have faith that if I do an imperfect piece of work, someone will read it and will be provoked to demonstrate this imperfection in a really interesting way. This will almost always lead to a greater understanding of the phenomenon under investigation. And maybe, after that other person does his research, the editor and the publisher and the world at large will say, "Hey, hats off." And that's OK.

Needless to say, I prefer the second way—as those of you who are familiar with my imperfect research can readily gather. And this "leap of faith" forms a vital aspect of my philosophy of science. Namely, that we don't have to make things perfect before we expose them to other people's thinking, criticism, and actions. It's more exciting, it's more yeasty, it's more provocative to treat science in big, broad brush strokes rather than meticulously polishing and honing and taking several years before coming out with a finished product. Since I believe that science is a self-correcting enterprise, I would prefer to be provocative than right. Of course, it goes without saying that I do not attempt to be wrong or sloppy. I attempt to do the best I can at the moment and share that less than perfect product with you— my colleagues and critics. William James maintains that there is an immense class of cases where faith creates its own verification. To use one of James' examples— if while climbing a mountain you must leap over a precipice, your belief that you can do it increases the probability of a successful outcome. To extrapolate our confidence that others will be excited by our research and provoked to carry it further or prove it wrong induces us to conduct research in a manner and publish it at a time that maximizes the probability of the research providing the yeast for this kind of empirical convocation.

It is now eight years since Carlsmith and I wrote that chapter in the *Handbook* in which we very blithely announced that experimentation in social psychology was fun. And as I read the journals, it seems to me that a lot of the fun has gone out of social psychology. One of the reasons for this, I think, is that we've gotten much too cautious, much too careful, much too afraid to be wrong, and it's taken a lot of the zest and a lot of the yeast out of the research in social psychology.

One of the by-products of excessive caution is excessive self-consciousness. Indeed, one of the most characteristic aspects of contemporary social psychology is a recent trend toward discipline-wide self-consciousness, handwringing, and "kvetching." By conservative estimate, in the past five or six years I must have been invited to at least a half dozen symposia with titles like "Where is Social Psychology Heading?" or "What Must We Do About Social Psychology?" or "Whither Social Psychology?" (with or without the "h" in whither). This kind of self-consciousness is a bore. Let me be clear. I'm not opposed to a certain degree of self-consciousness on an individual level. Indeed, I believe that any individual scientist should take stock of herself every few years and give herself the opportunity to re-order her priorities. Every few years it is probably quite useful to ask: What is important? What are my ethical and societal concerns? Although this process is important for an individual, when an entire discipline does it, the implication is that there is a particular place we should all be going, a particular methodology we should all be using, particular topics we should all be studying. I think that that kind of self-consciousness is deadly and stultifying. The proper question is "What might *individuals* be doing?", not "What should the *field* be doing?" I like to think of the discipline of social psychology as a really large circus tent where a lot of different acts are going on and the acts occasionally cross, intermingle, and overlap. In that way each individual is doing what that person thinks is the most interesting or useful thing to do *and* is being challenged all the time by the existence of other people in the science who are engaged in overlapping topics employing different methods or different topics employing overlapping methods.

Watch out—I feel another metaphor coming on. There is a short story by J. D. Salinger entitled *Seymour: An Introduction* in which Buddy Glass, who is a budding writer, presents his short stories as he writes them to his older brother, Seymour, for criticism. Seymour, who is a wise and good man, generally writes these criticisms in the form of a letter. After one story, Seymour writes Buddy a letter (which Buddy keeps for many years afterwards) in which he says, in effect:

> You really are a terrific craftsman. You really know how to write. You really know how to put a sentence together; you've mastered the technique. You know how to take these sentences and string them together into paragraphs. Your stories are constructed beautifully. The one thing you haven't learned yet is what to write about, and that's a really important problem. As I was thinking about it, I came up with a solution that was so simple and so direct and so "obvious" that it boggles the mind. The solution is this: Just remember that before you were a writer you were a reader. Then, all you have to do is think about the one story, the one thing that you always wanted to read, and then sit down and write it.

As a scientist, I continually find myself trying to make some use of Seymour's advice. In effect, I say to myself: "Hey, remember that before you were a researcher you were a consumer of research. If you want to know what to do research on, just think of the one experiment on human social behavior that you always wanted to read about, and then go out and do it." I have always tried to follow my translation of Seymour's advice, and occasionally I have succeeded. It would be very dangerous to say to a group like this that I always succeed—that every experiment I do involves the one question that I always wanted to know the answer to—because that would open me up to ridicule. I can just hear you saying, "You mean the one thing you always wanted to know about human nature is what happens when people spill coffee all over themselves?" (Aronson, Willerman, and Floyd, 1966). It's quite a dangerous statement even if it were true—and it's not true.

But I've had my moments and I can tell when I'm following Seymour's advice because I can feel my excitement rise, and I think I'm onto one of those questions now. My current research questions are: How can we turn the educational experience of millions of elementary school children into a less dehumanizing experience? How can we counteract the trend in American education toward lowering self-concepts of minority group people? How can we teach skills of cooperation painlessly and easily in the ordinary classroom situation? How can we make learning an exciting, interesting, social psychological, as well as educational, venture? Since public schools are institutions that 95 percent of us and 95 percent of our children and 95 percent of our grandchildren are going to be going through, I see these as important issues.

But I promised you I wouldn't dwell on the research itself. What I do want to dwell on are my failures as a scientist—the times when I don't follow Seymour's advice. Occasionally I do research that I'm not particularly excited about. How come? Sometimes I find myself between ideas. Or sometimes the research I'm interested in is too hard to do or is taking a lot of time to set up. When this happens, instead of doing nothing, I get scared. Scared of what? Let me backtrack. Yesterday when we had a symposium about the editorial policy of our journals, my friend and former student, Darwyn Linder, talked to us eloquently about what he thought the major function of journals should be. He presented a discussion of three separate functions: One is the archival function; the journal serves as a permanent body of knowledge so that 50 years from now when people want to know what was going on in social psychology in the 1970s, they can look in the back issues of the journals. A second function it serves is as a way of exchanging information; if you want to know what is going on in various labs across the country right now at this moment, or at least what went on three years ago when the researchers did the experiment that finally got published, you can look in the current journal. The third function that Linder mentioned was that for the younger social psychologists—people who are teaching at universities without tenure—it serves a pragmatic function. That is, publishing in journals convinces deans that the young social psychologist is doing his job by piling up vast numbers of publications. But what I want to say is that the third function is not one that is limited to younger people who are desirous of tenure. I think it is something that afflicts some of us old guys, too, in a very different way. Whereas our jobs aren't dependent on publications, something else is. And that something—for a lack of a better word—I would call collegial esteem. What keeps me from consistently doing research on the one thing that I always wanted to find out about is that I get scared once in a while that if I'm not always active, not always producing things—anything—then perhaps some of my colleagues might think that I've lost it, and they may begin to ask questions like ''Whatever became of Elliot Aronson?'' which was indeed the suggested topic of this talk. It is conceivable to me that I'm not the only person here who has experienced that fear. If my suspicion is correct, then perhaps by discussing that hobgoblin we can, if nothing else, get rid of what Harry Stack Sullivan called ''the fallacy of uniqueness.'' This may help us all to lay the hobgoblin to rest and get on with our proper business: for each of us in his own way to pursue as best as we can the answer to the one question he always wanted to know.

References

ARONSON, E., BLANEY, N., SIKES, J., STEPHAN, C., AND SNAPP, M. Busing and racial tension: The jig-saw route to learning and liking. *Psychology Today,* 1975, **8,** 43–50.

ARONSON, E., AND CARLSMITH, M. Experimentation in social psychology. In G. Lindzey and E. Aronson (eds.), *Handbook of social psychology.* Vol. 2 (2nd ed.). Reading, Mass.: Addison-Wesley, 1968.

ARONSON, E., WILLERMAN, B., AND FLOYD, J. The effect of a pratfall on increasing interpersonal attractiveness. *Psychonomic Science,* 1966, **4,** 227–228.

BLANEY, N., ROSENFIELD, R., STEPHAN, C., ARONSON, E., AND SIKES, J. Interdependence in the classroom: A field study. *Journal of Educational Psychology,* 1977, **69,** 139–146.

GERARD, H., AND MILLER, N. *School desegregation.* New York: Plenum Press, 1975.

JAMES, W. *The will to believe.* New York: Dover, 1956.

LUCKER, W., ROSENFIELD, D., SIKES, J., AND ARONSON, E. Performance in the interdependent classroom: A field study. *American Educational Research Review.* In press.

II

CONFORMITY AND OBEDIENCE

2

Opinions and Social Pressure

Solomon E. Asch

Exactly what is the effect of the opinions of others on our own? In other words, how strong is the urge toward social conformity? The question is approached by means of some unusual experiments.

That social influences shape every person's practices, judgments and beliefs is a truism to which anyone will readily assent. A child masters his "native" dialect down to the finest nuances; a member of a tribe of cannibals accepts cannibalism as altogether fitting and proper. All the social sciences take their departure from the observation of the profound effects that groups exert on their members. For psychologists, group pressure upon the minds of individuals raises a host of questions they would like to investigate in detail.

How, and to what extent, do social forces constrain people's opinions and attitudes? This question is especially pertinent in our day. The same epoch that has witnessed the unprecedented technical extension of communication has also brought into existence the deliberate manipulation of opinion and the "engineering of consent." There are many good reasons why, as citizens and as scientists, we should be concerned with studying the ways in which human beings form their opinions and the role that social conditions play.

Studies of these questions began with the interest in hypnosis aroused by the

French physician Jean Martin Charcot (a teacher of Sigmund Freud) toward the end of the nineteenth century. Charcot believed that only hysterical patients could be fully hypnotized, but this view was soon challenged by two other physicians, Hyppolyte Bernheim and A. A. Liébault, who demonstrated that they could put most people under the hypnotic spell. Bernheim proposed that hypnosis was but an extreme form of a normal psychological process which became known as "suggestibility." It was shown that monotonous reiteration of instructions could induce in normal persons in the waking state involuntary bodily changes such as swaying or rigidity of the arms, and sensations such as warmth and odor.

It was not long before social thinkers seized upon these discoveries as a basis for explaining numerous social phenomena, from the spread of opinion to the formation of crowds and the following of leaders. The sociologist Gabriel Tarde summed it all up in the aphorism: "Social man is a somnambulist."

When the new discipline of social psychology was born at the beginning of this century, its first experiments were essentially adaptations of the suggestion demonstration. The technique generally followed a simple plan. The subjects, usually college students, were asked to give their opinions or preferences concerning various matters; some time later they were again asked to state their choices, but now they were also informed of the opinions held by authorities or large groups of their peers on the same matters. (Often the alleged consensus was fictitious.) Most of these studies had substantially the same result: confronted with opinions contrary to their own, many subjects apparently shifted their judgments in the direction of the views of the majorities or the experts. The late psychologist Edward L. Thorndike reported that he had succeeded in modifying the esthetic preferences of adults by this procedure. Other psychologists reported that people's evaluations of the merit of a literary passage could be raised or lowered by ascribing the passage to different authors. Apparently the sheer weight of numbers or authority sufficed to change opinions, even when no arguments for the opinions themselves were provided.

Now the very ease of success in these experiments arouses suspicion. Did the subjects actually change their opinions, or were the experimental victories scored only on paper? On grounds of common sense, one must question whether opinions are generally as watery as these studies indicate. There is some reason to wonder whether it was not the investigators who, in their enthusiasm for a theory, were suggestible, and whether the ostensibly gullible subjects were not providing answers which they thought good subjects were expected to give.

The investigations were guided by certain underlying assumptions, which today are common currency and account for much that is thought and said about the operations of propaganda and public opinion. The assumptions are that people submit uncritically and painlessly to external manipulation by suggestion or prestige, and that any given idea or value can be "sold" or "unsold" without reference to its merits. We should be skeptical, however, of the supposition that the power of social pressure necessarily implies uncritical submission to it: independence and the capacity to rise above group passion are also open to human beings. Further, one may question on psychological grounds whether it is possible as a rule to change a person's judgment of a situation or an object without first changing his knowledge or assumptions about it.

In what follows I shall describe some experiments in an investigation of the effects of group pressure which was carried out recently with the help of a number of my associates. The tests not only demonstrate the operations of group pressure upon individuals but also illustrate a new kind of attack on the problem and some of the more subtle questions that it raises.

A group of seven to nine young men, all college students, are assembled in a classroom for a "psychological experiment" in visual judgment. The experimenter informs them that they will be comparing the lengths of lines. He shows two large white cards. On one is a single vertical black line—the standard whose length is to be matched. On the other card are three vertical lines of various lengths. The subjects are to choose the one that is of the same length as the line on the other card. One of the three actually is of the same length; the other two are substantially different, the difference ranging from three quarters of an inch to an inch and three quarters.

The experiment opens uneventfully. The subjects announce their answers in the order in which they have been seated in the room, and on the first round every person chooses the same matching line. Then a second set of cards is exposed; again the group is unanimous. The members appear ready to endure politely another boring experiment. On the third trial there is an unexpected disturbance. One person near the end of the group disagrees with all the others in his selection of

FIGURE 2.1
Experiment is repeated in the Laboratory of Social Relations at Harvard University. Seven student subjects are asked by the experimenter *(right)* to compare the length of lines *(see Fig. 2.2)*. Six of the subjects have been coached beforehand to give unanimously wrong answers. The seventh *(sixth from the left)* has merely been told that it is an experiment in perception. (Photograph by William Vandivert.)

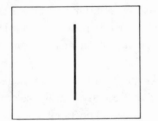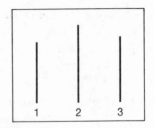

FIGURE 2.2
Subjects were shown two cards. One bore a standard line. The other
bore three lines, one of which was the same length as the standard.
The subjects were asked to choose this line.

the matching line. He looks surprised, indeed incredulous, about the disagreement. On the following trial he disagrees again, while the others remain unanimous in their choice. The dissenter becomes more and more worried and hesitant
as the disagreement continues in succeeding trials; he may pause before announcing his answer and speak in a low voice, or he may smile in an embarrassed way.

What the dissenter does not know is that all the other members of the group
were instructed by the experimenter beforehand to give incorrect answers in unanimity at certain points. The single individual who is not a party to this prear-
rangement is the focal subject of our experiment. He is placed in a position in
which, while he is actually giving the correct answers, he finds himself unexpectedly in a minority of one, opposed by a unanimous and arbitrary majority with
respect to a clear and simple fact. Upon him we have brought to bear two opposed
forces: the evidence of his senses and the unanimous opinion of a group of his
peers. Also, he must declare his judgments in public, before a majority which has
also stated its position publicly.

The instructed majority occasionally reports correctly in order to reduce the possibility that the naive subject will suspect collusion against him. (In only a few
cases did the subject actually show suspicion; when this happened, the experiment
was stopped and the results were not counted. There are 18 trials in each series,
and on 12 of these the majority responds erroneously. How do people respond to
group pressure in this situation? I shall report first the statistical results of a series
in which a total of 123 subjects from three institutions of higher learning (not including my own, Swarthmore College) were placed in the minority situation described above.

Two alternatives were open to the subject: he could act independently, repudiating the majority, or he could go along with the majority, repudiating the evidence of his senses. Of the 123 put to the test, a considerable percentage yielded
to the majority. Whereas in ordinary circumstances individuals matching the lines
will make mistakes less than 1 percent of the time, under group pressure the minority subjects swung to acceptance of the misleading majority's wrong judgments
in 36.8 percent of the selections.

Of course individuals differed in response. At one extreme, about one quarter of the subjects were completely independent and never agreed with the erroneous judgments of the majority. At the other extreme, some individuals went with the majority nearly all the time. The performances of individuals in this experiment tend to be highly consistent. Those who strike out on the path of independence do not, as a rule, succumb to the majority even over an extended series of trials, while those who choose the path of compliance are unable to free themselves as the ordeal is prolonged.

The reasons for the startling individual differences have not yet been investigated in detail. At this point we can only report some tentative generalizations from talks with the subjects, each of whom was interviewed at the end of the experiment. Among the independent individuals were many who held fast because of staunch confidence in their own judgment. The most significant fact about them was not absence of responsiveness to the majority but a capacity to recover from doubt and to reestablish their equilibrium. Others who acted independently came to believe that the majority was correct in its answers, but they continued their dissent on the simple ground that it was their obligation to call the play as they saw it.

Among the extremely yielding persons we found a group who quickly reached the conclusion: "I am wrong, they are right." Others yielded in order "not to spoil your results." Many of the individuals who went along suspected that the majority were "sheep" following the first responder, or that the majority were victims of an optical illusion; nevertheless, these suspicions failed to free them at the moment of decision. More disquieting were the reactions of subjects who construed their difference from the majority as a sign of some general deficiency in themselves, which at all costs they must hide. On this basis they desperately tried to merge with the majority, not realizing the longer-range consequences to themselves. All the yielding subjects underestimated the frequency with which they conformed.

Which aspect of the influence of a majority is more important—the size of the majority or its unanimity? The experiment was modified to examine this question. In one series the size of the opposition was varied from one to fifteen persons. The results showed a clear trend. When a subject was confronted with only a single individual who contradicted his answers, he was swayed little: he continued to answer independently and correctly in nearly all trials. When the opposition was increased to two, the pressure became substantial: minority subjects now accepted the wrong answer 13.6 percent of the time. Under the pressure of a majority of three, the subjects' errors jumped to 31.8 percent. But further increases in the size of the majority apparently did not increase the weight of the pressure substantially. Clearly the size of the opposition is important only up to a point.

Disturbance of the majority's unanimity had a striking effect. In this experiment the subject was given the support of a truthful partner—either another individual who did not know of the prearranged agreement among the rest of the group, or a person who was instructed to give correct answers throughout.

The presence of a supporting partner depleted the majority of much of its power.

Its pressure on the dissenting individual was reduced to one-fourth: that is, subjects answered incorrectly only one-fourth as often as under the pressure of a unanimous majority (Fig. 2.6). The weakest persons did not yield as readily. Most interesting were the reactions to the partner. Generally the feeling toward him was one of warmth and closeness; he was credited with inspiring confidence. However, the subjects repudiated the suggestion that the partner decided them to be independent.

Was the partner's effect a consequence of his dissent, or was it related to his accuracy? We now introduced into the experimental group a person who was instructed to dissent from the majority but also to disagree with the subject. In some experiments the majority was always to choose the worst of the comparison lines and the instructed dissenter to pick the line that was closer to the length of the standard one; in others the majority was consistently intermediate and the dissenter most in error. In this manner we were able to study the relative influence of "compromising" and "extremist" dissenters.

Again the results are clear. When a moderate dissenter is present, the effect of the majority on the subject decreases by approximately one-third, and extremes of yielding disappear. Moreover, most of the errors the subjects do make are moderate, rather than flagrant. In short, the dissenter largely controls the choice of errors. To this extent the subjects broke away from the majority even while bending to it.

On the other hand, when the dissenter always chose the line that was more flagrantly different from the standard, the results were of quite a different kind. The extremist dissenter produced a remarkable freeing of the subjects; their errors dropped to only 9 percent. Furthermore, all the errors were of the moderate variety. We were able to conclude that dissent *per se* increased independence and moderated the errors that occurred, and that the direction of dissent exerted consistent effects.

In all the foregoing experiments each subject was observed only in a single setting. We now turned to studying the effects upon a given individual of a change in the situation to which he was exposed. The first experiment examined the consequences of losing or gaining a partner. The instructed partner began by answering correctly on the first six trials. With his support the subject usually resisted pressure from the majority: eighteen of twenty-seven subjects were completely independent. But after six trials the partner joined the majority. As soon as he did so, there was an abrupt rise in the subjects' errors. Their submission to the ma-

FIGURE 2.3
Experiment proceeds as follows: In the top picture, the subject *(center)* hears rules of experiment for the first time. In the second picture, he makes his first judgment of a pair of cards, disagreeing with the unanimous judgment of othe others. In the third, he leans forward to look at another pair of cards. In the fourth, he shows the strain of repeatedly disagreeing with the majority. In the fifth, after twelve pairs of cards have been shown, he explains that "he has to call them as he sees them." This subject disagreed with the majority on all twelve trials. Seventy-five percent of experimental subjects agree with the majority in varying degrees. (Photographs by William Vandivert.)

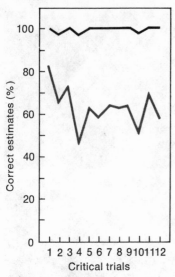

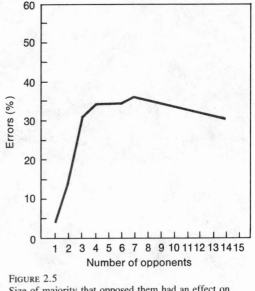

FIGURE 2.4
Error of 123 subjects, each of whom compared lines in the presence of six to eight opponents, is plotted in the gray curve. The accuracy of judgment not under pressure is indicated in black.

FIGURE 2.5
Size of majority that opposed them had an effect on the subjects. With a single opponent, the subject erred only 3.6 percent of the time; with two opponents he erred 13.6 percent; with three, 31.8 percent; with four, 35.1 percent; with six, 35.2 percent; with seven, 37.1 percent; with nine, 35.1 percent; with fifteen, 31.2 percent.

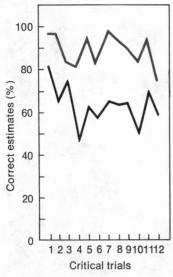

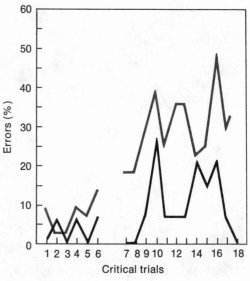

FIGURE 2.6
Two subjects supporting each other against a majority made fewer errors (gray curve) than one subject did against a majority (black curve).

FIGURE 2.7
Partner left subject after six trials in a single experiment. The gray curve shows the error of the subject when the partner "deserted" to the majority. The black curve shows error when partner merely left the room.

jority was just about as frequent as when the minority subject was opposed by a unanimous majority throughout.

It was surprising to find that the experience of having had a partner and of having braved the majority opposition with him had failed to strengthen the individuals' independence. Questioning at the conclusion of the experiment suggested that we had overlooked an important circumstance; namely, the strong specific effect of "desertion" by the partner to the other side. We therefore changed the conditions so that the partner would simply leave the group at the proper point. (To allay suspicion it was announced in advance that he had an appointment with the dean.) In this form of the experiment, the partner's effect outlasted his presence. The errors increased after his departure, but less markedly than after a partner switched to the majority.

In a variant of this procedure the trials began with the majority unanimously giving correct answers. Then they gradually broke away until on the sixth trial the naive subject was alone and the group unanimously against him. As long as the subject had anyone on his side, he was almost invariably independent, but as soon as he found himself alone, the tendency to conform to the majority rose abruptly.

As might be expected, an individual's resistance to group pressure in these experiments depends to a considerable degree on how wrong the majority is. We varied the discrepancy between the standard line and the other lines systematically, with the hope of reaching a point where the error of the majority would be so glaring that every subject would repudiate it and choose independently. In this we regretfully did not succeed. Even when the difference between the lines was seven inches, there were still some who yielded to the error of the majority.

The study provides clear answers to a few relatively simple questions, and it raises many others that await investigation. We would like to know the degree of consistency of persons in situations which differ in content and structure. If consistency of independence or conformity in behavior is shown to be a fact, how is it functionally related to qualities of character and personality? In what ways is independence related to sociological or cultural conditions? Are leaders more independent than other people, or are they adept at following their followers? These and many other questions may perhaps be answerable by investigations of the type described here.

Life in society requires consensus as an indispensable condition. But consensus, to be productive, requires that each individual contribute independently out of his experience and insight. When consensus comes under the dominance of conformity, the social process is polluted and the individual at the same time surrenders the powers on which his functioning as a feeling and thinking being depends. That we have found the tendency to conformity in our society so strong that reasonably intelligent and well-meaning young people are willing to call white black is a matter of concern. It raises questions about our ways of education and about the values that guide our conduct.

Yet anyone inclined to draw too pessimistic conclusions from this report would do well to remind himself that the capacities for independence are not to be under-estimated. He may also draw some consolation from a further observation: those who participated in this challenging experiment agreed nearly without exception that independence was preferable to conformity.

References

ASCH, S. E. Effects of group pressure upon the modification and distortion of judgments. *Groups, leadership, and men,* Harold Guetzdow (ed.). Carnegie Press, 1951.

ASCH, S. E. *Social psychology.* Prentice-Hall, Inc., 1952.

MILLER, N. E. AND DOLLARD, J. *Social learning and imitation.* Yale University Press, 1941.

3

Behavioral Study of Obedience

Stanley Milgram

This chapter describes a procedure for the study of destructive obedience in the laboratory. It consists of ordering a naive S to administer increasingly more severe punishment to a victim in the context of a learning experiment. Punishment is administered by means of a shock generator with thirty graded switches ranging from Slight Shock to Danger: Severe Shock. The victim is a confederate of the E. The primary dependent variable is the maximum shock the S is willing to administer before he refuses to continue further. Twenty-six Ss obeyed the experimental commands fully, and administered the highest shock on the generator. Fourteen Ss broke off the experiment at some point after the victim protested and refused to provide further answers. The procedure created extreme levels of nervous tension in some Ss. Profuse sweating, trembling and stuttering were typical expressions of this emotional disturbance. One unexpected sign of tension—yet to be explained—was the regular occurrence of nervous laughter, which in some Ss developed into uncontrollable seizures. The variety of interesting behavioral dynamics observed in the experiment, the reality of the situation for the S, and the possibility of parametric variation within the framework of the procedure, point to the fruitfulness of further study.

Obedience is as basic an element in the structure of social life as one can point to. Some system of authority is a requirement of all communal living, and it is only the man dwelling in isolation who is not forced to respond, through defiance or submission, to the commands of others. Obedience, as a determinant of behavior, is of particular relevance to our time. It has been reliably established that from 1933–1945 millions of innocent persons were systematically slaughtered on command. Gas chambers were built, death camps were guarded, daily quotas of corpses were produced with the same efficiency as the manufacture of appliances. These inhumane policies may have originated in the mind of a single person, but they could only be carried out on a massive scale if a very large number of persons obeyed orders.

Reprinted with permission from the author and *The Journal of Abnormal and Social Psychology,* Vol. 67, No. 4, 1963. Copyright 1963 by the American Psychological Association.

This research was supported by a grant (NSF G–17916) from the National Science Foundation. Exploratory studies conducted in 1960 were supported by a grant from the Higgins Fund at Yale University. The research assistance of Alan E. Elms and Jon Wayland is gratefully acknowledged.

Obedience is the psychological mechanism that links individual action to political purpose. It is the dispositional cement that binds men to systems of authority. Facts of recent history and observation in daily life suggest that for many persons obedience may be a deeply ingrained behavior tendency, indeed, a prepotent impulse overriding training in ethics, sympathy, and moral conduct. C. P. Snow (1961) points to its importance when he writes:

> When you think of the long and gloomy history of man, you will find more hideous crimes have been committed in the name of obedience than have ever been committed in the name of rebellion. If you doubt that, read William Shirer's "Rise and Fall of the Third Reich." The German Officer Corps were brought up in the most rigorous code of obedience . . . in the name of obedience they were party to, and assisted in, the most wicked large scale actions in the history of the world [p. 24].

While the particular form of obedience dealt with in the present study has its antecedents in these episodes, it must not be thought all obedience entails acts of aggression against others. Obedience serves numerous productive functions. Indeed, the very life of society is predicated on its existence. Obedience may be ennobling and educative and refer to acts of charity and kindness, as well as to destruction.

General Procedure

A procedure was devised which seems useful as a tool for studying obedience (Milgram, 1961). It consists of ordering a naive subject to administer electric shock to a victim. A simulated shock generator is used, with 30 clearly marked voltage levels that range from 15 to 450 volts. The instrument bears verbal designations that range from Slight Shock to Danger: Severe Shock. The responses of the victim, who is a trained confederate of the experimenter, are standardized. The orders to administer shocks are given to the naive subject in the context of a "learning experiment" ostensibly set up to study the effects of punishment on memory. As the experiment proceeds the naive subject is commanded to administer increasingly more intense shocks to the victim, even to the point of reaching the level marked Danger: Severe Shock. Internal resistances become stronger, and at a certain point the subject refuses to go on with the experiment. Behavior prior to this rupture is considered "obedience," in that the subject complies with the commands of the experimenter. The point of rupture is the act of disobedience. A quantitative value is assigned to the subject's performance based on the maximum intensity shock he is willing to administer before he refuses to participate further. Thus for any particular subject and for any particular experimental condition the degree of obedience may be specified with a numerical value. The crux of the study is to systematically vary the factors believed to alter the degree of obedience to the experimental commands.

The technique allows important variables to be manipulated at several points in the experiment. One may vary aspects of the source of command, content and form of command, instrumentalities for its execution, target object, general social setting, etc. The problem, therefore, is not one of designing increasingly more numerous experimental conditions, but of selecting those that best illuminate the *process* of obedience from the sociopsychological standpoint.

Related Studies

The inquiry bears an important relation to philosophic analyses of obedience and authority (Arendt, 1958; Friedrich, 1958; Weber, 1947), an early experimental study of obedience by Frank (1944), studies in "authoritarianism" (Adorno, Frenkel-Brunswik, Levinson, and Sanford, 1950; Rokeach, 1961), and a recent series of analytic and empirical studies in social power (Cartwright, 1959). It owes much to the long concern with *suggestion* in social psychology, both in its normal forms (e.g., Binet, 1900) and in its clinical manifestations (Charcot, 1881). But it derives, in the first instance, from direct observation of a social fact; the individual who is commanded by a legitimate authority ordinarily obeys. Obedience comes easily and often. It is a ubiquitous and indispensable feature of social life.

METHOD

Subjects

The subjects were 40 males between the ages of 20 and 50, drawn from New Haven and the surrounding communities. Subjects were obtained by a newspaper advertisement and direct mail solicitation. Those who responded to the appeal believed they were to participate in a study of memory and learning at Yale University. A wide range of occupations is represented in the sample. Typical subjects were postal clerks, high school teachers, salesmen, engineers, and laborers. Subjects ranged in educational level from one who had not finished elementary school, to those who had doctorate and other professional degrees. They were paid $4.50 for their participation in the experiment. However, subjects were told that payment was simply for coming to the laboratory, and that the money was theirs no matter what happened after they arrived. Table 3.1 shows the proportion of age and occupational types assigned to the experimental condition.

TABLE 3.1
Distribution of age and occupational types in the experiment

Occupations	20–29 years n	30–39 years n	40–50 years n	Percentage of total (occupations)
Workers, skilled and unskilled	4	5	6	37.5
Sales, business, and white-collar	3	6	7	40.0
Professional	1	5	3	22.5
Percentage of total (age)	20	40	40	

Note: Total $N = 40$.

Personnel and Locale

The experiment was conducted on the grounds of Yale University in the elegant interaction laboratory. (This detail is relevant to the perceived legitimacy of the experiment. In further variations, the experiment was dissociated from the university, with consequences for performance.) The role of experimenter was played by a 31-year-old high school teacher of biology. His manner was impassive, and his appearance somewhat stern throughout the experiment. He was dressed in a gray technician's coat. The victim was played by a 47-year-old accountant, trained for the role; he was of Irish-American stock, whom most observers found mild-mannered and likable.

Procedure

One naive subject and one victim (an accomplice) performed in each experiment. A pretext had to be devised that would justify the administration of electric shock by the naive subject. This was effectively accomplished by the cover story. After a general introduction on the presumed relation between punishment and learning, subjects were told:

> But actually, we know very *little* about the effect of punishment on learning, because almost no truly scientific studies have been made of it in human beings.
> For instance, we don't know how *much* punishment is best for learning—and we don't know how much difference it makes as to who is giving the punishment, whether an adult learns best from a younger or an older person than himself—or many things of that sort.
> So in this study we are bringing together a number of adults of different occupations and ages. And we're asking some of them to be teachers and some of them to be learners.

We want to find out just what effect different people have on each other as teachers and learners, and also what effect *punishment* will have on learning in this situation.

Therefore, I'm going to ask one of you to be the teacher here tonight and the other one to be the learner.

Does either of you have a preference?

Subjects then drew slips of paper from a hat to determine who would be the teacher and who would be the learner in the experiment. The drawing was rigged so that the naive subject was always the teacher and the accomplice always the learner. (Both slips contained the word "Teacher.") Immediately after the drawing the teacher and learner were taken to an adjacent room and the learner was strapped into an "electric chair" apparatus.

The experimenter explained that the straps were to prevent excessive movement while the learner was being shocked. The effect was to make it impossible for him to escape from the situation. An electrode was attached to the learner's wrist, and electrode paste was applied "to avoid blisters and burns." Subjects were told that the electrode was attached to the shock generator in the adjoining room.

In order to improve credibility the experimenter declared, in response to a question by the learner: "Although the shocks can be extremely painful, they cause no permanent tissue damage."

Learning task.   The lesson administered by the subject was a paired-associate learning task. The subject read a series of word pairs to the learner, and then read the first word of the pair along with four terms. The learner was to indicate which of the four terms had originally been paired with the first word. He communicated his answer by pressing one of four switches in front of him, which in turn lit up one of four numbered quadrants in an answer-box located atop the shock generator.

Shock generator.   The instrument panel consists of 30 lever switches set in a horizontal line. Each switch is clearly labeled with a voltage designation that ranges from 15 to 450 volts. There is a 15-volt increment from one switch to the next going from left to right. In addition, the following verbal designations are clearly indicated for groups of four switches going from left to right: Slight Shock, Moderate Shock, Strong Shock, Very Strong Shock, Intense Shock, Extreme Intensity Shock, Danger: Severe Shock. (Two switches after this last designation are simply marked XXX.)

Upon depressing a switch: a pilot light corresponding to each switch is illuminated in bright red; an electric buzzing is heard; an electric blue light, labeled "voltage energizer," flashes; the dial on the voltage meter swings to the right; various relay clicks are sounded.

The upper left-hand corner of the generator is labeled Shock Generator, Type ZLB, Dyson Instrument Company, Waltham, Mass. Output 15 Volts—450 Volts.

Details of the instrument were carefully handled to insure an appearance of authenticity. The panel was engraved by precision industrial engravers, and all components were of high quality. No subject in the experiment suspected that the instrument was merely a simulated shock generator.[1]

Sample shock.   Each naive subject is given a sample shock on the shock generator, prior to beginning his run as teacher. This shock is always 45 volts, and is applied by pressing the third switch of the generator. The shock is applied to the wrist of the naive subject, and has its source in a 45-volt battery wired into the generator. This further convinces the subject of the authenticity of the generator.

Shock instructions.   The subject is told to administer a shock to the learner each time he gives a wrong response. Moreover—and this is the key command—the subject is instructed to "move one lever higher on the shock generator each time the learner flashes a wrong answer." He is also instructed to announce the voltage level before administering a shock. This serves to continually remind subjects of the increasing intensity of shocks administered to the learner.

Preliminary and regular run.   Pretests revealed that the procedure of reading words and administering shocks requires some practice before it can be handled smoothly. Therefore, immediately preceding the regular run, the teacher is given a preliminary series of 10 words to read to the learner. There are 3 neutral words in the practice series (i.e., words that the learner answers correctly), so that shocks are administered for 7 of the words, with the maximum shock at 105 volts (moderate shock). Almost all subjects master the procedure by the time the preliminary run is over.

Subjects are then presented with a second list, and are told that the procedure is the same as for the first list; the experimenter adds, however:

> When you get to the bottom of the list, repeat it over again, and continue giving shocks, until the learner has learned all the pairs correctly.

The experimenter instructs the subject to

> start from 15 volts and increase the shock level one step each time the learner gives a wrong answer.

---

[1] A related technique, making use of a shock generator, was reported by Buss (1961) for the study of aggression in the laboratory. Despite the considerable similarity of technical detail in the experimental procedures, each investigator proceeded in ignorance of the other's work. Milgram provided plans and photographs of his shock generator, experimental procedure, and first results in a report to the National Science Foundation in January 1961. This report received only limited circulation. Buss reported his procedure six months later, but to a wider audience. Subsequently, technical information and reports were exchanged. The present article was first received in the editor's office on December 27, 1961; it was resubmitted with deletions on July 27, 1962.

Feedback from the victim.   In all conditions the learner gives a predetermined set of responses to the word pair test, based on a schedule of approximately three wrong answers to one correct answer. In the present experimental condition no vocal response or other sign of protest is heard from the learner until Shock Level 300 is reached. When the 300-volt shock is administered, the learner pounds on the wall of the room in which he is bound to the electric chair. The pounding can be heard by the subject. From this point on, the learner's answers no longer appear on the four-way panel.

At this juncture, subjects ordinarily turn to the experimenter for guidance. The experimenter instructs the subject to treat the absence of a response as a wrong answer, and to shock the subject according to the usual schedule. He advises the subjects to allow 5–10 seconds before considering no response as a wrong answer, and to increase the shock level one step each time the learner fails to respond correctly. The learner's pounding is repeated after the 315-volt shock is administered; afterwards he is not heard from, nor do his answers reappear on the four-way signal box.

Experimenter feedback.   At various points in the experiment the subject may turn to the experimenter for advice on whether he should continue to administer shocks. Or he may indicate that he does not wish to go on. It was important to standardize the experimenter's responses to these inquiries and reactions.

If the subject indicated his unwillingness to go on, the experimenter responded with a sequence of "prods," using as many as necessary to bring the subject into line.

> Prod 1: Please continue. *or* Please go on.
> Prod 2: The experiment requires that you continue.
> Prod 3: It is absolutely essential that you continue.
> Prod 4: You have no other choice, you *must* go on.

The prods were always made in sequence: Only if Prod 1 had been unsuccessful, could Prod 2 be used. If the subject refused to obey the experimenter after Prod 4, the experiment was terminated. The experimenter's tone of voice was at all times firm, but not impolite. The sequence was begun anew on each occasion that the subject balked or showed reluctance to follow orders.

Special prods. If the subject asked if the learner was liable to suffer permanent physical injury, the experimenter said:

> Although the shocks may be painful, there is no permanent tissue damage, so please go on. [Followed by Prods 2, 3, and 4 if necessary.]

If the subject said that the learner did not want to go on, the experimenter replied:

> Whether the learner likes it or not, you must go on until he has learned all the word pairs correctly. So please go on. [Followed by Prods 2, 3, and 4 if necessary.]

Dependent Measures

The primary dependent measure for any subject is the maximum shock he administers before he refuses to go any further. In principle this may vary from 0 (for a subject who refuses to administer even the first shock) to 30 (for a subject who administers the highest shock on the generator). A subject who breaks off the experiment at any point prior to administering the thirtieth shock level is termed a *defiant* subject. One who complies with experimental commands fully, and proceeds to administer all shock levels commanded, is termed an *obedient* subject.

Further records.   With few exceptions, experimental sessions were recorded on magnetic tape. Occasional photographs were taken through one-way mirrors. Notes were kept on any unusual behavior occurring during the course of the experiments. On occasion, additional observers were directed to write objective descriptions of the subjects' behavior. The latency and duration of shocks were measured by accurate timing devices.

Interview and dehoax.   Following the experiment, subjects were interviewed; open-ended questions, projective measures, and attitude scales were employed. After the interview, procedures were undertaken to assure that the subject would leave the laboratory in a state of well being. A friendly reconciliation was arranged between the subject and the victim, and an effort was made to reduce any tensions that arose as a result of the experiment.

RESULTS

Preliminary Notions

Before giving an account of the experimental results, it is instructive to consider how persons predict others will perform when commanded to administer powerful shocks to another person. Fourteen Yale seniors, all psychology majors, were provided with a detailed description of the experimental situation. They were asked to reflect carefully on it, and to predict the behavior of 100 hypothetical subjects. More specifically, they were instructed to plot the distribution of obedience of "100 Americans of diverse occupations, and ranging in age from 20 to 50 years," who were placed in the experimental situation.

There was considerable agreement among the respondents on the expected behavior of hypothetical subjects. All respondents predicted that only an insignificant minority would go through to the end of the shock series. (The estimates ranged from 0 to 3 percent; i.e., the most "pessimistic" member of the class predicted that of 100 persons, 3 would continue through to the most potent shock available on the shock generator—450 volts.) The class mean was 1.2 percent. The question was also posed informally to colleagues of the author, and the most general feeling was that few if any subjects would go beyond the designation Very Strong Shock.

Result of Experiment I

Subjects accept situation. With few exceptions subjects were convinced of the reality of the experimental situation, that they were administering shocks to another person, and the most potent shocks were extremely painful. In the post-experimental interview subjects were asked: "How painful to the learner were the last few shocks that you administered to him?" Subjects were instructed to indicate their answers on a printed 14-point scale ranging from "Not at all painful" to "Extremely painful." The modal response was 14 (Extremely painful) with the mean at 13.42.

Signs of extreme tension. Many subjects showed signs of nervousness in the experimental situation, and especially upon administering the more powerful shocks. In a large number of cases the degree of tension reached extremes that are rarely seen in sociopsychological laboratory studies. Subjects were observed to sweat, tremble, stutter, bite their lips, groan, and dig their fingernails into their flesh. These were characteristic rather than exceptional responses to the experiment.

One sign of tension was the regular occurrence of nervous laughing fits. Fourteen of the 40 subjects showed definite signs of nervous laughter and smiling. The laughter seemed entirely out of place, even bizarre. Full-blown, uncontrollable seizures were observed for 3 subjects. On one occasion we observed a seizure so violently convulsive that it was necessary to call a halt to the experiment. The subject, a 46-year-old encyclopedia salesman, was seriously embarrassed by his untoward and uncontrollable behavior. In the post-experimental interviews subjects took pains to point out that they were not sadistic types, and that the laughter did not mean they enjoyed shocking the victim.

Distribution of scores. It had been conjectured that persons would not, in general, agree to administer shocks beyond the level designated as Very Strong Shock. The distribution of maximum shocks for this experiment is shown in Table 3.2. The obtained distribution of scores deviates radically from the prediction.

Upon command of the experimenter, each of the 40 subjects went beyond the expected breakoff point. No subject stopped prior to administering Shock Level 20. (At this level—300 volts—the victim kicks on the wall and no longer provides answers to the teacher's multiple-choice questions.)

Of the 40 subjects, 5 refused to obey the experimental commands beyond the 300-volt level. Four more subjects administered one further shock, and then refused to go on. Two broke off at the 330-volt level, and 1 each at 345, 360, and

TABLE 3.2
Distribution of breakoff points

Verbal designation and voltage indication	Number of subjects for whom this was maximum shock
Slight Shock	
15	0
30	0
45	0
60	0
Moderate Shock	
75	0
90	0
105	0
120	0
Strong Shock	
135	0
150	0
165	0
180	0
Very Strong Shock	
195	0
210	0
225	0
240	0
Intense Shock	
255	0
270	0
285	0
300	5
Extreme Intensity Shock	
315	4
330	2
345	1
360	1
Danger: Severe Shock	
375	1
390	0
405	0
420	0
XXX	
435	0
450	26

375 volts. Thus a total of 14 subjects defied the experimenter. It is instructive to reprint their remarks at the point of defiance, as transcribed from the tape recordings:

[0124] I think he's trying to communicate, he's knocking . . . Well it's not fair to shock the guy . . . these are terrific volts. I don't think this is very humane . . . Oh, I can't go on with this; no, this isn't right. It's a hell of an experiment. The guy is suffering in there. No, I don't want to go on. This is crazy. [Subject refused to administer more shocks.]

[0123] He's banging in there. I'm gonna chicken out. I'd like to continue, but I can't do that to a man . . . I'm sorry I can't do that to a man. I'll hurt his heart. You take your check . . . No really, I couldn't do it.

These subjects were frequently in a highly agitated and even angered state. Sometimes, verbal protest was at a minimum, and the subject simply got up from his chair in front of the shock generator, and indicated that he wished to leave the laboratory.

Of the 40 subjects, 26 obeyed the orders of the experimenter to the end, proceeding to punish the victim until they reached the most potent shock available on the shock generator. At that point, the experimenter called a halt to the session. (The maximum shock is labeled 450 volts, and is two steps beyond the designation: Danger: Severe Shock.) Although obedient subjects continued to administer shocks, they often did so under extreme stress. Some expressed reluctance to administer shocks beyond the 300-volt level, and displayed fears similar to those who defied the experimenter; yet they obeyed.

After the maximum shocks had been delivered, and the experimenter called a halt to the proceedings, many obedient subjects heaved sighs of relief, mopped their brows, rubbed their fingers over their eyes, or nervously fumbled cigarettes. Some shook their heads, apparently in regret. Some subjects had remained calm throughout the experiment, and displayed only minimal signs of tension from beginning to end.

DISCUSSION

The experiment yielded two findings that were surprising. The first finding concerns the sheer strength of obedient tendencies manifested in this situation. Subjects have learned from childhood that it is a fundamental breach of moral conduct to hurt another person against his will. Yet, 26 subjects abandon this tenet in following the instructions of an authority who has no special powers to enforce his commands. To disobey would bring no material loss to the subject; no punishment would ensue. It is clear from the remarks and outward behavior of many participants that in punishing the victim they are often acting against their own values. Subjects often expressed deep disapproval of shocking a man in the face of his objections, and others denounced it as stupid and senseless. Yet the majority complied with the experimental commands. This outcome was surprising from two perspectives: first, from the standpoint of predictions made in the questionnaire described earlier. (Here, however, it is possible that the remoteness of the respondents from the actual situation, and the difficulty of conveying to them the concrete details of the experiment, could account for the serious underestimation of obedience.)

But the results were also unexpected to persons who observed the experiment in progress, through one-way mirrors. Observers often uttered expressions of disbelief upon seeing a subject administer more powerful shocks to the victim. These persons had a full acquaintance with the details of the situation, and yet systematically underestimated the amount of obedience that subjects would display.

The second unanticipated effect was the extraordinary tension generated by the procedures. One might suppose that a subject would simply break off or continue as his conscience dictated. Yet, this is very far from what happened. There were striking reactions of tension and emotional strain. One observer related:

> I observed a mature and initially poised businessman enter the laboratory smiling and confident. Within 20 minutes he was reduced to a twitching, stuttering wreck, who was rapidly approaching a point of nervous collapse. He constantly pulled on his earlobe, and twisted his hands. At one point he pushed his fist into his forehead and muttered: "Oh God, let's stop it." And yet he continued to respond to every word of the experimenter, and obeyed to the end.

Any understanding of the phenomenon of obedience must rest on an analysis of the particular conditions in which it occurs. The following features of the experiment go some distance in explaining the high amount of obedience observed in the situation.

1. The experiment is sponsored by and takes place on the grounds of an institution of unimpeachable reputation, Yale University. It may be reasonably presumed that the personnel are competent and reputable. The importance of this background authority is now being studied by conducting a series of experiments outside of New Haven, and without any visible ties to the university.

2. The experiment is, on the face of it, designed to attain a worthy purpose—advancement of knowledge about learning and memory. Obedience occurs not as an end in itself, but as an instrumental element in a situation that the subject construes as significant, and meaningful. He may not be able to see its full significance, but he may properly assume that the experimenter does.

3. The subject perceives that the victim has voluntarily submitted to the authority system of the experimenter. He is not (at first) an unwilling captive impressed for involuntary service. He has taken the trouble to come to the laboratory presumably to aid the experimental research. That he later becomes an involuntary subject does not alter the fact that, initially, he consented to participate without qualification. Thus he has in some degree incurred an obligation toward the experimenter.

4. The subject, too, has entered the experiment voluntarily, and perceives himself under obligation to aid the experimenter. He has made a commitment, and to disrupt the experiment is a repudiation of this initial promise of aid.

5. Certain features of the procedure strengthen the subject's sense of obligation

to the experimenter. For one, he has been paid for coming to the laboratory. In part this is canceled out by the experimenter's statement that:

> Of course, as in all experiments, the money is yours simply for coming to the laboratory. From this point on, no matter what happens, the money is yours.[2]

6. From the subject's standpoint, the fact that he is the teacher and the other man the learner is purely a chance consequence (it is determined by drawing lots) and he, the subject, ran the same risk as the other man in being assigned the role of learner. Since the assignment of positions in the experiment was achieved by fair means, the learner is deprived of any basis of complaint on this count. (A similar situation obtains in Army units, in which—in the absence of volunteers— a particularly dangerous mission may be assigned by drawing lots, and the unlucky soldier is expected to bear his misfortune with sportsmanship.)

7. There is, at best, ambiguity with regard to the prerogatives of a psychologist and the corresponding rights of his subject. There is a vagueness of expectation concerning what a psychologist may require of his subject, and when he is overstepping acceptable limits. Moreover, the experiment occurs in a closed setting, and thus provides no opportunity for the subject to remove these ambiguities by discussion with others. There are few standards that seem directly applicable to the situation, which is a novel one for most subjects.

8. The subjects are assured that the shocks administered to the subject are "painful but not dangerous." Thus they assume that the discomfort caused the victim is momentary, while the scientific gains resulting from the experiment are enduring.

9. Through Shock Level 20 the victim continues to provide answers on the signal box. The subject may construe this as a sign that the victim is still willing to "play the game." It is only after Shock Level 20 that the victim repudiates the rules completely, refusing to answer further.

These features help to explain the high amount of obedience obtained in this experiment. Many of the arguments raised need not remain matters of speculation, but can be reduced to testable propositions to be confirmed or disproved by further experiments.[3]

The following features of the experiment concern the nature of the conflict which the subject faces.

10. The subject is placed in a position in which he must respond to the competing demands of two persons: the experimenter and the victim. The conflict must be resolved by meeting the demands of one or the other; satisfaction of the victim and the experimenter are mutually exclusive. Moreover, the resolution must take the form of a highly visible action, that of continuing to shock the victim or break-

---

[2] Forty-three subjects, undergraduates at Yale University, were run in the experiment without payment. The results are very similar to those obtained with paid subjects.

[3] A series of recently completed experiments employing the obedience paradigm is reported in Milgram (1965).

ing off the experiment. Thus the subject is forced into a public conflict that does not permit any completely satisfactory solution.

11. While the demands of the experimenter carry the weight of scientific authority, the demands of the victim spring from his personal experience of pain and suffering. The two claims need not be regarded as equally pressing and legitimate. The experimenter seeks an abstract scientific datum; the victim cries out for relief from physical suffering caused by the subject's actions.

12. The experiment gives the subject little time for reflection. The conflict comes on rapidly. It is only minutes after the subject has been seated before the shock generator that the victim begins his protests. Moreover, the subject perceives that he has gone through but two-thirds of the shock levels at the time the subject's first protests are heard. Thus he understands that the conflict will have a persistent aspect to it, and may well become more intense as increasingly more powerful shocks are required. The rapidity with which the conflict descends on the subject, and his realization that it is predictably recurrent may well be sources of tension to him.

13. At a more general level, the conflict stems from the opposition of two deeply ingrained behavior dispositions: first, the disposition not to harm other people, and second, the tendency to obey those whom we perceive to be legitimate authorities.

References

ADORNO, T., FRENKEL-BRUNSWIK, ELSE, LEVINSON, D. J., AND SANFORD, R. N. *The authoritarian personality.* New York: Harper, 1950.

ARENDT, H. What was authority? In C. J. Friedrich (ed.), *Authority.* Cambridge: Harvard Univer. Press, 1958. Pp. 81–112.

BINET, A. *La suggestibilité.* Paris: Schleicher, 1900.

BUSS, A. H. *The psychology of aggression.* New York: Wiley, 1961.

CARTWRIGHT, S. (ed.) *Studies in social power.* Ann Arbor: University of Michigan Institute for Social Research, 1959.

CHARCOT, J. M. *Oeuvres complètes.* Paris: Bureaux du Progrès Médical, 1881.

FRANK, J. D. Experimental studies of personal pressure and resistance. *J. gen. Psychol.,* 1944, **30,** 23–64.

FRIEDRICH, C. J. (ed.) *Authority.* Cambridge: Harvard Univer. Press, 1958.

MILGRAM, S. Dynamics of obedience. Washington: National Science Foundation, 25 January 1961. (Mimeo)

MILGRAM, S. Some conditions of obedience and disobedience to authority. *Hum. Relat.,* 1965, **18,** 57–76.

ROKEACH, M. Authority, authoritarianism, and conformity. In I. A. Berg and B. M. Bass (eds.), *Conformity and deviation.* New York: Harper, 1961. Pp. 230–257.

SNOW, C. P. Either-or. *Progressive,* 1961 (Feb.), 24.

WEBER, M. *The theory of social and economic organization.* Oxford: Oxford Univer. Press, 1947.

4

"From Jerusalem to Jericho": A Study of Situational and Dispositional Variables in Helping Behavior

John M. Darley and C. Daniel Batson

The influence of several situational and personality variables on helping behavior was examined in an emergency situation suggested by the parable of the Good Samaritan. People going between two buildings encountered a shabbily dressed person slumped by the side of the road. Subjects in a hurry to reach their destination were more likely to pass by without stopping. Some subjects were going to give a short talk on the parable of the Good Samaritan, others on a nonhelping relevant topic; this made no significant difference in the likelihood of their giving the victim help. Religious personality variables did not predict whether an individual would help the victim or not. However, if a subject did stop to offer help, the character of the helping response was related to his type of religiosity.

Helping other people in distress is, among other things, an ethical act. That is, it is an act governed by ethical norms and precepts taught to children at home, in school, and in church. From Freudian and other personality theories, one would expect individual differences in internalization of these standards that would lead to differences between individuals in the likelihood with which they would help others. But recent research on bystander intervention in emergency situations (Bickman, 1969; Darley & Lantané, 1968; Korte, 1969; but see also Schwartz & Clausen, 1970) has had bad luck in finding personality determinants of helping behavior. Although personality variables that one might expect to correlate with helping behavior have been measured (Machiavellianism, authoritarianism, social desirability, alienation, and social responsibility), these were not predictive of helping. Nor was this due to a generalized lack of predictability in the helping

Reprinted with permission of the authors and *The Journal of Personality and Social Psychology*, Vol. 27, No. 1, 1973. Copyright 1973 by The American Psychological Association.

For assistance in conducting this research thanks are due Robert Wells, Beverly Fisher, Mike Shafto, Peter Sheras, Richard Detweiler, and Karen Glasser. The research was funded by National Science Foundation Grant GS-2293.

situation examined, since variations in the experimental situation, such as the availability of other people who might also help, produced marked changes in rates of helping behavior. These findings are reminiscent of Hartshorne and May's (1928) discovery that resistance to temptation, another ethically relevant act, did not seem to be a fixed characteristic of an individual. That is, a person who was likely to be honest in one situation was not particularly likely to be honest in the next (but see also Burton, 1963).

The rather disappointing correlation between the social psychologist's traditional set of personality variables and helping behavior in emergency situations suggests the need for a fresh perspective on possible predictors of helping and possible situations in which to test them. Therefore, for inspiration we turned to the Bible, to what is perhaps the classical helping story in the Judeo-Christian tradition, the parable of the Good Samaritan. The parable proved of value in suggesting both personality and situational variables relevant to helping.

> "And who is my neighbor?" Jesus replied, "A man was going down from Jerusalem to Jericho, and he fell among robbers, who stripped him and beat him, and departed, leaving him half dead. Now by chance a priest was going down the road; and when he saw him he passed by on the other side. So likewise a Levite, when he came to the place and saw him, passed by on the other side. But a Samaritan, as he journeyed, came to where he was; and when he saw him, he had compassion, and went to him and bound his wounds, pouring on oil and wine; then he set him on his own beast and brought him to an inn, and took care of him. And the next day he took out two dennarii and gave them to the innkeeper, saying, "Take care of him; and whatever more you spend, I will repay you when I come back." Which of these three, do you think, proved neighbor to him who fell among the robbers? He said, "The one who showed mercy on him." And Jesus said to him, "Go and do likewise." [Luke 10: 29–37 RSV]

To psychologists who reflect on the parable, it seems to suggest situational and personality differences between the nonhelpful priest and Levite and the helpful Samaritan. What might each have been thinking and doing when he came upon the robbery victim on that desolate road? What sort of persons were they?

One can speculate on differences in thought. Both the priest and the Levite were religious functionaries who could be expected to have their minds occupied with religious matters. The priest's role in religious activities is obvious. The Levite's role, although less obvious, is equally important: The Levites were necessary participants in temple ceremonies. Much less can be said with any confidence about what the Samaritan might have been thinking, but, in contrast to the others, it was most likely not of a religious nature, for Samaritans were religious outcasts.

Not only was the Samaritan most likely thinking about more mundane matters than the priest and Levite, but, because he was socially less important, it seems likely that he was operating on a quite different time schedule. One can imagine the priest and Levite, prominent public figures, hurrying along with little black books full of meetings and appointments, glancing furtively at their sundials. In contrast, the Samaritan would likely have far fewer and less important people counting on him to be at a particular place at a particular time, and therefore might be expected to be in less of a hurry than the prominent priest or Levite.

In addition to these situational variables, one finds personality factors suggested as well. Central among these, and apparently basic to the point that Jesus was trying to make, is a distinction between types of religiosity. Both the priest and Levite are extremely "religious." But it seems to be precisely their type of religiosity that the parable challenges. At issue is the motivation for one's religion and ethical behavior. Jesus seems to feel that the religious leaders of his time, though certainly respected and upstanding citizens, may be "virtuous" for what it will get them, both in terms of the admiration of their fellowmen and in the eyes of God. New Testament scholar R. W. Funk (1966) noted that the Samaritan is at the other end of the spectrum:

> The Samaritan does not love with side glances at God. The need of neighbor alone is made self-evident, and the Samaritan responds without other motivation [pp. 218–219].

That is, the Samaritan is interpreted as responding spontaneously to the situation, not as being preoccupied with the abstract ethical or organizational do's and don'ts of religion as the priest and Levite would seem to be. This is not to say that the Samaritan is portrayed as irreligious. A major intent of the parable would seem to be to present the Samaritan as a religious and ethical example, but at the same time to contrast his type of religiosity with the more common conception of religiosity that the priest and Levite represent.

To summarize the variables suggested as affecting helping behavior by the parable, the situational variables include the content of one's thinking and the amount of hurry in one's journey. The major dispositional variable seems to be differing types of religiosity. Certainly these variables do not exhaust the list that could be elicited from the parable, but they do suggest several research hypotheses.

Hypothesis 1. The parable implies that people who encounter a situation possibly calling for a helping response while thinking religious and ethical thoughts will be no more likely to offer aid than persons thinking about something else. Such a hypothesis seems to run counter to a theory that focuses on norms as determining helping behavior because a normative account would predict that the increased salience of helping norms produced by thinking about religious and ethical examples would increase helping behavior.

Hypothesis 2. Persons encountering a possible helping situation when they are in a hurry will be less likely to offer aid than persons not in a hurry.

Hypothesis 3. Concerning types of religiosity, persons who are religious in a Samaritan-like fashion will help more frequently than those religious in a priest or Levite fashion.

Obviously, this last hypothesis is hardly operationalized as stated. Prior research by one of the investigators on types of religiosity (Batson, 1971), however, led us to differentiate three distinct ways of being religious: *(a)* for what it will gain one (cf. Freud, 1927, and perhaps the priest and Levite), *(b)* for its own intrinsic value (cf. Allport and Ross, 1967), and *(c)* as a response to and quest for meaning in one's everyday life (cf. Batson, 1971). Both of the latter conceptions would be proposed by their exponents as related to the more Samaritanlike "true" religiosity. Therefore, depending on the theorist one follows, the third hypothesis may be stated like this: People *(a)* who are religious for intrinsic reasons (Allport and Ross, 1967) or *(b)* whose religion emerges out of questioning the meaning of their everyday lives (Batson, 1971) will be more likely to stop to offer help to the victim.

The parable of the Good Samaritan also suggested how we would measure people's helping behavior—their response to a stranger slumped by the side of one's path. The victim should appear somewhat ambiguous—ill-dressed, possibly in need of help, but also possibly drunk or even potentially dangerous.

Further, the parable suggests a means by which the incident could be perceived as a real one rather than part of a psychological experiment in which one's behavior was under surveillance and might be shaped by demand characteristics (Orne, 1962), evaluation apprehension (Rosenberg, 1965), or other potentially artifactual determinants of helping behavior. The victim should be encountered not in the experimental context but on the road between various tasks.

METHOD

In order to examine the influence of these variables on helping behavior, seminary students were asked to participate in a study on religious education and vocations. In the first testing session, personality questionnaires concerning types of religiosity were administered. In a second individual session, the subject began experimental procedures in one building and was asked to report to another building for later procedures. While in transit, the subject passed a slumped "victim" planted in an alleyway. The dependent variable was whether and how the subject helped the victim. The independent variables were the degree to which the subject was told to hurry in reaching the other building and the talk he was to give when he arrived there. Some subjects were to give a talk on the jobs in which seminary students would be most effective, others, on the parable of the Good Samaritan.

Subjects

The subjects for the questionnaire administration were 67 students at Princeton Theological Seminary. Forty-seven of them, those who could be reached by telephone, were scheduled for the experiment. Of the 47, 7 subjects' data were not included in the analyses—3 because of contamination of the experimental procedures during their testing and 4 due to suspicion of the experimental situation. Each subject was paid $1 for the questionnaire session and $1.50 for the experimental session.

Personality Measures

Detailed discussion of the personality scales used may be found elsewhere (Batson, 1971), so the present discussion will be brief. The general personality construct under examination was religiosity. Various conceptions of religiosity have been offered in recent years based on different psychometric scales. The conception seeming to generate the most interest is the Allport and Ross (1967) distinction between "intrinsic" versus "extrinsic" religiosity (cf. also Allen and Spilka, 1967, on "committed" versus "consensual" religion). This bipolar conception of religiosity has been questioned by Brown (1964) and Batson (1971), who suggested three-dimensional analyses instead. Therefore, in the present research, types of religiosity were measured with three instruments which together provided six separate scales: *(a)* a *doctrinal orthodoxy* (D-O) scale patterned after that used by Glock and Stark (1966), scaling agreement with classic doctrines of Protestant theology; *(b)* the Allport-Ross *extrinsic* (AR-E) scale, measuring the use of religion as a means to an end rather than as an end in itself; *(c)* the Allport-Ross *intrinsic* (AR-I) scale, measuring the use of religion as an end in itself; *(d)* the *extrinsic external* scale of Batson's Religious Life Inventory (RELI-EE), designed to measure the influence of significant others and situations in generating one's religiosity; *(e)* the *extrinsic internal* scale of the Religious Life Inventory (RELI-EI), designed to measure the degree of "driveness" in one's religiosity; and *(f)* the *intrinsic* scale of the Religious Life Inventory (RELI-I), designed to measure the degree to which one's religiosity involves a questioning of the meaning of life arising out of one's interactions with his social environment. The order of presentation of the scales in the questionnaire was RELI, AR, D-O.

Consistent with prior research (Batson, 1971), a principal-component analysis of the total scale scores and individual items for the 67 seminarians produced a theoretically meaningful, orthogonally rotated three-component structure with the following loadings:

Religion as means received a single very high loading from AR-E (.903) and therefore was defined by Allport and Ross's (1967) conception of this scale as measuring religiosity as a means to other ends. This component also received moderate negative loadings from D-O ($-.400$) and AR-I ($-.372$) and a moderate positive loading from RELI-EE (.301).

Religion as end received high loadings from RELI-EI (.874), RELI-EE (.725), AR-I (.768), and D-O (.704). Given this configuration, and again following Allport and Ross's conceptualization, this component seemed to involve religiosity as an end in itself with some intrinsic value.

Religion as quest received a single very high loading from RELI-I (.945) and a moderate loading from RELI-EE (.75). Following Batson, this component was conceived to involve religiosity emerging out of an individual's search for meaning in his personal and social world.

The three religious personality scales examined in the experimental research were constructed through the use of complete-estimation factor score coefficients from these three components.

Scheduling of Experimental Study

Since the incident requiring a helping response was staged outdoors, the entire experimental study was run in 3 days, December 14–16, 1970, between 10 A.M. and 4 P.M. A tight schedule was used in an attempt to maintain reasonably consistent weather and light conditions. Temperature fluctuation according to the *New York Times* for the 3 days during these hours was not more than 5 degrees Fahrenheit. No rain or snow fell, although the third day was cloudy, whereas the first two were sunny. Within days the subjects were randomly assigned to experimental conditions.[1]

Procedure

When a subject appeared for the experiment, an assistant (who was blind with respect to the personality scores) asked him to read a brief statement which explained that he was participating in a study of the vocational careers of seminary students. After developing the rationale for the study, the statement read:

> What we have called you in for today is to provide us with some additional material which will give us a clearer picture of how you think than does the questionnaire material we have gathered thus far. Questionnaires are helpful, but tend to be somewhat oversimplified. Therefore, we would like to record a 3–5-minute talk you give based on the following passage. . . .

Variable 1: Message.   In the task-relevant condition the passage read,

> With increasing frequency the question is being asked: What jobs or professions do seminary students subsequently enjoy most, and in what jobs are they most effective? The answer to this question used to be so obvious that the question was not

---

[1]An error was made in randomizing that increased the number of subjects in the intermediate-hurry conditions. This worked against the prediction that was most highly confirmed (the hurry prediction) and made no difference to the message variable tests.

even asked. Seminary students were being trained for the ministry, and since both society at large and the seminary student himself had a relatively clear understanding of what made a "good" minister, there was no need even to raise the question of for what other jobs seminary experience seems to be an asset. Today, however, neither society nor many seminaries have a very clearly defined conception of what a "good" minister is or of what sorts of jobs and professions are the best context in which to minister. Many seminary students, apparently genuinely concerned with "ministering," seem to feel that it is impossible to minister in the professional clergy. Other students, no less concerned, find the clergy the most viable profession for ministry. But are there other jobs and/or professions for which seminary experience is an asset? And, indeed, how much of an asset is it for the professional ministry? Or, even more broadly, can one minister through an "establishment" job at all?

In the helping-relevant condition, the subject was given the parable of the Good Samaritan exactly as printed earlier in this article. Next, regardless of condition, all subjects were told,

> You can say whatever you wish based on the passage. Because we are interested in how you think on your feet, you will not be allowed to use notes in giving the talk. Do you understand what you are to do? If not, the assistant will be glad to answer questions.

After a few minutes the assistant returned, asked if there were any questions, and then said:

> Since they're rather tight on space in this building, we're using a free office in the building next door for recording the talks. Let me show you how to get there [draws and explains map on 3×5 card]. This is where Professor Steiner's laboratory is. If you go in this door [points at map], there's a secretary right here, and she'll direct you to the office we're using for recording. Another of Professor Steiner's assistants will set you up for recording your talk. Is the map clear?

Variable 2: Hurry. In the high-hurry condition the assistant then looked at his watch and said, "Oh, you're late. They were expecting you a few minutes ago. We'd better get moving. The assistant should be waiting for you so you'd better hurry. It shouldn't take but just a minute." In the intermediate-hurry condition he said, "The assistant is ready for you, so please go right over." In the low-hurry condition he said, "It'll be a few minutes before they're ready for you, but you might as well head on over. If you have to wait over there, it shouldn't be long."

The incident. When the subject passed through the alley, the victim was sitting slumped in a doorway, head down, eyes closed, not moving. As the subject went by, the victim coughed twice and groaned, keeping his head down. If the subject stopped and asked if something was wrong or offered to help, the victim, startled and somewhat groggy, said, "Oh, thank you [cough]. . . . No, it's all right. [Pause] I've got this respiratory condition [cough]. . . . The doctor's given me these pills to take, and I just took one. . . . If I just sit and rest for a few minutes I'll be O.K. . . . . Thanks very much for stopping though [smiles weakly]." If the subject persisted, insisting on taking the victim inside the building, the victim allowed him to do so and thanked him.

Helping ratings.   The victim rates each subject on a scale of helping behavior as follows:

> 0 = failed to notice the victim as possibly in need at all; 1 = perceived the victim as possibly in need but did not offer aid; 2 = did not stop but helped indirectly (e.g., by telling Steiner's assistant about the victim); 3 = stopped and asked if victim needed help; 4 = after stopping, insisted on taking the victim inside and then left him.

The victim was blind to the personality scale scores and experimental conditions of all subjects. At the suggestion of the victim, another category was added to the rating scales, based on his observations of pilot subjects' behavior:

> 5 = after stopping, refused to leave the victim (after 3–5 minutes) and/or insisted on taking him somewhere outside experimental context (e.g., for coffee or to the infirmary).

(In some cases it was necessary to distinguish Category 0 from Category 1 by the postexperimental questionnaire and Category 2 from Category 1 on the report of the experimental assistant.)

This 6-point scale of helping behavior and a description of the victim were given to a panel of 10 judges (unacquainted with the research) who were asked to rank order the (unnumbered) categories in terms of "the amount of helping behavior displayed toward the person in the doorway." Of the 10, 1 judge reversed the order of Categories 0 and 1. Otherwise there was complete agreement with the ranking implied in the presentation of the scale above.

The speech.   After passing through the alley and entering the door marked on the map, the subject entered a secretary's office. She introduced him to the assistant who gave the subject time to prepare and privately record his talk.

Helping behavior questionnaire.   After recording the talk, the subject was sent to another experimenter, who administered "an exploratory questionnaire on personal and social ethics." The questionnaire contained several initial questions about the interrelationship between social and personal ethics, and then asked three key questions: *(a)* "When was the last time you saw a person who seemed to be in need of help?" *(b)* "When was the last time you stopped to help someone in need?" *(c)* "Have you had experience helping persons in need? If so, outline briefly." These data were collected as a check on the victim's ratings of whether subjects who did not stop perceived the situation in the alley as one possibly involving need or not.

When he returned, the experimenter reviewed the subject's questionnaire, and, if no mention was made of the situation in the alley, probed for reactions to it and then phased into an elaborate debriefing and discussion session.

Debriefing

In the debriefing, the subject was told the exact nature of the study, including the deception involved, and the reasons for the deception were explained. The subject's reactions to the victim and to the study in general were discussed. The role of situational determinants of helping behavior was explained in relation to this particular incident and to other experiences of the subject. All subjects seemed readily to understand the necessity for the deception, and none indicated any resentment of it. After debriefing, the subject was thanked for his time and paid, then he left.

RESULTS AND DISCUSSION

Overall Helping Behavior

The average amount of help that a subject offered the victim, by condition, is shown in Table 4.1. The unequal-N analysis of variance indicates that while the hurry variable was significantly ($F = 3.56$, $df = 2/34$, $p < .05$) related to helping behavior, the message variable was not. Subjects in a hurry were likely to offer less help than were subjects not in a hurry. Whether the subject was going to give a speech on the parable of the Good Samaritan or not did not significantly affect his helping behavior on this analysis.

TABLE 4.1
Means and analysis of variance of graded helping responses

	M			
		Hurry		
Message	Low	Medium	High	Summary
Helping relevant	3.800	2.000	1.000	2.263
Task relevant	1.667	1.667	.500	1.333
Summary	3.000	1.818	.700	

Analysis of variance				
Source	SS	df	MS	F
Message (A)	7.766	1	7.766	2.65
Hurry (B)	20.884	2	10.442	3.56*
A × B	5.237	2	2.619	.89
Error	99.633	34	2.930	

Note: $N = 40$.
*$p < .05$.

TABLE 4.2
Stepwise multiple regression analysis

	Help vs. no help				Graded helping				
	Individual variable		Overall equation			Individual variable		Variable equation	
Step	r^a	F	R	F	Step	r	F	R	F
1. Hurry[b]	−.37	4.537*	.37	5.884*	1. Hurry	−.42	6.665*	.42	8.196**
2. Message[c]	.25	1.495	.41	3.834*	2. Message	.25	1.719	.46	5.083*
3. Religion as quest	−.03	.081	.42	2.521	3. Religion as quest	−.16	1.297	.50	3.897*
4. Religion as means	−.03	.003	.42	1.838*	4. Religion as means	−.08	.018	.50	2.848*
5. Religion as end	.06	.000	.42	1.430	5. Religion as end	−.07	.001	.50	2.213

Note: N=40. Helping is the dependent variable. $df=1/34$.
[a]Individual variable correlation coefficient is a point biserial where appropriate.
[b]Variables are listed in order of entry into stepwise regression equations.
[c]Helping-relevant message is positive.
* $p<.05$.
** $p<.01$.

Other studies have focused on the question of whether a person initiates helping action or not, rather than on scaled kinds of helping. The data from the present study can also be analyzed on the following terms: Of the 40 subjects, 16 (40%) offered some form of direct or indirect aid to the victim (Coding Categories 2–5), 24 (60%) did not (Coding Categories 0 and 1). The percentages of subjects who offered aid by situational variable were, for low hurry, 63% offered help, intermediate hurry 45%, and high hurry 10%; for helping-relevant message 53%, task-relevant message 29%. With regard to this more general question of whether help was offered or not, an unequal-N analysis of variance (arc sine transformation of percentages of helpers, with low- and intermediate-hurry conditions pooled) indicated that again only the hurry main effect was significantly ($F = 5.22$, $p < .05$) related to helping behavior; the subjects in a hurry were more likely to pass by the victim than were those in less of a hurry.

Reviewing the predictions in the light of these results, the second hypothesis, that the degree of hurry a person is in determines his helping behavior, was supported. The prediction involved in the first hypothesis concerning the message content was based on the parable. The parable itself seemed to suggest that thinking pious thoughts would not increase helping. Another and conflicting prediction might be produced by a norm salience theory. Thinking about the parable should make norms for helping salient and therefore produce more helping. The data, as hypothesized, are more congruent with the prediction drawn from the parable. A person going to speak on the parable of the Good Samaritan is not significantly more likely to stop to help a person by the side of the road than is a person going to talk about possible occupations for seminary graduates.

Since both situational hypotheses are confirmed, it is tempting to stop the analysis of these variables at this point. However, multiple regression analysis procedures were also used to analyze the relationship of all of the independent variables of the study and the helping behavior. In addition to often being more statistically powerful due to the use of more data information, multiple regression analysis has an advantage over analysis of variance in that it allows for a comparison of the relative effect of the various independent variables in accounting for variance in the dependent variable. Also, multiple regression analysis can compare the effects of continuous as well as nominal independent variables on both continuous and nominal dependent variables (through the use of point biserial correlations, r_{pb}) and shows considerable robustness to violation of normality assumptions (Cohen, 1965, 1968). Table 4.2 reports the results of the multiple regression analysis using both help versus no help and the graded helping scale as dependent measures. In this table the overall equation Fs show the F value of the entire regression equation as a particular row variable enters the equation. Individual variable Fs were computed with all five independent variables in the equation. Although the two situational variables, hurry and message condition, correlated more highly with the dependent measure than any of the religious dispositional variables, only hurry was a significant predictor of whether one will help or not (column 1) or of the overall amount of help given (column 2). These results corroborate the findings of the analysis of variance.[2]

Notice also that neither form of the third hypothesis, that types of religiosity will predict helping, received support from these data. No correlation between the various measures of religiosity and any form of the dependent measure ever came near statistical significance, even though the multiple regression analysis procedure is a powerful and not particularly conservative statistical test.

Personality Difference among Subjects Who Helped

To further investigate the possible influence of personality variables, analyses were carried out using only the data from subjects who offered some kind of help to the victim. Surprisingly (since the number of these subjects was small, only 16) when this was done, one religiosity variable seemed to be significantly related to the kind of helping behavior offered. (The situational variables had no significant effect.) Subjects high on the religion as quest dimension appear likely, when they stop for the victim, to offer help of a more tentative or incomplete nature than are subjects scoring low on this dimension ($r = -.53$, $p < .05$).

This result seemed unsettling for the thinking behind either form of Hypothesis 3. Not only do the data suggest that the Allport-Ross–based conception of religion as *end* does not predict the degree of helping, but the religion as quest component is a significant predictor of offering less help. This latter result seems counterintuitive and out of keeping with previous research (Batson, 1971), which found that this type of religiosity correlated positively with other socially valued characteristics. Further data analysis, however, seemed to suggest a different interpretation of this result.

It will be remembered that one helping coding category was added at the suggestion of the victim after his observation of pilot subjects. The correlation of religious personality variables with helping behavior dichotomized between the added category (1) and all of the others (0) was examined. The correlation between religion as quest and this dichotomous helping scale was essentially unchanged ($r_{pb} = -.54$, $p < .05$). Thus, the previously found correlation between the helping scale and religion as quest seems to reflect the tendency of those who score low on the quest dimension to offer help in the added helping category.

What does help in this added category represent? Within the context of the experiment, it represented an embarrassment. The victim's response to persistent offers of help was to assure the helper he was all right, had taken his medicine, just needed to rest for a minute or so, and, if ultimately necessary, to request the helper

---

[2] To check the legitimacy of the use of both analysis of variance and multiple regression analysis, parametric analyses, on this ordinal data, Kendall rank correlation coefficients were calculated between the helping scale and the five indepedent variables. As expected τ approximated the correlation quite closely in each case and was significant for hurry only (hurry, $\tau = -.38$, $p < .001$).

to leave. But the *super* helpers in this added category often would not leave until the final appeal was repeated several times by the victim (who was growing increasingly panicky at the possibility of the arrival of the next subject). Since it usually involved the subject's attemping to carry through a preset plan (e.g., taking the subject for a cup of coffee or revealing to him the strength to be found in Christ), and did not allow information from the victim to change that plan, we originally labeled this kind of helping as rigid—an interpretation supported by its increased likelihood among highly doctrinal orthodox subjects ($r = .63$, $p < .01$). It also seemed to have an inappropriate character. If this more extreme form of helping behavior is indeed effectively less helpful, then the second form of Hypothesis 3 does seem to gain support.

But perhaps it is the experimenters rather than the super helpers who are doing the inappropriate thing; perhaps the best characterization of this kind of helping is as different rather than as inappropriate. This kind of helper seems quickly to place a particular interpretation on the situation, and the helping response seems to follow naturally from this interpretation. All that can safely be said is that one style of helping that emerged in this experiment was directed toward the presumed underlying needs of the victim and was little modified by the victim's comments about his own needs. In contrast, another style was more tentative and seemed more responsive to the victim's statements of his need.

The former kind of helping was likely to be displayed by subjects who expressed strong doctrinal orthodoxy. Conversely, this fixed kind of helping was unlikely among subjects high on the religion as quest dimension. These latter subjects, who conceived their religion as involving an ongoing search for meaning in their personal and social world, seemed more responsive to the victim's immediate needs and more open to the victim's definitions of his own needs.

CONCLUSION AND IMPLICATIONS

A person not in a hurry may stop and offer help to a person in distress. A person in a hurry is likely to keep going. Ironically, he is likely to keep going even if he is hurrying to speak on the parable of the Good Samaritan, thus inadvertently confirming the point of the parable. (Indeed, on several occasions, a seminary student going to give his talk on the parable of the Good Samaritan literally stepped over the victim as he hurried on his way!)

Although the degree to which a person was in a hurry had a clearly significant effect on his likelihood of offering the victim help, whether he was going to give a sermon on the parable or on possible vocational roles of ministers did not. This lack of effect of sermon topic raises certain difficulties for an explanation of helping behavior involving helping norms and their salience. It is hard to think of a context in which norms concerning helping those in distress are more salient than for a person thinking about the Good Samaritan, and yet it did not significantly increase helping behavior. The results were in the direction suggested by the norm salience hypothesis, but they were not significant. The most accurate conclusion seems to be that salience of helping norms is a less strong determinant of helping behavior in the present situation than many, including the present authors, would expect.

Thinking about the Good Samaritan did not increase helping behavior, but being in a hurry decreased it. It is difficult not to conclude from this that the frequently cited explanation that ethics becomes a luxury as the speed of our daily lives increases is at least an accurate description. The picture that this explanation conveys is of a person seeing another, consciously noting his distress, and consciously choosing to leave him in distress. But perhaps this is not entirely accurate, for, when a person is in a hurry, something seems to happen that is akin to Tolman's (1948) concept of the "narrowing of the cognitive map." Our seminarians in a hurry noticed the victim in that in the postexperiment interview almost all mentioned him as, on reflection, possibly in need of help. But it seems that they often had not worked this out when they were near the victim. Either the interpretation of their visual picture as a person in distress or the empathic reactions usually associated with that interpretation had been deferred because they were hurrying. According to the reflections of some of the subjects, it would be inaccurate to say that they realized the victim's possible distress, then chose to ignore it; instead, because of the time pressures, they did not perceive the scene in the alley as an occasion for an ethical decision.

For other subjects it seems more accurate to conclude that they decided not to stop. They appeared aroused and anxious after the encounter in the alley. For these subjects, what were the elements of the choice that they were making? Why were the seminarians hurrying? Because the experimenter, *whom the subject was helping,* was depending on him to get to a particular place quickly. In other words, he was in conflict between stopping to help the victim and continuing on his way to help the experimenter. And this is often true of people in a hurry; they hurry because somebody depends on their being somewhere. Conflict, rather than callousness, can explain their failure to stop.

Finally, as in other studies, personality variables were not useful in predicting whether a person helped or not. But in this study, unlike many previous ones, considerable variations were possible in the kinds of help given, and these variations did relate to personality measures—specifically to religiosity of the quest sort. The clear light of hindsight suggests that the dimension of kinds of helping would have been the appropriate place to look for personality differences all along; *whether* a person helps or not is an instant decision likely to be situationally controlled. How a person helps involves a more complex and considered number of decisions, including the time and scope to permit personality characteristics to shape them.

References

ALLEN, R. O., AND SPILKA, B. Committed and consensual religion. A specification of religion-prejudice relationships. *Journal for the Scientific Study of Religion,* 1967, **6**, 191–206.

ALLPORT, G. W. AND ROSS, J. M. Personal religious orientation and prejudice. *Journal of Personality and Social Psychology,* 1967, **5**, 432–443.

BATSON, C. D. Creativity and religious development: Toward a structural-functional psychology of religion. Unpublished doctoral dissertation, Princeton Theological Seminary, 1971.

BICKMAN, L. B. The effect of the presence of others on bystander intervention in an emergency. Unpublished doctoral dissertation, City College of the City University of New York, 1969.

BROWN, L. B. Classifications of religious orientation. *Journal for the Scientific Study of Religion,* 1964, **4**, 91–99.

BURTON, R. V. The generality of honesty reconsidered. *Psychological Review,* 1963, **70**, 481–499.

COHEN, J. Multiple regression as a general data-analytic system. *Psychological Bulletin,* 1968, **70**, 426–443.

COHEN, J. Some statistical issues in psychological research. In B. B. Wolman (Ed.), *Handbook of clinical psychology.* New York: McGraw-Hill, 1965.

DARLEY, J. M., AND LATANÉ, B. Bystander intervention in emergencies: Diffusion of responsibility. *Journal of Personality and Social Psychology,* 1968, **8**, 377–383.

FREUD, S. *The future of an illusion.* New York: Liveright, 1953.

FUNK, R. W. *Language, hermeneutic, and word of God.* New York: Harper & Row, 1966.

GLOCK, C. Y., AND STARK, R. *Christian beliefs and anti-Semitism.* New York: Harper & Row, 1966.

HARTSHORNE, H., AND MAY, M. A. *Studies in the nature of character.* Vol. 1. *Studies in deceit.* New York: Macmillan, 1928.

KORTE, C. Group effects on help-giving in an emergency. *Proceedings of the 77th Annual Convention of the American Psychological Association,* 1969, **4**, 383–384. (Summary)

ORNE, M. T. On the social psychology of the psychological experiment: With particular reference to demand characteristics and their implications. *American Psychologist,* 1962, **17**, 776–783.

ROSENBERG, M. J. When dissonance fails: On eliminating evaluation apprehension from attitude measurement. *Journal of Personality and Social Psychology,* 1965, **1**, 28–42.

SCHWARTZ, S. H., AND CLAUSEN, G. T. Responsibility, norms, and helping in an emergency. *Journal of Personality and Social Psychology,* 1970, **16**, 299–310.

TOLMAN, E. C. Cognitive maps in rats and men. *Psychological Review,* 1948, **55**, 189–208.

5

A Study of Prisoners and Guards
in a Simulated Prison

Craig Haney, Curtis Banks, and Philip Zimbardo

After he had spent four years in a Siberian prison the great Russian novelist Dostoevsky commented surprisingly that his time in prison had created in him a deep
optimism about the ultimate future of mankind because, as he put it, if man could
survive the horrors of prison life he must surely be a "creature who could with-
stand anything." The cruel irony which Dostoevsky overlooked is that the reality
of prison bears witness not only to the resiliency and adaptiveness of the men who
tolerate life within its walls, but as well to the "ingenuity" and tenacity of those
who devised and still maintain our correctional and reformatory systems.

Nevertheless, in the century which has passed since Dostoevsky's imprisonment, little has changed to render the main thrust of his statement less relevant.
Although we have passed through periods of enlightened humanitarian reform, in
which physical conditions within prison have improved somewhat, and the rheto-
ric of rehabilitation has replaced the language of punitive incarceration, the social
institution of prison has continued to fail. On purely pragmatic grounds, there is
substantial evidence that prisons really neither "rehabilitate" nor act as a deterrent

Reprinted with permission from the authors and *Naval Research Reviews*, Sept. 1973, Dept. of the
Navy.

to future crime—in America, recidivism rates upwards of 75 percent speak quite decisively to these criteria. And, to perpetuate what is also an economic failure, American taxpayers alone must provide an expenditure for "corrections" of 1.5 billion dollars annually. On humanitarian grounds as well, prisons have failed: our mass media are increasingly filled with accounts of atrocities committed daily, man against man, in reaction to the penal system or in the name of it. The experience of prison creates undeniably, almost to the point of cliche, an intense hatred and disrespect in most inmates for the authority and the established order of society into which they will eventually return. And the toll it takes in the deterioration of human spirit for those who must administer it, as well as for those upon whom it is inflicted, is incalculable.

Attempts to provide an explanation of the deplorable condition of our penal system and its dehumanizing effects upon prisoners and guards, often focus upon what might be called the *dispositional hypothesis*. While this explanation is rarely expressed explicitly, it is central to a prevalent nonconscious ideology: that the state of the social institution of prison is due to the "nature" of the people who administrate it, or the "nature" of the people who populate it, or both. That is, a major contributing cause to despicable conditions, violence, brutality, dehumanization and degradation existing within any prison can be traced to some innate or acquired characteristic of the correctional and inmate population. Thus on the one hand, there is the contention that violence and brutality exist within prison because guards are sadistic, uneducated, and insensitive people. It is the "guard mentality," a unique syndrome of negative traits which they bring into the situation, that engenders the inhumane treatment of prisoners. On the other hand, there is the argument that prison violence and brutality are the logical and predictable results of the involuntary confinement of a collective of individuals whose life histories are, by definition, characterized by disregard for law, order and social convention and a concurrent propensity for impulsivity and aggression. In seeming logic, it follows that these individuals, having proven themselves incapable of functioning satisfactorily within the "normal" structure of society, cannot do so either inside the structure provided by prisons. To control such men, the argument continues, whose basic orientation to any conflict situation is to react with physical power or deception, force must be met with force, and a certain number of violent encounters must be expected and tolerated by the public.

The dispositional hypothesis has been embraced by the proponents of the prison *status quo* (blaming conditions on the evil in the prisoners), as well as by its critics (attributing the evil to guards and staff with their evil motives and deficient personality structures). The appealing simplicity of this proposition localizes the source of prison riots, recidivism and corruption in these "bad seeds" and not in the conditions of the "prison soil." Such an analysis directs attention away from the complex matrix of social, economic and political forces that combine to make prisons what they are—and what would require complex, expensive, revolutionary actions to bring about any meaningful change. Instead, rioting prisoners are identified, punished, transferred to maximum security institutions or shot, outside agitators sought, and corrupt officials suspended—while the system itself goes on essentially unchanged, its basic structure unexamined and unchallenged.

However, the dispositional hypothesis cannot be critically evaluated directly through observation in existing prison settings, because such naturalistic observation necessarily confounds the acute effects of the environment with the chronic characteristics of the inmate and guard populations. To separate the effects of the prison environment *per se* from those attributable to *a priori* dispositions of its inhabitants requires a research strategy in which a "new" prison is constructed, comparable in its fundamental social psychological milieu to existing prison systems, but entirely populated by individuals who are undifferentiated in all essential dimensions from the rest of society.

Such was the approach taken in the present empirical study, namely, to create a prison-like situation in which the guards and inmates were initially comparable and characterized as being "normal-average," and then to observe the patterns of behavior which resulted, as well as the cognitive, emotional and attitudinal reactions which emerged. Thus, we began our experiment with a sample of individuals who were in the normal range of the general population on a variety of dimensions we were able to measure. Half were randomly assigned to the role of "prisoner," the others to that of "guard," neither group having any history of crime, emotional disability, physical handicap or even intellectual or social disadvantage.

The environment created was that of a "mock" prison which physically constrained the prisoners in barred cells and psychologically conveyed the sense of imprisonment to all participants. Our intention was not to create a *literal* simulation of an American prison, but rather a functional representation of one. For ethical, moral and pragmatic reasons we could not exercise the threat and promise of severe physical punishment, we could not allow homosexual or racist practices to flourish, nor could we duplicate certain other specific aspects of prison life. Nevertheless, we believed that we could create a situation with sufficient mundane realism to allow the role-playing participants to go beyond the superficial demands of their assignment into the deep structure of the characters they represented. To do so, we established functional equivalents for the activities and experiences of actual prison life which were expected to produce qualitatively similar psychological reactions in our subjects—feelings of power and powerlessness, of control and oppression, of satisfaction and frustration, of arbitrary rule and resistance to authority, of status and anonymity, of machismo and emasculation. In the conventional terminology of experimental social psychology, we first identified a number of relevant conceptual variables through analysis of existing prison situations, then designed a setting in which these variables were operationalized. No specific hypotheses were advanced other than the general one that assignment to the treatment of "guard" or "prisoner" would result in significantly different reactions on behavioral measures of interaction, emotional measures of mood state and pathology, attitudes toward self, as well as other indices of coping and adaptation to this novel situation. What follows is a discussion of how we created and peopled our prison, what we observed, what our subjects reported, and finally, what we can conclude about the nature of the prison environment and the psychology of imprisonment which can account for the failure of our prisons.

METHOD

Overview

The effects of playing the role of "guard" or "prisoner" were studied in the context of an experimental simulation of a prison environment. The research design was a relatively simple one, involving as it did only a single treatment variable, the random assignment to either a "guard" or "prisoner" condition. These roles were enacted over an extended period of time (nearly one week) within an environment that was physically constructed to resemble a prison. Central to the methodology of creating and maintaining a psychological state of imprisonment was the functional simulation of significant properties of "real prison life" (established through information from former inmates, correctional personnel and texts).

The "guards" were free within certain limits to implement the procedures of induction into the prison setting and maintenance of custodial retention of the "prisoners." These inmates, having voluntarily submitted to the conditions of this total institution in which they now lived, coped in various ways with its stresses and its challenges. The behavior of both groups of subjects was observed, recorded, and analyzed. The dependent measures were of two general types: (1) transactions between and within each group of subjects, recorded on video and audio tape as well as directly observed; (2) individual reactions on questionnaires, mood inventories, personality tests, daily guard shift reports, and post-experimental interviews.

Subjects

The 22 subjects who participated in the experiment were selected from an initial pool of 75 respondents, who answered a newspaper ad asking for male volunteers to participate in a psychological study of "prison life" in return for payment of $15 per day. Each respondent completed an extensive questionnaire concerning his family background, physical and mental health history, prior experience and attitudinal propensities with respect to sources of psychopathology (including their involvements in crime). Each respondent also was interviewed by one of two experimenters. Finally, the 24 subjects who were judged to be most stable (physically and mentally), most mature, and least involved in anti-social behaviors were selected to participate in the study. On a random basis, half of the subjects were assigned the role of "guard," half were assigned the role of "prisoner."

The subjects were normal, healthy, male college students who were in the Stanford area during the summer. They were largely of middle class socioeconomic status and Caucasians (with the exception of one Oriental subject). Initially they were strangers to each other, a selection precaution taken to avoid the disruption of any pre-existing friendship patterns and to mitigate any transfer into the experimental situation of previously established relationships or patterns of behavior.

This final sample of subjects was administered a battery of psychological tests on the day prior to the start of the simulation, but to avoid any selective bias on the part of the experimenter-observers, scores were not tabulated until the study was completed.

Two subjects who were assigned to be a "stand-by" in case an additional "prisoner" was needed were not called, and one assigned to be a "stand-by" guard decided against participating just before the simulation phase began—thus, our data analysis is based upon ten prisoners and eleven guards in our experimental conditions.

PROCEDURE

Physical Aspects of the Prison

The prison was built in a 35-foot section of a basement corridor in the psychology building at Stanford University. It was partitioned by two fabricated walls; one was fitted with the only entrance door to the cell block and the other contained a small observation screen. Three small cells (6 × 9 ft.) were made from converted laboratory rooms by replacing the usual doors with steel barred, black painted ones, and removing all furniture.

A cot (with mattress, sheet and pillow) for each prisoner was the only furniture in the cells. A small closet across from the cells served as a solitary confinement facility; its dimensions were extremely small (2 × 2 × 7 ft.), and it was unlighted.

In addition, several rooms in an adjacent wing of the building were used as guards' quarters (to change in and out of uniform or for rest and relaxation), a bedroom for the "warden" and "superintendent," and an interview-testing room. Behind the observation screen at one end of the "yard" (small enclosed room representing the fenced prison grounds) was video recording equipment and sufficient space for several observers.

Operational Details

The "prisoner" subjects remained in the mock-prison 24 hours per day for the duration of the study. Three were arbitrarily assigned to each of the three cells; the others were on stand-by call at their homes. The "guard" subjects worked on three-man, eight-hour shifts; remaining in the prison environment only during their work shift and going about their usual lives at other times.

Role Instructions

All subjects had been told that they would be assigned either the guard or the prisoner role on a completely random basis and all had voluntarily agreed to play either role for $15.00 per day for up to two weeks. They signed a contract guaranteeing a minimally adequate diet, clothing, housing and medical care as well as the financial remuneration in return for their stated "intention" of serving in the assigned role for the duration of the study.

It was made explicit in the contract that those assigned to be prisoners should expect to be under surveillance (have little or no privacy) and to have some of their basic civil rights suspended during their imprisonment, excluding physical abuse. They were given no other information about what to expect nor instructions about behavior appropriate for a prisoner role. Those actually assigned to this treatment were informed by phone to be available at their place of residence on a given Sunday when we would start the experiment.

The subjects assigned to be guards attended an orientation meeting on the day prior to the induction of the prisoners. At this time they were introduced to the principal investigators, the "Superintendent" of the prison (the author) and an undergraduate research assistant who assumed the administrative role of "Warden." They were told that we wanted to try to simulate a prison environment within the limits imposed by pragmatic and ethical considerations. Their assigned task was to "maintain the reasonable degree of order within the prison necessary for its effective functioning," although the specifics of how this duty might be implemented were not explicitly detailed. They were made aware of the fact that, while many of the contingencies with which they might be confronted were essentially unpredictable (e.g., prisoner escape attempts), part of their task was to be prepared for such eventualities and to be able to deal appropriately with the variety of situations that might arise. The "Warden" instructed the guards in the administrative details, including: the work-shifts, the mandatory daily completion of "critical incident" reports which detailed unusual occurrences, and the administration of meals, work and recreation programs for the prisoners. In order to begin to involve these subjects in their roles even before the first prisoner was incarcerated, the guards assisted in the final phases of completing the prison complex— putting the cots in the cells, signs on the walls, setting up the guards' quarters, moving furniture, water coolers, refrigerators, etc.

The guards generally believed that we were primarily interested in studying the behavior of the prisoners. Of course, we were as interested in the effects which enacting the role of guard in this environment would have on their behavior and subjective states.

To optimize the extent to which their behavior would reflect their genuine reactions to the experimental prison situation and not simply their ability to follow instructions, they were intentionally given only minimal guidelines for what it meant to be a guard. An explicit and categorical prohibition against the use of physical punishment or physical aggression was, however, emphasized by the experimen-

ters. Thus, with this single notable exception, their roles were relatively unstructured initially, requiring each "guard" to carry out activities necessary for interacting with a group of "prisoners" as well as with other "guards" and the
"correctional staff."

Uniforms

In order to promote feelings of anonymity in the subjects each group was issued
identical uniforms. For the guards, the uniform consisted of plain khaki shirts and
trousers, a whistle, a police night-stick (wooden baton), and reflecting sunglasses
which made eye contact impossible. The prisoners' uniform consisted of a loose
fitting muslin smock with an identification number on front and back, no underclothes, a light chain and lock around one ankle, rubber sandals and a cap made
from a nylon stocking. Each prisoner also was issued a toothbrush, soap, soap-
dish, towel and bed linen. No personal belongings were allowed in the cells.

The outfitting of both prisoners and guards in this manner served to enhance
group identity and reduce individual uniqueness within the two groups. The khaki
uniforms were intended to convey a military attitude, while the whistle and night-
stick were carried as symbols of control and power. The prisoners' uniforms were
designed not only to deindividuate the prisoners but to be humiliating and serve as
symbols of their dependence and subservience. The ankle chain was a constant
reminder (even during their sleep when it hit the other ankle) of the oppressiveness
of the environment. The stocking cap removed any distinctiveness associated with
hair length, color or style (as does shaving of heads in some "real" prisons and
the military). The ill-fitting uniforms made the prisoners feel awkward in their
movements; since these "dresses" were worn without undergarments, the uniforms forced them to assume unfamiliar postures, more like those of a woman
than a man—another part of the emasculating process of becoming a prisoner.

Induction Procedure

With the cooperation of the Palo Alto City Police Department all of the subjects assigned to the prisoner treatment were unexpectedly "arrested" at their res-
idences. A police officer charged them with suspicion of burglary or armed robbery, advised them of their legal rights, handcuffed them, thoroughly searched them
(often as curious neighbors looked on) and carried them off to the police station
in the rear of the police car. At the station they went through the standard routines
of being fingerprinted, having an identification file prepared and then being placed
in a detention cell. Each prisoner was blindfolded and subsequently driven by one
of the experimenters and a subject-guard to our mock prison. Throughout the entire arrest procedure, the police officers involved maintained a formal, serious at-

titude, avoiding answering any questions of clarification as to the relation of this "arrest" to the mock prison study.

Upon arrival at our experimental prison, each prisoner was stripped, sprayed with a delousing preparation (a deodorant spray) and made to stand alone naked for a while in the cell yard. After being given the uniform described previously and having an I.D. picture taken ("mug shot"), the prisoner was put in his cell and ordered to remain silent.

Administrative Routine

When all the cells were occupied, the warden greeted the prisoners and read them the rules of the institution (developed by the guards and the warden). They were to be memorized and to be followed. Prisoners were to be referred to only by the number on their unfiroms, also in an effort to depersonalize them.

The prisoners were to be served three bland meals per day, were allowed three supervised toilet visits, and given two hours daily for the privilege of reading or letterwriting. Work assignments were issued for which the prisoners were to receive an hourly wage to constitute their $15 daily payment. Two visiting periods per week were scheduled, as were movie rights and exercise periods. Three times a day prisoners were lined up for a "count" (one on each guard work-shift). The initial purpose of the "count" was to ascertain that all prisoners were present, and to test them on their knowledge of the rules and their I.D. numbers. The first perfunctory counts lasted only about ten minutes, but on each successive day (or night) they were spontaneously increased in duration until some lasted several hours. Many of the preestablished features of administrative routine were modified or abandoned by the guards, and some privileges were forgotten by the staff over the course of the study.

RESULTS

Overview

Although it is difficult to anticipate exactly what the influence of incarceration will be upon the individuals who are subjected to it and those charged with its maintenance, especially in a simulated reproduction, the results of the present experiment support many commonly held conceptions of prison life and validate anecdotal evidence supplied by articulate ex-convicts. The environment of arbitrary custody had great impact upon the affective states of both guards and prisoners as well as upon the interpersonal processes taking place between and within those role-groups.

In general, guards and prisoners showed a marked tendency toward increased negativity of affect, and their overall outlook became increasingly negative. As the experiment progressed, prisoners expressed intentions to do harm to others more frequently. For both prisoners and guards, self-evaluations were more deprecating as the experience of the prison environment became internalized.

Overt behavior was generally consistent with the subjective self-reports and affective expressions of the subjects. Despite the fact that guards and prisoners were essentially free to engage in any form of interaction (positive or negative, supportive or affrontive, etc.), the characteristic nature of their encounters tended to be negative, hostile, affrontive and dehumanizing. Prisoners immediately adopted a generally passive response mode while guards assumed a very active initiative role in all interactions. Throughout the experiment, commands were the most frequent form of verbal behavior and, generally, verbal exchanges were strikingly impersonal, with few references to individual identity. Although it was clear to all subjects that the experimenters would not permit physical violence to take place, varieties of less direct aggressive behavior were observed frequently (especially on the part of guards). In lieu of physical violence, verbal affronts were used as one of the most frequent forms of interpersonal contact between guards and prisoners.

The most dramatic evidence of the impact of this situation upon the participants was seen in the gross reactions of five prisoners who had to be released because of extreme emotional depression, crying, rage and acute anxiety. The pattern of symptoms was quite similar in four of the subjects and began as early as the second day of imprisonment. The fifth subject was released after being treated for a psychosomatic rash which covered portions of his body. Of the remaining prisoners, only two said they were not willing to forfeit the money they had earned in return for being "paroled." When the experiment was terminated prematurely after only six days, all the remaining prisoners were delighted by their unexpected good fortune. In contrast, most of the guards seemed to be distressed by the decision to stop the experiment and it appeared to us that they had become sufficiently involved in their roles that they now enjoyed the extreme control and power which they exercised and were reluctant to give it up. One guard did report being personally upset at the suffering of the prisoners, and claimed to have considered asking to change his role to become one of them—but never did so. None of the guards ever failed to come to work on time for their shift, and indeed, on several occasions guards remained on duty voluntarily and uncomplaining for extra hours— without additional pay.

The extremely pathological reactions which emerged in both groups of subjects testify to the power of the social forces operating, but still there were individual differences seen in styles of coping with this novel experience and in degrees of successful adaptation to it. Half the prisoners did endure the oppressive atmosphere, and not all the guards resorted to hostility. Some guards were tough but fair ("played by the rules"), some went far beyond their roles to engage in creative cruelty and harassment, while a few were passive and rarely instigated any coercive control over the prisoners.

REALITY OF THE SIMULATION

At this point it seems necessary to confront the critical question of "reality" in the simulated prison environment: were the behaviors observed more than the mere acting out of assigned roles convincingly? To be sure, ethical, legal and practical considerations set limits upon the degree to which this situation could approach the conditions existing in actual prisons and penitentiaries. Necessarily absent were some of the most salient aspects of prison life reported by criminologists and documented in the writing of prisoners. There was no involuntary homosexuality, no racism, no physical beatings, no threat to life by prisoners against each other or the guards. Moreover, the maximum anticipated "sentence" was only two weeks and, unlike some prison systems, could not be extended indefinitely for infractions of the internal operating rules of the prison.

In one sense, the profound psychological effects we observed under the relatively minimal prison-like conditions which existed in our mock prison made the results even more significant, and force us to wonder about the devastating impact of chronic incarceration in real prisons. Nevertheless, we must contend with the criticism that our conditions were too minimal to provide a meaningful analogue to existing prisons. It is necessary to demonstrate that the participants in this experiment transcended the conscious limits of their preconceived stereotyped roles and their awareness of the artificiality and limited duration of imprisonment. We feel there is abundant evidence that virtually all of the subjects at one time or another experienced reactions which went well beyond the surface demands of role-playing and penetrated the deep structure of the psychology of imprisonment.

Although instructions about how to behave in the roles of guard or prisoner were not explicitly defined, demand characteristics in the experiment obviously exerted some directing influence. Therefore, it is enlightening to look to circumstances where role demands were minimal, where the subjects believed they were not being observed, or where they should not have been behaving under the constraints imposed by their roles (as in "private" situations), in order to assess whether the role behaviors reflected anything more than public conformity or good acting.

When the private conversations of the prisoners were monitored, we learned that almost all (a full 90 percent) of what they talked about was directly related to immediate prison conditions, that is, food, privileges, punishment, guard harassment, etc. Only one-tenth of the time did their conversations deal with their life outside the prison. Consequently, although they had lived together under such intense conditions, the prisoners knew surprisingly little about each other's past history or future plans. This excessive concentration on the vicissitudes of their current situation helped to make the prison experience more oppressive for the prisoners because, instead of escaping from it when they had a chance to do so in the privacy of their cells, the prisoners continued to allow it to dominate their thoughts and social relations. The guards too, rarely exchanged personal information during their relaxation breaks. They either talked about "problem prisoners," other prison topics, or did not talk at all. There were few instances of any personal communication across the two role groups. Moreover, when prisoners referred to other prisoners during interviews, they typically deprecated each other, seemingly adopting the guards' negative attitude.

From post-experimental data, we discovered that when individual guards were alone with solitary prisoners and out of range of any recording equipment, as on the way to or in the toilet, harassment often was greater than it was on the "Yard." Similarly, video-taped analyses of total guard aggression showed a daily escalation even after most prisoners had ceased resisting and prisoner deterioration had become visibly obvious to them. Thus, guard aggression was no longer elicited as it was initially in response to perceived threats, but was emitted simply as a "natural" consequence of being in the uniform of a "guard" and asserting the power inherent in that role. In specific instances we noted cases of a guard (who did not know he was being observed) in the early morning hours pacing the Yard as the prisoners slept—vigorously pounding his night stick into his hand while he "kept watch" over his captives. Or another guard who detained an "incorrigible" prisoner in solitary confinement beyond the duration set by the guards' own rules and then conspired to keep him in the hole all night while attempting to conceal this information from the experimenters who were thought to be too soft on the prisoners.

In passing we may note an additional point about the nature of role-playing and the extent to which actual behavior is "explained away" by reference to it. It will be recalled that many guards continued to intensify their harassment and aggressive behavior even after the second day of the study, when prisoner deterioration became marked and visible and emotional breakdowns began to occur (in the presence of the guards). When questioned after the study about their persistent affrontive and harassing behavior in the face of prisoner emotional trauma, most guards replied that they were "just playing the role" of a tough guard, although none ever doubted the magnitude or validity of the prisoners' emotional response. The reader may wish to consider to what extremes an individual may go, how great must be the consequences of his behavior for others, before he can no longer rightfully attribute his actions to "playing a role" and thereby abdicate responsibility.

When introduced to a Catholic priest, many of the role-playing prisoners referred to themselves by their prison numbers rather than their Christian names. Some even asked him to get a lawyer to help them get out. When a public defender was summoned to interview those prisoners who had not yet been released, almost all of them strenuously demanded that he "bail" them out immediately.

One of the most remarkable incidents of the study occurred during a parole board hearing when each of five prisoners eligible for parole was asked by the senior author whether he would be willing to forfeit all the money earned as a prisoner if he were to be paroled (released from the study). Three of the five prisoners said, "yes," they would be willing to do this. Notice that the original incentive for participating in the study had been the promise of money, and they were, after only four days, prepared to give this up completely. And, more suprisingly, when told that this possibility would have to be discussed with the members of the staff before a decision could be made, each prisoner got up quietly and was escorted by a guard back to his cell. If they regarded themselves simply as "subjects" participating in an experiment for money, there was no longer any incentive to remain in the study and they could have easily escaped this situation which had so clearly become aversive for them by quitting. Yet, so powerful was the control which the situation had come to have over them, so much a reality had this simulated environment become, that they were unable to see that their original and singular motive for remaining no longer obtained, and they returned to their cells to await a "parole" decision by their captors.

The reality of the prison was also attested to by our prison consultant who had spent over 16 years in prison, as well as the priest who had been a prison chaplain and the public defender, all of whom were brought into direct contact with our simulated prison environment. Further, the depressed affect of the prisoners, the guards' willingness to work overtime for no additional pay, the spontaneous use of prison titles and I.D. numbers in non-role-related situations all point to a level of reality as real as any other in the lives of all those who shared this experience.

To understand how an illusion of imprisonment could have become so real, we need now to consider the uses of power by the guards as well as the effects of such power in shaping the prisoner mentality.

PATHOLOGY OF POWER

Being a guard carried with it social status within the prison, a group identity (when wearing the uniform), and above all, the freedom to exercise an unprecedented degree of control over the lives of other human beings. This control was invariably expressed in terms of sanctions, punishment, demands, and with the threat of manifest physical power. There was no need for the guards to rationally justify a request as they did their ordinary life, and merely to make a demand was sufficient to have it carried out. Many of the guards showed in their behavior and revealed in post-experimental statements that this sense of power was exhilarating.

The use of power was self-aggrandizing and self-perpetuating. The guard power, derived initially from an arbitrary and randomly assigned label, was intensified whenever there was any perceived threat by the prisoners and this new level subsequently became the baseline from which further hostility and harassment would begin. The most hostile guards on each shift moved spontaneously into the leadership roles of giving orders and deciding on punishments. They became role models whose behavior was emulated by other members of the shift. Despite minimal contact between the three separate guard shifts and nearly 16 hours a day spent away from the prison, the absolute level of aggression, as well as more subtle and "creative" forms of aggression manifested, increased in a spiralling function. Not to be tough and arrogant was to be seen as a sign of weakness by the guards, and even those "good" guards who did not get as drawn into the power syndrome as the others respected the implicit norm of *never* contradicting or even interfering with an action of a more hostile guard on their shift.

After the first day of study, practically all prisoner rights (even such things as the time and conditions of sleeping and eating) came to be redefined by the guards as "privileges" which were to be earned by obedient behavior. Constructive activities such as watching movies or reading (previously planned and suggested by the experimenters) were arbitrarily cancelled until further notice by the guards— and were subsequently never allowed. "Reward" then became granting approval for prisoners to eat, sleep, go to the toilet, talk, smoke a cigarette, wear eyeglasses, or the temporary diminution of harassment. One wonders about the conceptual nature of "positive" reinforcement when subjects are in such conditions of deprivation, and the extent to which even minimally acceptable conditions become rewarding when experienced in the context of such an impoverished environment.

We might also question whether there are meaningful non-violent alternatives as models for behavior modification in real prisons. In a world where men are either powerful or powerless, everyone learns to despise the lack of power in others and in oneself. It seems to us, that prisoners learn to admire power for its own sake—power becoming the ultimate reward. Real prisoners soon learn the means to gain power whether through ingratiation, informing, sexual control of other prisoners or development of powerful cliques. When they are released from prison, it is likely they will never want to feel so powerless again and will take action to establish and assert a sense of power.

THE PATHOLOGICAL PRISONER SYNDROME

Various coping strategies were employed by our prisoners as they began to react to their perceived loss of personal identity and the arbitrary control of their lives. At first they exhibited disbelief at the total invasion of their privacy, constant surveillance, and atmosphere of oppression in which they were living. Their next response was rebellion, first by the use of direct force, and later by subtle divisive tactics designed to foster distrust among the prisoners. They then tried to work within the system by setting up an elected grievance committee. When that collective action failed to produce meaningful changes in their existence, individual self-interests emerged. The breakdown in prisoner cohesion was the start of social disintegration which gave rise not only to feelings of isolation, but deprecation of other prisoners as well. As noted before, half the prisoners coped with the prison situation by becoming "sick"—extremely disturbed emotionally—as a passive way of demanding attention and help. Others became excessively obedient in trying to be "good" prisoners. They sided with the guards against a solitary fellow prisoner who coped with his situation by refusing to eat. Instead of supporting this final and major act of rebellion, the prisoners treated him as a troublemaker who deserved to be punished for his disobedience. It is likely that the negative self-regard among the prisoners noted by the end of the study was the product of their coming to believe that the continued hostility toward all of them was justified because they "deserved it" (following Walster, 1966). As the days wore on, the model prisoner reaction was one of passivity, dependence, and flattened affect.

Let us briefly consider some of the relevant processes involved in bringing about these reactions.

Loss of personal identity.   For most people identity is conferred by social recognition of one's uniqueness and established through one's name, dress, appearance, behavior style, and history. Living among strangers who do not know your name or history (who refer to you only by number), dressed in a uniform exactly like all other prisoners, not wanting to call attention to oneself because of the unpredictable consequences it might provoke—all led to a weakening of self-identity among the prisoners. As they began to lose initiative and emotional responsivity, while acting ever more compliantly, indeed, the prisoners became deindividuated not only to the guards and the observers, but also to themselves.

Arbitrary control.   On post-experimental questionnaires, the most frequently mentioned aversive aspect of the prison experience was that of being subjugated to the patently arbitrary, capricious decisions and rules of the guards. A question by a prisoner as often elicited derogation and aggression as it did a rational answer. Smiling at a joke could be punished in the same way that failing to smile might be. An individual acting in defiance of the rules could bring punishment to innocent cell partners (who became, in effect, "mutually yoked controls"), to himself, or to all.

As the environment became more unpredictable, and previously learned assumptions about a just and orderly world were no longer functional, prisoners ceased to initiate any action. They moved about on orders and when in their cells rarely engaged in any purposeful activity. Their zombie-like reaction was the functional equivalent of the learned helplessness phenomenon reported by Seligman & Groves (1970). Since their behavior did not seem to have any contingent relationship to environmental consequences, the prisoners essentially gave up and stopped behaving. Thus the subjective magnitude of aversiveness was manipulated by the guards not in terms of physical punishment but rather by controlling the psychological dimension of environmental predictability (Glass & Singer, 1972).

Dependency and emasculation.   The network of dependency relations established by the guards not only promoted helplessness in the prisoners but served to emasculate them as well. The arbitrary control by the guards put the prisoners at their mercy for even the daily, commonplace functions like going to the toilet. To do so, required publicly obtained permission (not always granted) and then a personal escort to the toilet while blindfolded and handcuffed. The same was true for many other activities ordinarily practiced spontaneously without thought, such as lighting a cigarette, reading a novel, writing a letter, drinking a glass of water, or brushing one's teeth. These were all privileged activities requiring permission and necessitating a prior show of good behavior. These low level dependencies engendered a regressive orientation in the prisoners. Their dependency was defined in terms of the extent of the domain of control over all aspects of their lives which they allowed other individuals (the guards and prison staff) to exercise.

As in real prisons, the assertive, independent, aggressive nature of male prisoners posed a threat which was overcome by a variety of tactics. The prisoner uniforms resembled smocks or dresses, which made them look silly and enabled the guards to refer to them as ''sissies'' or ''girls.'' Wearing these uniforms without any underclothes forced the prisoners to move and sit in unfamiliar, feminine postures. Any sign of individual rebellion was labelled as indicative of ''incorrigibility'' and resulted in loss of privileges, solitary confinement, humiliation, or punishment of cell mates. Physically smaller guards were able to induct stronger prisoners to act foolishly and obediently. Prisoners were encouraged to belittle each other publicly during the counts. These and other tactics all served to engender in the prisoners a lessened sense of their masculinity (as defined by their external culture). It followed then, that although the prisoners usually outnumbered the guards during line-ups and counts (nine vs. three) there never was an attempt to directly overpower them. (Interestingly, after the study was terminated, the prisoners expressed the belief that the basis for assignment to guard and prisoner groups was physical size. They perceived the guards were ''bigger,'' when, in fact, there was no difference in average height or weight between these randomly determined groups.)

In conclusion, we believe this demonstration reveals new dimensions in the social psychology of imprisonment worth pursuing in future research. In addition, this research provides a paradigm and information base for studying alternatives to existing guard training, as well as for questioning the basic operating principles on which penal institutions rest. If our mock prison could generate the extent of pathology it did in such a short time, then the punishment of being imprisoned in a real prison does not ''fit the crime'' for most prisoners—indeed, it far exceeds it! Moreover, since both prisoners and guards are locked into a dynamic, symbiotic relationship which is destructive to their human nature, guards are also society's prisoners.

References

ADORNO, T. W., FRENKEL-BRUNSWIK, E., LEVINSON, D. J., AND SANFORD, R. N. *The authoritarian personality*. New York: Harper, 1950.

CHARRIERE, H. *Papillon*. Robert Laffont, 1969.

CRISTIE, R., AND GEIS, F. L. (Eds.) *Studies in machiavellianism*. New York: Academic Press, 1970.

COMREY, A. L. *Comrey personality scales*. San Diego: Educational and Industrial Testing Service, 1970.

GLASS, D. C., AND SINGER, J. E. Behavioral after-effects of unpredictable and uncontrollable aversive events. *American Scientist*, 1972, **6**, No. 4, 457–465.

JACKSON, G. *Soledad brother: The prison letters of George Jackson*. New York: Bantam Books, 1970.

MILGRAM, S. Some conditions of obedience and disobedience to authority. *Human Relations*, 1965, **18**, No. 1, 57–76.

MISCHEL, W. *Personality and assessment*. New York: Wiley, 1968.

SCHEIN, E. *Coercive persuasion*. New York: Norton, 1961.

SELIGMAN, M. E. AND GROVES, D. P. Nontransient learned helplessness. *Psychonomic Science*, 1970, **19**, No. 3, 191–192.

WALSTER, E. Assignment of responsibility for an accident. *Journal of Personality and Social Psychology*, 1966, **3**, No. 1, 73–79.

6

Making Sense of the Nonsensical: An Analysis of Jonestown

Neal Osherow

Those who do not remember the past are condemned to repeat it.

—quotation on placard over Jim Jones's rostrum
at Jonestown

Close to one thousand people died at Jonestown. The members of the Peoples Temple settlement in Guyana, under the direction of the Reverend Jim Jones, fed a poison-laced drink to their children, administered the potion to their infants, and drank it themselves. Their bodies were found lying together, arm in arm; over 900 perished.

How could such a tragedy occur? The image of an entire community destroying itself, of parents killing their own children, appears incredible. The media stories about the event and full-color pictures of the scene documented some of its horror but did little to illuminate the causes or to explain the processes that led to the deaths. Even a year afterwards, a CBS Evening News broadcast asserted that "it was widely assumed that time would offer some explanation for the ritualistic suicide/murder of over 900 people. . . . One year later, it does not appear that any lessons have been uncovered" (CBS News, 1979).

The story of the Peoples Temple is not enshrouded in mystery, however. Jim Jones had founded his church over twenty years before, in Indiana. His preaching stressed the need for racial brotherhood and integration, and his group helped feed

I am very grateful to Elliot Aronson for his assistance with this essay. His insights, suggestions, and criticism were most valuable to its development. Also, my thanks to Elise Bean for her helpful editing.

the poor and find them jobs. As his congregation grew, Jim Jones gradually increased the discipline and dedication that he required from the members. In 1965, he moved to northern California; about 100 of his faithful relocated with him. The membership began to multiply, new congregations were formed, and the headquarters was established in San Francisco.

Behind his public image as a beloved leader espousing interracial harmony, "Father," as Jones was called, assumed a messiah-like presence in the Peoples Temple. Increasingly, he became the personal object of the members' devotion, and he used their numbers and obedience to gain political influence and power. Within the Temple, Jones demanded absolute loyalty, enforced a taxing regimen, and delivered sermons forecasting nuclear holocaust and an apocalyptic destruction of the world, promising his followers that they alone would emerge as survivors. Many of his harangues attacked racism and capitalism, but his most vehement anger focused on the "enemies" of the Peoples Temple—its detractors and especially its defectors. In mid-1977, publication of unfavorable magazine articles, coupled with the impending custody battle over a six-year-old Jones claimed as a "son," prompted emigration of the bulk of Temple membership to a jungle outpost in Guyana.

In November, 1978, Congressman Leo Ryan responded to charges that the Peoples Temple was holding people against their will at Jonestown. He organized a trip to the South American settlement; a small party of journalists and "Concerned Relatives" of Peoples Temple members accompanied him on his investigation. They were in Jonestown for one evening and part of the following day. They heard most residents praise the settlement, expressing their joy at being there and indicating their desire to stay. Two families, however, slipped messages to Ryan that they wanted to leave with him. After the visit, as Ryan's party and these defectors tried to board planes to depart, the group was ambushed and fired upon by Temple gunmen—five people, including Ryan, were murdered.

As the shootings were taking place at the jungle airstrip, Jim Jones gathered the community at Jonestown. He informed them that the Congressman's party would be killed and then initiated the final ritual: the "revolutionary suicide" that the membership had rehearsed on prior occasions. The poison was brought out. It was taken.

Jonestown's remoteness caused reports of the event to reach the public in stages. First came bulletins announcing the assassination of Congressman Ryan along with several members of his party. Then came rumors of mass-deaths at Jonestown, then confirmations. The initial estimates put the number of dead near 400, bringing the hope that substantial numbers of people had escaped into the jungle. But as the bodies were counted, many smaller victims were discovered under the corpses of larger ones—virtually none of the inhabitants of Jonestown survived. The public was shocked, then horrified, then incredulous.

Amid the early stories about the tragedy, along with the lurid descriptions and sensational photographs, came some attempts at analysis. Most discussed the charisma of Jim Jones and the power of "cults." Jones was described as "a character Joseph Conrad might have dreamt up" (Krause, 1978), a "self-appointed messiah" whose "lust for dominion" led hundreds of "fanatic" followers to their demise (Special Report: The Cult of Death, *Newsweek*, 1978a).

While a description in terms of the personality of the perpetrator and the vulnerability of the victims provides some explanation, it relegates the event to the category of being an aberration, a product of unique forces and dispositions. Assuming such a perspective distances us from the phenomenon. This might be comforting, but I believe that it limits our understanding and is potentially dangerous. My aim in this analysis is not to blunt the emotional impact of a tragedy of this magnitude by subjecting it to academic examination. At the same time, applying social psychological theory and research makes it more conceivable and comprehensible, thus bringing it closer (in kind rather than in degree) to processes each of us encounters. Social psychological concepts can facilitate our understanding: The killings themselves, and many of the occurrences leading up to them, can be viewed in terms of obedience and compliance. The processes that induced people to join and to believe in the Peoples Temple made use of strategies involved in propaganda and persuasion. In grappling with the most perplexing questions—Why didn't more people leave the Temple? How could they actually kill their children and themselves?—the psychology of self-justification provides some insight.

CONFORMITY

The character of a church . . . can be seen in its attitude toward its detractors.

—Hugh Prather, *Notes to Myself*

At one level, the deaths at Jonestown can be viewed as the product of obedience, of people complying with the orders of a leader and reacting to the threat of force. In the Peoples Temple, whatever Jim Jones commanded, the members did. When he gathered the community at the pavilion and the poison was brought out, the populace was surrounded by armed guards who were trusted lieutenants of Jones. There are reports that some people did not drink voluntarily but had the poison forced down their throats or injected (Winfrey, 1979). While there were isolated acts of resistance and suggestions of opposition to the suicides, excerpts from a tape, recorded as the final ritual was being enacted, reveal that such dissent was quickly dismissed or shouted down:

> JONES: I've tried my best to give you a good life. In spite of all I've tried, a handful of people, with their lies, have made our life impossible. If we can't live

in peace then let's die in peace. (Applause) . . . We have been so terribly betrayed. . . .

What's going to happen here in the matter of a few minutes is that one of the people on that plane is going to shoot the pilot—I know that. I didn't plan it, but I know it's going to happen. . . . So my opinion is that you be kind to children, and be kind to seniors, and take the potion like they used to in ancient Greece, and step over quietly, because we are not committing suicide—it's a revolutionary act. . . . We can't go back. They're now going back to tell more lies. . . .

FIRST WOMAN: I feel like that as long as there's life, there's hope.

JONES: Well, someday everybody dies.

CROWD: That's right, that's right!

JONES: What those people gone and done, and what they get through will make our lives worse than hell. . . . But to me, death is not a fearful thing. It's living that's cursed. . . . Not worth living like this.

FIRST WOMAN: But I'm afraid to die.

JONES: I don't think you are. I don't think you are.

FIRST WOMAN: I think there were too few who left for 1,200 people to give them their lives for those people who left. . . . I look at all the babies and I think they deserve to live.

JONES: But don't they deserve much more—they deserve peace. The best testimony we can give is to leave this goddam world. (Applause)

FIRST MAN: It's over, sister. . . . We've made a beautiful day. (Applause)

SECOND MAN: If you tell us we have to give our lives now, we're ready. (Applause) [*Baltimore Sun,* 1979.]

Above the cries of babies wailing, the tape continues, with Jones insisting upon the need for suicide and urging the people to complete the act:

JONES: Please get some medication. Simple. It's simple. There's no convulsions with it. . . . Don't be afraid to die. You'll see people land out here. They'll torture our people. . . .

SECOND WOMAN: There's nothing to worry about. Everybody keep calm and try to keep your children calm. . . . They're not crying from pain; it's just a little bitter tasting . . .

THIRD WOMAN: This is nothing to cry about. This is something we could all rejoice about. (Applause)

JONES: Please, for God's sake, let's get on with it. . . . This is a revolutionary suicide. This is not a self-destructive suicide. (Voices praise "Dad." Applause)

THIRD MAN: Dad has brought us this far. My vote is to go with Dad. . . .

JONES: We must die with dignity. Hurry, hurry, hurry. We must hurry. . . . Stop this hysterics. Death is a million times more perferable to spending more days in this life. . . . If you knew what was ahead, you'd be glad to be stepping over tonight . . .

FOURTH WOMAN: It's been a pleasure walking with all of you in this revolutionary struggle. . . . No other way I would rather go than to give my life for socialism. Communism, and I thank Dad very much.

JONES: Take our life from us. . . . We didn't commit suicide. We committed an act of revolutionary suicide protesting against the conditions of an inhuman world [*Newsweek,* 1978b, 1979].

If you hold a gun at someone's head, you can get that person to do just about anything. As many accounts have attested,[1] by the early 1970s the members of the Peoples Temple lived in constant fear of severe punishment—brutal beatings coupled with public humiliation—for commiting trivial or even inadvertent offenses. But the power of an authority need not be so explicitly threatening in order to induce compliance with its demands, as demonstrated by social psychological research. In Milgram's experiments (1963), a surprisingly high proportion of subjects obeyed the instructions of an experimenter to administer what they thought were very strong electric shocks to another person. Nor does the consensus of a group need be so blatantly coercive to induce agreement with its opinion, as Asch's experiments (1955) on conformity to the incorrect judgments of a majority indicate.

Jim Jones utilized the threat of severe punishment to impose the strict discipline and absolute devotion that he demanded, and he also took measures to eliminate those factors that might encourage resistance or rebellion among his followers. Research showed that the presence of a "disobedient" partner greatly reduced the extent to which most subjects in the Milgram situation (1965) obeyed the instructions to shock the person designated the "learner." Similarly, by including just one confederate who expressed an opinion different from the majority's, Asch (1955) showed that the subject would also agree far less, even when the "other dissenter's" judgment was also incorrect and differed from the subject's. In the Peoples Temple, Jones tolerated no dissent, made sure that members had no allegiance more powerful than to himself, and tried to make the alternative of leaving the Temple an unthinkable option.

Jeanne Mills, who spent six years as a high-ranking member before becoming one of the few who left the Peoples Temple, writes: "There was an unwritten but perfectly understood law in the church that was very important: 'No one is to criticize Father, his wife, or his children' " (Mills, 1979). Deborah Blakey, another long-time member who managed to defect, testified:

> Any disagreement with [Jim Jones's] dictates came to be regarded as "treason."
> . . . Although I felt terrible about what was happening, I was afraid to say anything because I knew that anyone with a differing opinion gained the wrath of Jones and other members. [Blakey, June 15, 1978.]

Conditions in the Peoples Temple became so oppressive, the discrepancy between Jim Jones's stated aims and his practices so pronounced, that it is almost inconceivable that members failed to entertain questions about the church. But these

---

[1] The reports of ex-Peoples Temple members who defected create a very consistent picture of the tactics Jim Jones employed in his church. Jeanne Mills (1979) provides the most comprehensive personal account, and there are affidavits about the Peoples Temple sworn to by Deborah Blakey (May 12, 1978 and June 15, 1978) and Yolanda Crawford (April 10, 1978). Media stories about the Peoples Temple, which usually rely on interviews with defectors, and about Jonestown, which are based on interviews with survivors, also corroborate one another. (See especially Kilduff and Tracy (1977), *Newsweek* (1978a), Lifton (1979), and Cahill (1979).

doubts went unreinforced. There were no allies to support one's disobedience of the leader's commands and no fellow dissenters to encourage the expression of disagreement with the majority. Public disobedience or dissent was quickly punished. Questioning Jones's word, even in the company of family or friends, was dangerous—informers and "counselors" were quick to report indiscretions, even by relatives.

The use of informers went further than to stifle dissent; it also diminished the solidarity and loyalty that individuals felt toward their families and friends. While Jones preached that a spirit of brotherhood should pervade his church, he made it clear that each member's personal dedication should be directed to "Father." Families were split: First, children were seated away from parents during services; then, many were assigned to another member's care as they grew up; and ultimately, parents were forced to sign documents surrendering custody rights. "Families are part of the enemy system," Jones stated, because they hurt one's total dedication to the "Cause" (Mills, 1979). Thus, a person called before the membership to be punished could expect his or her family to be among the first and most forceful critics (Cahill, 1979).

Besides splitting parent and child, Jones sought to loosen the bonds between wife and husband. He forced spouses into extramarital sexual relations, which were often of a homosexual or humiliating nature, or with Jones himself. Sexual partnerships and activities not under his direction and control were discouraged and publicly ridiculed.

Thus, expressing any doubts or criticism of Jones—even to a friend, child, or partner—became risky for the individual. As a consequence, such thoughts were kept to oneself, and with the resulting impression that nobody else shared them. In addition to limiting one's access to information, this "fallacy of uniqueness" precluded the sharing of support. It is interesting that among the few who successfully defected from the Peoples Temple were couples such as Jeanne and Al Mills, who kept together, shared their doubts, and gave each other support.

Why didn't more people leave? Once inside the Peoples Temple, getting out was discouraged; defectors were hated. Nothing upset Jim Jones so much; people who left became the targets of his most vitriolic attacks and were blamed for any problems that occurred. One member recalled that after several teen-age members left the Temple, "We hated those eight with such a passion because we knew any day they were going to try bombing us. I mean Jim Jones had us totally convinced of this" (Winfrey, 1979).

Defectors were threatened: Immediately after she left, Grace Stoen headed for the beach at Lake Tahoe, where she found herself looking over her shoulder, checking to make sure that she hadn't been tracked down (Kilduff and Tracy, 1977). Jeanne Mills reports that she and her family were followed by men in cars, their home was burglarized, and they were threatened with the use of confessions they had signed while still members. When a friend from the Temple paid a visit, she quickly examined Mills' ears—Jim Jones had vowed to have one of them cut off

(Mills, 1979). He had made ominous predictions concerning other defectors as well: Indeed, several ex-members suffered puzzling deaths or committed very questionable "suicides" shortly after leaving the Peoples Temple (Reiterman, 1977; Tracy, 1978).

Defecting became quite a risky enterprise, and, for most members, the potential benefits were very uncertain. They had little to hope for outside of the Peoples Temple; what they had, they had committed to the church. Jim Jones had vilified previous defectors as "the enemy" and had instilled the fear that, once outside of the Peoples Temple, members' stories would not be believed by the "racist, fascist" society, and they would be subjected to torture, concentration camps, and execution. Finally, in Guyana, Jonestown was surrounded by dense jungle, the few trails patrolled by armed security guards (Cahill, 1979). Escape was not a viable option. Resistance was too costly. With no other alternatives apparent, compliance became the most reasonable course of action.

The power that Jim Jones wielded kept the membership of the Peoples Temple in line, and the difficulty of defecting helped to keep them in. But what attracted them to join Jones's church in the first place?

PERSUASION

Nothing is so unbelievable that oratory cannot make it acceptable.

—Cicero

Jim Jones was a charismatic figure, adept at oratory. He sought people for his church who would be receptive to his messages and vulnerable to his promises, and he carefully honed his presentation to appeal to each specific audience.

The bulk of the Peoples Temple membership was comprised of society's needy and neglected: the urban poor, the black, the elderly, and a sprinkling of ex-addicts and ex-convicts (Winfrey, 1979). To attract new members, Jones held public services in various cities. Leaflets would be distributed:

> PASTOR JIM JONES . . . Incredible! . . . Miraculous! . . . Amazing! . . . The Most Unique Prophetic Healing Service You've Ever Witnessed! Behold the Word Made Incarnate In Your Midst!
>
> God works as tumorous masses are passed in every service. . . . Before your eyes, the cripple walk, the blind see! [Kilduff and Javers, 1978.]

Potential members first confronted an almost idyllic scene of blacks and whites living, working, and worshipping together. Guests were greeted and treated most warmly and were invited to share in the group's meal. As advertised, Jim Jones also gave them miracles. A number of members would recount how Jones had cured them of cancer or other dread diseases; during the service Jones or one of his nurses would reach into the member's throat and emerge with a vile mass of tissue—the "cancer" that had been passed as the person gagged. Sometimes Jim Jones would make predictions that would occur with uncanny frequency. He also received revelations about members or visitors that nobody but those individuals could know—what they had eaten for dinner the night before, for instance, or news about a far-off relative. Occasionally, he performed miracles similar to more well-established religious figures:

> There were more people than usual at the Sunday service, and for some reason the church members hadn't brought enough food to feed everyone. It became apparent that the last fifty people in line weren't going to get any meat. Jim announced, "Even though there isn't enough food to feed this multitude, I am blessing the food that we have and multiplying it—just as Jesus did in biblical times."
>
> Sure enough, a few minutes after he made this startling announcement, Eva Pugh came out of the kitchen beaming, carrying two platters filled with fried chicken. A big cheer came from the people assembled in the room, especially from the people who were at the end of the line.
>
> The "blessed chicken" was extraordinarily delicious, and several of the people mentioned that Jim had produced the best-tasting chicken they had ever eaten. [Mills, 1979.]

These demonstrations were dramatic and impressive; most members were convinced of their authenticity and believed in Jones's "powers." They didn't know that the "cancers" were actually rancid chicken gizzards, that the occurrences Jones "forecast" were staged, or that sending people to sift through a person's garbage could reveal packages of certain foods or letters of out-of-town relatives to serve as grist for Jones' "revelations" (Kilduff and Tracy, 1977; Mills, 1979). Members were motivated to believe in Jones; they appreciated the racial harmony, sense of purpose, and relief from feelings of worthlessness that the Peoples Temple provided them (Winfrey, 1979; Lifton, 1979). Even when suspecting that something was wrong, they learned that it was unwise to voice their doubts:

> One of the men, Chuck Beikman . . . jokingly mentioned to a few people standing near him that he had seen Eva drive up a few moments earlier with buckets from the Kentucky Fried Chicken stand. He smiled as he said, "The person that blessed this chicken was Colonel Sanders."
>
> During the evening meeting Jim mentioned the fact that Chuck had made fun of his gift. "He lied to some of the members here, telling them that the chicken had come from a local shop," Jim stormed. "But the Spirit of Justice has prevailed. Because of his lie Chuck is in the men's room right now, wishing that he was dead. He is vomiting and has diarrhea so bad he can't talk!"
>
> An hour later a pale and shaken Chuck Beikman walked out of the men's room and up to the front, being supported by one of the guards. Jim asked him, "Do you have anything you'd like to say?"
>
> Chuck looked up weakly and answered, "Jim, I apologize for what I said. Please forgive me."

> As we looked at Chuck, we vowed in our hearts that we would never question any of Jim's "miracles"—at least not out loud. Years later, we learned that Jim had put a mild poison in a piece of cake and given it to Chuck. [Mills, 1979.]

While most members responded to presentations that were emotional, one-sided, and almost sensational in tone, those who eventually assumed positions of responsibility in the upper echelons of the Peoples Temples were attracted by different considerations. Most of these people were white and came from upper-middle-class backgrounds—they included lawyers, a medical student, nurses, and people representing other occupations that demanded education and reflected a strong social consciousness. Jones lured these members by stressing the social and political aspects of the church, its potential as an idealistic experiment with integration and socialism. Tim Stoen, who was the Temple's lawyer, stated later, "I wanted utopia so damn bad I could die" (Winfrey, 1979). These members had the information and intelligence to see through many of Jones's ploys, but, as Jeanne Mills explains repeatedly in her book, they dismissed their qualms and dismissed Jones's deception as being necessary to achieve a more important aim—furthering the Cause: "For the thousandth time, I rationalized my doubts. 'If Jim feels it's necessary for the Cause, who am I to question his wisdom?' " (Mills, 1979).

It turned out to be remarkably easy to overcome their hesitancy and calm their doubts. Mills recalls that she and her husband initially were skeptical about Jones and the Peoples Temple. After attending their first meeting, they remained unimpressed by the many members who proclaimed that Jones had healed their cancers or cured their drug habits. They were annoyed by Jones' arrogance, and they were bored by most of the long service. But in the weeks following their visit, they received numerous letters containing testimonials and gifts from the Peoples Temple, they had dreams about Jones, and they were attracted by the friendship and love they had felt from both the black and the white members. When they went back for their second visit, they took their children with them. After the long drive, the Mills's were greeted warmly by many members and by Jones himself. "This time . . . my mind was open to hear his message because my own beliefs had become very shaky" (Mills, 1979). As they were driving home afterwards, the children begged their parents to join the church:

> We had to admit that we enjoyed the service more this time and we told the children that we'd think it over. Somehow, though, we knew that it was only a matter of time before we were going to become members of the Peoples Temple. [Mills, 1979.]

Jim Jones skillfully manipulated the impression that his church would convey to newcomers. He carefully managed its public image. He used the letter-writing and political clout of hundreds of members to praise and impress the politicians and press that supported the Peoples Temple, as well as to criticize and intimidate its opponents (Kasindorf, 1978). Most importantly, Jones severely restricted the information that was available to the members. In addition to indoctrinating members into his own belief system through extensive sermons and lectures, he incul-

cated a distrust of any contradictory messages, labelling them the product of enemies. By destroying the credibility of their sources, he inoculated the membership
against being persuaded by outside criticism. Similarly, any contradictory thoughts
that might arise within each member were to be discredited. Instead of seeing them
as having any basis in reality, members interpreted them as indications of their
own shortcomings or lack of faith. Members learned to attribute the apparent discrepancies between Jones's lofty pronouncements and the rigors of life in the Peoples Temple to their personal inadequacies rather than blaming them on any fault
of Jones. As ex-member Neva Sly was quoted: "We always blamed ourselves for
things that didn't seem right" (Winfrey, 1979). A unique and distorting language
developed within the church, in which "the Cause" became anything that Jim Jones
said (Mills, 1979). It was spoken at Jonestown, where a guard tower was called
the "playground" (Cahill, 1979). Ultimately, through the clever use of oratory,
deception, and language, Jones could speak of death as "stepping over," thereby
camouflaging a hopeless act of self-destruction as a noble and brave act of "revolutionary suicide," and the members accepted his words.

SELF-JUSTIFICATION

Both salvation and punishment for man lie in the fact
that if he lives wrongly he can befog himself so as not
to see the misery of his position.

—Tolstoy, "The Kreutzer Sonata"

Analyzing Jonestown in terms of obedience and the power of the situation can
help to explain why the people *acted* as they did. Once the Peoples Temple had
moved to Jonestown, there was little the members could do other than follow Jim
Jones's dictates. They were comforted by an authority of absolute power. They
were left with few options, being surrounded by armed guards and by the jungle,
having given their passports and various documents and confessions to Jones, and
believing that conditions in the outside world were even more threatening. The
members' poor diet, heavy workload, lack of sleep, and constant exposure to Jones's
diatribes exacerbated the coerciveness of their predicament; tremendous pressures
encouraged them to obey.

By the time of the final ritual, opposition or escape had become almost impossible for most of the members. Yet even then, it is doubtful that many *wanted* to
resist or to leave. Most had come to believe in Jones—one woman's body was
found with a message scribbled on her arm during the final hours: "Jim Jones is
the only one" (Cahill, 1979). They seemed to have accepted the necessity, and
even the beauty, of dying—just before the ritual began, a guard approached Charles

Garry, one of the Temple's hired attorneys, and exclaimed, "It's a great moment . . . we all die" (Lifton, 1979). A survivor of Jonestown, who happened to be away at the dentist, was interviewed a year following the deaths:

> If I had been there, I would have been the first one to stand in that line and take that poison and I would have been proud to take it. The thing I'm sad about is this; that I missed the ending. [Gallagher, 1979.]

It is this aspect of Jonestown that is perhaps the most troubling. To the end, and even beyond, the vast majority of the Peoples Temple members *believed* in Jim Jones. External forces, in the form of power or persuasion, can exact compliance. But one must examine a different set of processes to account for the members' internalizing those beliefs.

Although Jones's statements were often inconsistent and his methods cruel, most members maintained their faith in his leadership. Once they were isolated at Jonestown, there was little opportunity or motivation to think otherwise—resistance or escape was out of the question. In such a situation, the individual is motivated to rationalize his or her predicament; a person confronted with the inevitable tends to regard it more positively. For example, social psychological research has shown that when children believe that they will be served more of a vegetable they dislike, they will convince themselves that it is not so noxious (Brehm, 1959), and when a person thinks that she will be interacting with someone, she tends to judge a description of that individual more favorably (Darley and Berscheid, 1967).

A member's involvement in the Temple did not begin at Jonestown—it started much earlier, closer to home, and less dramatically. At first, the potential member would attend meetings voluntarily and might put in a few hours each week working for the church. Though the established members would urge the recruit to join, he or she felt free to choose whether to stay or to leave. Upon deciding to join, a member expended more effort and became more committed to the Peoples Temple. In small increments, Jones increased the demands made on the member, and only after a long sequence did he escalate the oppressiveness of his rule and the desperation of his message. Little by little, the individual's alternatives became more limited. Step by step, the person was motivated to rationalize his or her commitment and to justify his or her behavior.

Jeanne Mills, who managed to defect two years before the Temple relocated in Guyana, begins her account, *Six Years With God* (1979), by writing: "Every time I tell someone about the six years we spent as members of the Peoples Temple, I am faced with an unanswerable question: 'If the church was so bad, why did you and your family stay in for so long?" Several classic studies from social psychological research investigating processes of self-justification and the theory of cognitive dissonance (see Aronson, 1980, chapter 4; Aronson, 1969) can point to explanations for such seemingly irrational behavior.

According to dissonance theory, when a person commits an act or holds a cognition that is psychologically inconsistent with his or her self-concept, the inconsistency arouses an unpleasant state of tension. The individual tries to reduce this "dissonance," usually by altering his or her attitudes to bring them more into line with the previously discrepant action or belief. A number of occurrences in the Peoples Temple can be illuminated by viewing them in light of this process. The horrifying events of Jonestown were not due merely to the threat of force, nor did they erupt instantaneously. That is, it was *not* the case that something "snapped" in people's minds, suddenly causing them to behave in bizarre ways. Rather, as the theory of cognitive dissonance spells out, people seek to *justify* their choices and commitments.

Just as a towering waterfall can begin as a trickle, so too can the impetus for doing extreme or calamitous actions be provided by the consequences of agreeing to do seemingly trivial ones. In the Peoples Temple, the process started with the effects of undergoing a severe initiation to join the church, was reinforced by the tendency to justify one's commitments, and was strengthened by the need to rationalize one's behavior.

Consider the prospective member's initial visit to the People's Temple, for example. When a person undergoes a severe initiation in order to gain entrance into a group, he or she is apt to judge that group as being more attractive, in order to justify expending the effort or enduring the pain. Aronson and Mills (1959) demonstrated that students who suffered greater embarrassment as a prerequisite for being allowed to participate in a discussion group rated its conversation (which actually was quite boring) to be significantly more interesting than did those students who experienced little or no embarrassment in order to be admitted. Not only is there a tendency to justify undergoing the experience by raising one's estimation of the goal—in some circumstances, choosing to experience a hardship can go so far as to affect a person's perception of the discomfort or pain he or she felt. Zimbardo (1969) and his colleagues showed that when subjects volunteered for a procedure that involves their being given electric shocks, those thinking that they had more choice in the matter reported feeling less pain from the shocks. More specifically, those who experienced greater dissonance, having little external justification to account for their choosing to endure the pain, described it as being less intense. This extended beyond their impressions and verbal reports; their performance on a task was hindered less, and they even recorded somewhat lower readings on a physiological instrument measuring galvanic skin responses. Thus the dissonance-reducing process can be double-edged: Under proper guidance, a person who voluntarily experiences a severe initiation not only comes to regard its ends more positively, but may also begin to see the means as less aversive: "We begin to appreciate the long meetings, because we were told that spiritual growth comes from self-sacrifice" (Mills, 1979).

Once involved, a member found ever-increasing portions of his or her time and energy devoted to the Peoples Temple. The services and meetings occupied weekends and several evenings each week. Working on Temple projects and writing the required letters to politicians and the press took much of one's "spare" time.

Expected monetary contributions changed from "voluntary" donations (though they were recorded) to the required contribution of a quarter of one's income. Eventually, a member was supposed to sign over all personal property, savings, social security checks, and the like to the Peoples Temple. Before entering the meeting room for each service, a member stopped at a table and wrote self-incriminating letters or signed blank documents that were turned over to the church. If anyone objected, the refusal was interpreted as denoting a "lack of faith" in Jones. Finally, members were asked to live at Temple facilities to save money and to be able to work more efficiently, and many of their children were raised under the care of other families. Acceding to each new demand had two repercussions: In practical terms, it enmeshed the person further into the Peoples Temple web and made leaving more difficult; on an attitudinal level, it set the aforementioned processes of self-justification into motion. As Mills (1979) describes:

> We had to face painful reality. Our life savings were gone. Jim had demanded that we sell the life insurance policy and turn the equity over to the church, so that was gone. Our property had all been taken from us. Our dream of going to an overseas mission was gone. We thought that we had alienated our parents when we told them we were leaving the country. Even the children whom we had left in the care of Carol and Bill were openly hostile toward us. Jim had accomplished all this in such a short time! All we had left now was Jim and the Cause, so we decided to buckle under and give our energies to these two.

Ultimately, Jim Jones and the Cause would require the members to give their lives.

What could cause people to kill their children and themselves? From a detached perspective, the image seems unbelievable. In fact, at first glance, so does the idea of so many individuals committing so much of their time, giving all of their money, and even sacrificing the control of their children to the Peoples Temple. Jones took advantage of rationalization processes that allow people to justify their commitments by raising their estimations of the goal and minimizing its costs. Much as he gradually increased his demands, Jones carefully orchestrated the members' exposure to the concept of a "final ritual." He utilized the leverage provided by their previous commitments to push them closer and closer to its enactment. Gaining a "foot in the door" by getting a person to agree to a moderate request makes it more probable that he or she will agree to do a much larger deed later, as social psychologists—and salespeople—have found (Freedman and Fraser, 1966). Doing the initial task causes something that might have seemed unreasonable at first appear less extreme in comparison, and it also motivates a person to make his or her behavior appear more consistent by consenting to the larger request as well.

After indoctrinating the members with the workings of the Peoples Temple itself, Jones began to focus on broader and more basic attitudes. He started by undermining the members' belief that death was to be fought and feared and set the stage by introducing the possibility of a cataclysmic ending for the church. As several accounts corroborate (see Mills, 1979; Lifton, 1979; Cahill, 1979), Jones directed several "fake" suicide drills, first with the elite Planning Commission of the Peoples Temple and later with the general membership. He would give them wine and then announce that it had been poisoned and that they would soon die. These became tests of faith, of the members' willingness to follow Jones even to death. Jones would ask people if they were ready to die and on occasion would have the membership "decide" its own fate by voting whether to carry out his wishes. An ex-member recounted that one time, after a while

> Jones smiled and said, "Well, it was a good lesson. I see you're not dead." He made it sound like we needed the 30 minutes to do very strong, introspective type of thinking. We all felt strongly dedicated, proud of ourselves. . . . [Jones] taught that it was a privilege to die for what you believed in, which is exactly what I would have been doing. [Winfrey, 1979.]

After the Temple moved to Jonestown, the "White Nights," as the suicide drills were called, occurred repeatedly. An exercise that appears crazy to the observer was a regular, justifiable occurrence for the Peoples Temple participant. The reader might ask whether this caused the members to think that the actual suicides were merely another practice, but there were many indications that they knew 'that the poison was truly deadly on that final occasion. The Ryan visit had been climatic, there were several new defectors, the cooks—who had been excused from the prior drills in order to prepare the upcoming meal—were included, Jones had been growing increasingly angry, desperate, and unpredictable, and, finally, everyone could see the first babies die. The membership was manipulated, but they were not unaware that this time the ritual was for real.

A dramatic example of the impact of self-justification concerns the physical punishment that was meted out in the Peoples Temple. As discussed earlier, the threat of being beaten or humiliated forced the member to comply with Jones's orders: A person will obey as long as he or she is being threatened and supervised. To affect a person's *attitudes,* however, a mild threat has been demonstrated to be more effective than a severe threat (Aronson and Carlsmith, 1963) and its influence has been shown to be far longer lasting (Freedman, 1965). Under a mild threat, the individual has more difficulty attributing his or her behavior to such a minor external restraint, forcing the person to alter his or her attitudes in order to justify the action. Severe threats elicit compliance, but, imposed from the outside, they usually fail to cause the behavior to be internalized. Quite a different dynamic ensues when it is not so clear that the action is being imposed upon the person. When an individual feels that he or she played an active role in carrying out an action that hurts someone, there comes a motivation to justify one's part in the cruelty by rationalizing it as necessary or by derogating the victim by thinking that the punishment was deserved (Davis and Jones, 1960).

Let's step back for a moment. The processes going on at Jonestown obviously were not as simple as those in a well-controlled laboratory experiment; several themes were going on simultaneously. For example, Jim Jones had the power to impose any punishments that he wished in the Peoples Temple, and, especially towards the end, brutality and terror at Jonestown were rampant. But Jones carefully controlled how the punishments were carried out. He often called upon the members themselves to agree to the imposition of beatings. They were instructed to testify against fellow members, bigger members told to beat up smaller ones, wives or lovers forced to sexually humiliate their partners, and parents asked to consent to and assist in the beatings of their children (Mills, 1979; Kilduff and Javers, 1978). The punishments grew more and more sadistic, the beatings so severe as to knock the victim unconscious and cause bruises that lasted for weeks. As Donald Lunde, a psychiatrist who has investigated acts of extreme violence, explains:

> Once you've done something that major, it's very hard to admit even to yourself that you've made a mistake, and subconsciously you will go to great lengths to rationalize what you did. It's very tricky defense mechanism exploited to the hilt by the charismatic leader. [*Newsweek,* 1978a.]

A more personal account of the impact of this process is provided by Jeanne Mills. At one meeting, she and her husband were forced to consent to the beating of their daughter as punishment for a very minor transgression. She relates the effect this had on her daughter, the victim, as well as on herself, one of the perpetrators:

> As we drove home, everyone in the car was silent. We were all afraid that our words would be considered treasonous. The only sounds came from Linda, sobbing quietly in the back seat. When we got into our house, Al and I sat down to talk with Linda. She was in too much pain to sit. She stood quietly while we talked with her. "How do you feel about what happened tonight?" Al asked her.
>
> "Father was right to have me whipped," Linda answered. "I've been so rebellious lately, and I've done a lot of things that were wrong. . . . I'm sure Father knew about those things, and that's why he had me hit so many times."
>
> As we kissed our daughter goodnight, our heads were spinning. It was hard to think clearly when things were so confusing. Linda had been the victim, and yet we were the only people angry about it. She should have been hostile and angry. Instead, she said that Jim had actually helped her. We knew Jim had done a cruel thing, and yet everyone acted as if he were doing a loving thing in whipping our disobedient child. Unlike a cruel person hurting a child, Jim had seemed calm, almost loving, as he observed the beating and counted off the whacks. Our minds were not able to comprehend the atrocity of the situation because none of the feedback we were receiving was accurate. [Mills, 1979.]

The feedback one received from the outside was limited, and the feedback from inside the Temple member was distorted. By justifying the previous actions and commitments, the groundwork for accepting the ultimate commitment was established.

CONCLUSION

Only months after we defected from Temple did we realize the full extent of the cocoon in which we'd lived.
And only then did we understand the fraud, sadism, and
emotional blackmail of the master manipulator.

—Jeanne Mills, *Six Years with God*

Immediately following the Jonestown tragedy, there came a proliferation of articles about "cults" and calls for their investigation and control. From Synanon
to Transcendental Meditation, groups and practices were examined by the press,
which had a difficult time determining what constituted a "cult" or differentiating
between those that might be safe and beneficial and those that could be dangerous.
The Peoples Temple and the events at Jonestown make such a definition all the
more problematic. A few hours before his murder, Congressman Ryan addressed
the membership: "I can tell you right now that by the few conversations I've had
with some of the folks . . . there are some people who believe this is the best
thing that ever happened in their whole lives" (Krause, 1978). The acquiescence
of so many and the letters they left behind indicate that this feeling was widely
shared—or at least expressed—by the members.

Many "untraditional"—to mainstream American culture—groups or practices,
such as Eastern religions or meditation techniques, have proven valuable for the
people who experience them but may be seen as very strange and frightening to
others. How can people determine whether they are being exposed to a potentially
useful alternative way of living their lives or if they are being drawn to a dangerous one?

The distinction is a difficult one. Three questions suggested by the previous
analysis, however, can provide important clues: Are alternatives being provided
or taken away? Is one's access to new and different information being broadened
or denied? Finally, does the individual assume personal responsibility and control
or is it usurped by the group or by its leader?

The Peoples Temple attracted many of its members because it provided them an
alternative way of viewing their lives; it gave many people who were downtrodden
a sense of purpose, and even transcendence. But it did so at a cost, forcing them
to disown their former friendships and beliefs and teaching them to fear anything
outside of the Temple as "the enemy." Following Jones became the *only* alternative.

Indeed, most of the members grew increasingly unaware of the possibility of
any other course. Within the Peoples Temple, and especially at Jonestown, Jim
Jones controlled the information to which members would be exposed. He effectively stifled any dissent that might arise within the church and instilled a distrust

in each member for contradictory messages from outside. After all, what credibility could be carried by information supplied by "the enemy" that was out to destroy the Peoples Temple with "lies"?

Seeing no alternatives and having no information, a member's capacity for dissent or resistance was minimized. Moreover, for most members, part of the Temple's attraction resulted from their willingness to relinquish much of the responsibility and control over their lives. These were primarily the poor, the minorities, the elderly, and the unsuccessful—they were happy to exchange personal autonomy (with its implicit assumption of personal responsibility for their plights) for security, brotherhood, the illusion of miracles, and the promise of salvation. Stanley Cath, a psychiatrist who has studied the conversion techniques used by cults, generalizes: "Converts have to believe only what they are told. They don't have to think, and this relieves tremendous tensions" (*Newsweek*, 1978a). Even Jeanne Mills, one of the better-educated Temple members, commented:

> I was amazed at how little disagreement there was between the members of this church. Before we joined the church, Al and I couldn't even agree on whom to vote for in the presidential election. Now that we all belonged to a group, family arguments were becoming a thing of the past. There was never a question of who was right, because Jim was always right. When our large household met to discuss family problems, we didn't ask for opinions. Instead, we put the question to the children, "What would Jim do?" It took the difficulty out of life. There was a type of "manifest destiny" which said the Cause was right and would succeed. Jim was right and those who agreed with him were right. If you disagreed with Jim, you were wrong. It was as simple as that. [Mills, 1979.]

Though it is unlikely that he had any formal exposure to the social psychological literature, Jim Jones utilized several very powerful and effective techniques for controlling people's behavior and altering their attitudes. Some analyses have compared his tactics to those involved in "brainwashing," for both include the control of communication, the manipulation of guilt, and dispensing power over people's existence (Lifton, 1979), as well as isolation, an exacting regimen, physical pressure, and the use of confessions (Cahill, 1979). But using the term brainwashing makes the process sound too esoteric and unusual. There *were* some unique and scary elements in Jones' personality—paranoia, delusions of grandeur, sadism, and a preoccupation with suicide. Whatever his personal motivation, however, having formulated his plans and fantasies, he took advantage of well-established social psychological tactics to carry them out. The decision to have a community destroy itself was crazy, but those who performed the deed were "normal" people who were subjected to a tremendously impactful situation, the victims of powerful internal forces as well as external pressures.

POSTSCRIPT

Within a few weeks of the deaths at Jonestown, the bodies had been transported back to the United States, the remnants of the Peoples Temple membership were said to have disbanded, and the spate of stories and books about the sui-cide/murders had begun to lose the public's attention. Three months afterwards, Michael Prokes, who had escaped from Jonestown because he was assigned to carry away a box of Peoples Temple funds, called a press conference in a California motel room. After claiming that Jones had been misunderstood and de-manding the release of a tape recording of the final minutes [quoted earlier], he stepped into the bathroom and shot himself in the head. He left behind a note, saying that if his death inspired another book about Jonestown, it was worthwhile (*Newsweek,* 1979).

POSTSCRIPT

Jeanne and Al Mills were among the most vocal of the Peoples Temples critics following their defection, and they topped an alleged "death list" of its enemies. Even after Jonestown, the Mills's had repeatedly expressed fear for their lives. Well over a year after the Peoples Temple deaths, they and their daughter were murdered in their Berkeley home. Their teen-aged son, himself an ex-Peoples Temple member, has testified that he was in another part of the large house at the time. At this writing, no suspect has been charged. There are indications that the Mills's knew their killer—there were no signs of forced entry, and they were shot at close range. Jeanne Mills had been quoted as saying, "It's going to happen. If not to-day, then tomorrow." On the final tape of Jonestown, Jim Jones had blamed Jeanne Mills by name, and had promised that his followers in San Francisco "will not take our death in vain" (*Newsweek,* 1980).

References

ARONSON, E. *The social animal* (3rd ed.) San Francisco: W. H. Freeman and Company, 1980.

ARONSON, E. The theory of cognitive dissonance: A current perspective. In L. Berkowitz (ed.), *Advances in experimental social psychology.* Vol. 4. New York: Academic Press, 1969.

ARONSON E., AND CARLSMITH, J. M. Effect of the severity of threat on the devaluation of forbidden behavior. *Journal of Abnormal and Social Psychology,* 1963, **66,** 584–588.

ARONSON, E., AND MILLS, J. The effects of severity of initiation on liking for a group. *Journal of Abnormal and Social Psychology.* 1959, 59, 177–181.

ASCH, S. Opinions and social pressure. *Scientific American,* 1955, 193. (Also reprinted in this volume.)

Tape hints early decision by Jones on mass suicide. *Baltimore Sun,* March 15, 1979.

BLAKEY, D. Affidavit: Georgetown, Guyana. May 12, 1978.

BLAKEY, D. Affidavit: San Francisco. June 15, 1978.

BREHM, J. Increasing cognitive dissonance by a *fait-accompli. Journal of Abnormal and Social Psychology,* 1959, 58, 379–382.

CAHILL, T. In the valley of the shadow of death. *Rolling Stone.* January 25, 1979.

Committee on Foreign Affairs, U.S. House of Representatives, Report of a staff investigative group. *The assassination of Representative Leo J. Ryan and the Jonestown, Guyana tragedy.* Washington, D.C.: Government Printing Office, May 15, 1979. (Many of the other press reports are reprinted in this volume.)

CRAWFORD, Y. Affidavit: San Francisco. April 10, 1978.

DARLEY, J., AND BERSCHEID, E. Increased liking as a result of the anticipation of personal contact. *Human Relations,* 1967, **20,** 29–40.

DAVIS, K., AND JONES, E. Changes in interpersonal perception as a means of reducing cognitive dissonance. *Journal of Abnormal and Social Psychology,* 1960, **61,** 402–410.

"Don't be afraid to die" and another victim of Jonestown. *Newsweek,* March 10, 1980.

A fatal prophecy is fulfilled. *Newsweek,* March 10, 1980.

FREEDMAN, J. Long-term behavioral effects of cognitive dissonance. *Journal of Experimental Social Psychology,* 1965, **1,** 145–155.

FREEDMAN, J., AND FRASER, S. Compliance without pressure: The foot-in-the-door technique. *Journal of Personality and Social Psychology,* 1966, **4,** 195–202.

GALLAGHER, N. Jonestown: The survivors' story. *New York Times Magazine,* November 18, 1979.

KASINDORF, J. Jim Jones: The seduction of San Francisco. *New West,* December 18, 1978.

KILDUFF, M., AND JAVERS, R. *The suicide cult.* New York: Bantam, 1978, and *San Francisco Chronicle.*

KILDUFF, M., AND TRACY, P. Inside Peoples Temple. *New West,* August 1, 1977.

KRAUSE, C. *Guyana massacre.* New York: Berkley, 1978, and *Washington Post.*

LIFTON, R. J. Appeal of the death trip. *New York Times Magazine,* January 7, 1979.

MILGRAM, S. Behavioral study of obedience. *Journal of Abnormal and Social Psychology,* 1963, 67, 371–378. (Also reprinted in E. Aronson (ed.), *Readings about the social animal.*)

MILGRAM, S. Liberating effects of group pressure. *Journal of Personality and Social Psychology,* 1965, **1,** 127–134.

MILLS, J. *Six years with God.* New York: A & W Publishers, 1979.

REITERMAN, T. Scared too long. *San Francisco Examiner,* November 13, 1977.

The sounds of death. *Newsweek,* December 18, 1978b.

Special report: The cult of death. *Newsweek,* December 4, 1978a.

TRACY, P. Jim Jones: The making of a madman. *New West,* December 18, 1978.

WINFREY, C. Why 900 died in Guyana. *New York Times Magazine,* February 25, 1979.

YOUNG, S. Report on Jonestown. *The CBS Evening News with Walter Cronkite.* November 16, 1979.

ZIMBARDO, P. *The cognitive control of motivation.* Glenview, Ill.: Scott Foresman, 1969.

ZIMBARDO, P., EBBESON, E., AND MASLACH, C. *Influencing attitudes and changing behavior* (2nd ed.). Reading, Mass.: Addison-Wesley, 1977.

III

MASS COMMUNICATION, PROPAGANDA, AND PERSUASION

7

Effects of Varying the Recommendations in a Fear-Arousing Communication

James M. Dabbs, Jr. and Howard Leventhal

It has been suggested that divergent effects of fear arousal on attitude change can be caused by variations in the recommendations in a persuasive communication. In a three-way factorial design Ss were presented with communications manipulating fear of tetanus and the perceived effectiveness and painfulness of inoculation against tetanus. Inoculation was recommended for all Ss. It was expected that more Ss would take shots described as highly effective and not painful, and that this tendency would change as level of fear was increased. The manipulations of effectiveness and painfulness were perceived as intended, but they did not affect intentions to take shots or shot-taking behavior. The fear manipulation influenced both intentions and behavior, with higher fear producing greater compliance with the recommendations.

A number of studies have investigated the effects of fear arousal on persuasion. Although the majority report that fear increases persuasion, the picture is not completely clear. Facilitating effects of fear on persuasion have been reported in studies of dental hygiene practices (Haefner, 1965; Leventhal and Singer, 1966; Singer, 1965), tetanus inoculations (Leventhal, Jones, and Trembly, 1966; Leventhal, Singer, and Jones, 1965), safe driving practices (Berkowitz and Cottingham, 1960; Leventhal and Niles, 1965), and cigarette smoking (Insko, Arkoff, and Insko, 1965; Leventhal and Watts, 1966; Niles, 1964). However, under increasing levels of fear Janis and Feshbach (1953) observed no increase in acceptance of beliefs about the proper type of toothbrush to use, and Leventhal and Niles (1964) observed some decrease in acceptance of a recommendation to stop smoking. All these results are based on verbal measures of *attitude* change.

Reprinted with permission from the authors and *The Journal of Personality and Social Psychology,* Vol. 4, No, 1966. Copyright 1966 by the American Psychological Association.

The present study was supported in part by United States Public Health Service Grant CH00077-04 to Howard Leventhal. The authors wish to thank John S. Hathaway and James S. Davie of the Yale University Department of University Health for their many contributions to the research.

The picture is even less clear when one considers actual *behavior* change. The study on tetanus by Leventhal et al. (1965) reports that some minimal amount of fear is necessary for behavior change, but that further increases in fear do not affect change. The later study by Leventhal et al. (1966) reports a slight tendency for increases in fear to increase behavior change. In the studies of dental hygiene practices, Janis and Feshbach (1953) reported decreased behavior change under high fear, while Singer (1965) found no main effect of fear. In the study on smoking by Leventhal and Watts (1966) high fear simultaneously increased compliance with a recommendation to cut down on smoking and decreased compliance with a recommendation to take an X-ray.

The present study attempted to account for some of these divergent findings. It is reasonable to expect that the behavior being recommended by a persuasive communication is of critical importance (Leventhal, 1965). Recommendations that are seen as effective in controlling danger may be accepted more readily as fear is increased, while ineffective recommendations may be rejected rationally or may produce reactions of denial (Janis and Feshbach, 1953) or aggression (Janis and Terwilliger, 1962). Additionally, rejection of a recommended behavior may occur if subjects have become afraid of the behavior itself (Leventhal and Watts, 1966).

Recommendations presented as part of fear-arousing communications vary in their effectiveness in controlling danger and in the unpleasantness associated with them. For example, brushing the teeth offers no guarantee of preventing decay, while taking a chest X-ray can lead to the unpleasant discovery of lung cancer. An audience might well reject recommendations which are ineffective in warding off danger or which are difficult, painful, or apt to bring unpleasant consequences. Recommendation factors have been invoked post hoc to explain research findings, but have not been manipulated and studied directly.

In the present study fear was manipulated by presenting differing discussions of the danger of tetanus. Inoculations and booster shots were recommended for protection against tetanus. Under high and low levels of fear, inoculation was portrayed so that it would be seen as more or less *effective* in preventing tetanus and more or less *painful* to take (these manipulations were orthogonal). Subjects' intentions to take shots and their actual shot-taking behavior were used to measure compliance with the recommendations.

It was expected that compliance would be greater when shots were highly effective or not painful. These factors might produce simple main effects or they might interact with level of fear. It seemed more likely that the latter would be the case—that increased fear would make subjects either more or less sensitive to differences in the recommendations.

METHOD

Subjects and Design

Letters were sent to all Yale College seniors asking them to participate in a study to be conducted jointly by the John Slade Ely Center, a local research organization, and the Department of University Health. The study was presented as a survey of student health practices at Yale and an evaluation of some health-education materials. An attempt was then made to contact all seniors by telephone for scheduling.

Of approximately 1000 students who received letters, 274 were scheduled and run in the experiment. Seventy-seven of these were excluded because they had been inoculated since the preceding academic year, and 15 were excluded because of suspicion, involvement in compulsory inoculation programs, allergic reactions to inoculation, or religious convictions against inoculation. The final usable N was 182.

Each subject received a communication which was intended to manipulate perceived effectiveness and painfulness of inoculation. Three levels of fear (including a no-fear control level), two levels of effectiveness, and two levels of pain were combined in a $3 \times 2 \times 2$ factorial design. The n's for the resulting 12 conditions ranged from 11 to 20, with smaller n's in the no-fear control conditions.

Procedure

Experimental sessions were conducted in a classroom with groups ranging in size from 1 to 12. Subjects within each session were randomly assigned to conditions. Control (no-fear) conditions were run separately because of the brevity of the control communications.

Questionnaires containing medical items and personality premeasures were administered at the beginning of the session. Subjects then read a communication on tetanus and gave their reactions to it in a second questionnaire. They were assured that all their responses would be kept confidential.

Communications

Communications were ten-page pamphlets which discussed the danger of tetanus and the effectiveness and painfulness of inoculation.[1] All pamphlets gave spe-

---

[1] Communications have been deposited with the American Documentation Institute. Order Document No. 9011 from ADI Auxiliary Publications Project, Photoduplication Service, Library of Congress, Washington, D.C. 20540. Remit in advance $2.00 for microfilm or $3.75 for photocopies and make checks payable to: Chief, Photoduplication Service, Library of Congress.

cific instructions on how to become inoculated and were similar in style and content to those used by Leventhal et al. (1965) and Leventhal et al. (1966).

Fear. Low-fear material described the very low incidence of tetanus and indicated that bleeding from a wound usually flushes the poison-producing bacilli out of the body. A case history was included which reported recovery from tetanus following mild medication and throat-suction procedures. High-fear material indicated that tetanus can be contracted through seemingly trivial means and that if contracted the chances of death are high. A high-fear case history was included which reported death from tetanus despite heavy medication and surgery to relieve throat congestion. Black-and-white photographs were included in the low-fear material and color photographs in the high-fear material. The discussion of tetanus and case history were omitted entirely from control (no-fear) communications.

Effectiveness. The effectiveness manipulation stressed either the imperfections or the unusual effectiveness of inoculation. Low-effectiveness material stated that inoculation is generally effective and about as adequate as the measures available to deal with other kinds of danger. It pointed out, however, that no protection is perfect and that there is a possibility that even an inoculated person will contract tetanus. High-effectiveness material described inoculation as almost perfect and as far superior to methods available to deal with other kinds of danger. It emphasized that inoculation reduces the chances of contracting tetanus, for all practical purposes, to zero. All communications reported that a new type of inoculation was available at the Department of University Health which would provide protection against tetanus for a period of 10 years.

Pain. To produce fear of the recommended behavior, it was pointed out that inoculation against tetanus has always been painful. Subjects were told that the new inoculation requires a deep intramuscular injection of tetanus toxoid and alum precipitate, making the injection even more painful than before and the local reaction longer lasting. The discussion of pain was presented to subjects as a forewarning so that the discomfort would not take them by surprise. This discussion was omitted from pamphlets in the no-pain conditions.

Specific instructions on how to get a shot and a map showing the location of the Department of University Health were included in all pamphlets. Subjects were encouraged to get a shot or at least to check on whether or not they needed one.

Measures

Most questions on the pre- and postcommunication questionnaires were answered on 7-point rating scales. The precommunication questionnaire contained medical questions and four personality measures (susceptibility, coping, anxiety, and self-esteem). The susceptibility measure was made up of three items which asked subjects how susceptible they felt toward common illnesses, toward unusual diseases, and toward illness and disease in general. Three items used to measure coping asked subjects whether they tended to tackle problems actively or to postpone dealing with them. The problems concerned subjects' health habits, their everyday lives as students, and their decisions regarding summer activities. The anxiety scale was made up of 10 true-false items from the Taylor (1953) Manifest Anxiety scale.

The self-esteem measure was similar to that used by Dabbs (1964). Subjects rated themselves on 20 adjectives and descriptive phrases and then rated each of the 20 items as to its desirability. Eight items were classified as desirable and 12 as undersirable on the basis of mean ratings from the entire sample of subjects. Using this group criterion of desirability, each subject's self-esteem score was defined as the sum of his ratings on the desirable items. (It was subsequently discovered that 12 subjects in the present study had participated in the earlier study by Dabbs, and the correlation between their self-esteem scores in the two studies was .62, $p < .01$. This correlation, despite a lapse of 3 years and changes in the measuring instrument, suggests this type of measure is reasonably stable.)

The postcommunication questionnaire included checks on each of the experimental manipulations, a 10-item mood adjective check list, and questions on intentions to take shots, the importance of shots, and the likelihood of contracting tetanus. The subject's evaluation of the pamphlet and the date and place of his last tetanus shot were obtained on this questionnaire.

A measure of behavioral compliance with the recommendations was obtained from shot records of the Department of University Health. Subjects were counted as complying if they took tetanus shots between the experimental sessions and the end of the semester, about one month later. When contacted by letter and phone, no subjects reported receiving shots at places other than the Department of University Health. A few reported that they had tried to take shots and had been told they did not need any. These subjects were counted as having taken shots but their data are presented separately in footnote 2.

TABLE 7.1
Compliance with recommendations

	Control (no fear)	Low fear	High fear
Mean intentions to take shots	4.12	4.73	5.17
Proportion of Ss taking shots	.06	.13	.22
N	48	62	72

RESULTS

Main Effects of the Manipulations

Compliance with the recommendations was unaffected by the manipulations of effectiveness and pain. The manipulation of fear, however, influenced both intentions to take shots ($F_{2,179} = 4.85$, $p < .01$) and actual shot-taking behavior[2] ($F = 3.39$, $p < .05$). The effects were linear (Table 6.1), with compliance being greatest under high fear. The consistency of subjects' responses is indicated by high biserial correlations between intention and behavior measures within high-fear ($r_b = .62$, $p < .01$) and low-fear ($r_b = .68$, $p < .01$) conditions; no correlation was computed for the control conditions since only three control subjects took shots.

Table 7.2 shows that the fear manipulation increased feelings of fear, as it was intended to do. It also increased feelings of interest and nausea, belief in the severity of tetanus and the importance of taking shots, and desire to have additional information. None of these measures were affected by the manipulations of effectiveness or pain.

TABLE 7.2
Other reactions to fear manipulation

	Control	Low fear	High fear	F^a
Fear[b]	7.74	9.25	12.19	21.50**
Feelings of nausea	1.27	1.24	1.90	7.95**
Feelings of interest	4.48	4.70	5.39	6.12**
Evaluation of the severity of tetanus	4.64	4.83	5.34	3.86*
Feelings of susceptibility to tetanus	3.46	3.99	4.25	3.31*
Evaluation of the importance of shots	5.85	6.16	6.52	6.44**
Desire for more information about tetanus	3.08	2.68	3.57	3.90*

[a] F ratios are computed from three-way analyses of variance. In each analysis $df = 2/170$ (approximately).
[b] "Fear" represents the sum of three items: feelings of fear, fear of contracting tetanus, and fear produced by the pamphlet.
*$p < .05$.
**$p < .01$.

[2] "Shot-taking behavior" combines 20 subjects who took shots and 7 who reported trying to take shots. These two categories are distributed similarly across the fear treatment conditions: 2, 6, and 12 subjects took shots and 1, 2, and 4 subjects tried to do so.
An arc sine transformation of the proportions in Table 7.1 was used (Winer, 1962). This made it possible to test the significance of the differences between groups against the *baseline variance* of the transformation (see Gilson and Abelson, 1965). Base-line variance has a theoretical value which does not depend on computations from observed data. In the present case this value is given by the reciprocal of the harmonic mean number of cases on which each proportion is based, or $1/59 = .0169$.

Both effectiveness and pain were successfully manipulated. Check questions showed that subjects in the high-effectiveness conditions felt inoculation was more effective than did those in the low-effectiveness conditions ($\overline{X}_{high} = 6.7$, $\overline{X}_{low} = 5.8$, $F_{1,170} = 105.85$, $p < .01$). Subjects in the pain conditions felt shots would be more painful ($\overline{X}_{pain} = 4.1$, $\overline{X}_{no\ pain} = 2.3$, $F_{1,170} = 74.56$, $p < .01$) and reported more "mixed feelings" about taking shots ($F_{1,170} = 7.20$, $p < .01$) than did subjects in the no-pain conditions. But the clear perception of differences in effectiveness did not affect subjects' intentions to take shots, nor did increasing the anticipated painfulness of shots decrease intentions to take them. In fact, there was a slight tendency for painfulness to strengthen intentions to be inoculated ($F_{1,170} = 2.72$, $p = .10$).

Correlations Among Responses

Only within the high-fear condition did reported fear correlate with intentions to take shots ($r = .23$, $p < .01$). A scatterplot of scores revealed that the high-fear condition increased the range of reported fear and that the positive correlation could be attributed to subjects in the extended portion of the range (those scoring higher than 13 on a 19-point composite scale). These extreme subjects all showed strong intentions to take shots, while other subjects in the high-fear condition sometimes did and sometimes did not intend to take shots. This pattern suggests that fear and acceptance of a recommendation are more closely associated when fear is relatively high, though acceptance may occur at any level of reported fear.

Unlike fear, anger was negatively associated with intent to take shots. The overall within-class correlation between anger and intentions was $-.18$ ($p < .05$). This correlation remained essentially the same within different levels of the fear treatment, but became increasingly negative as the recommendation was portrayed as less effective and more painful (Table 7.3). Both the row and column differences

TABLE 7.3
Correlations between anger and intentions to take shots under varying portrayal of recommendations

	Effectiveness	
	Low	High
Pain		
Low	−.29	.15
High	−.52	−.18

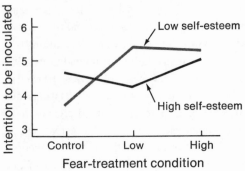

FIGURE 7.1
Self-esteem differences in reactions to the
communications on tetanus.

in Table 7.3 are significant[3] (for rows, $CR = 2.02$, $p < .05$; for columns, $CR = 2.81$, $p < .01$). It should be emphasized that these are correlational differences only. Low effectiveness and high pain did not increase the mean level of anger or decrease intentions (or decrease actual shot taking). The ranges of anger and intention scores also did not differ among the four conditions.

Personality Differences

None of the premeasures on personality (susceptibility, coping, anxiety, self-esteem) were significantly correlated with intentions to take shots, nor did the correlations vary systematically across the 12 experimental conditions. However, differences were observed when subjects were split at the median into groups high and low on self-esteem subjects. Figure 7.1 shows intentions to take shots among high- and low-self-esteem subjects. The only significant effect within this data (other than the main effect of fear) is the interaction between self-esteem and fear level ($F_{2,166} = 4.74$, $p < .01$). Subjects low in self-esteem increased their intentions to take shots from control to low-fear conditions, then showed no further increase under high fear. Subjects high in self-esteem, on the other hand, showed increased intentions only from low- to high-fear conditions.

DISCUSSION

A positive relationship between fear arousal and persuasion was observed. Increases in the intensity of the fear manipulation were associated with increases in

---

[3] Significance of these differences was tested after applying Fisher's z' transformation to the correlation coefficients.

attitude and behavior change, with high correlations between intentions to take shots and actual shot taking. These findings are similar to those of Leventhal et al. (1966), who reported a slight tendency for shot taking to increase as fear was raised from low to high levels.

Subjects' beliefs about the effectiveness of inoculation did not affect their compliance; they responded equally well to recommendations portrayed as low and high in effectiveness. This may be because even the low-effective recommendation was rated relatively high in effectiveness (5.8 on a 7-point scale). However, the manipulation was sufficient for subjects to perceive significant differences between low and high effectiveness, and the failure of this variable to influence compliance suggests caution in using it to reconcile divergent results of studies on fear arousal and persuasion (Janis and Leventhal, 1967; Leventhal, 1965; Leventhal and Singer, 1966).

The description of pain produced mixed feelings about shots, but did not prevent subjects from taking them. Perhaps this is because the discomfort of inoculation is negligible in comparison with the pain of tetanus itself. A stronger manipulation of anticipated "painfulness" was unintentionally introduced in the study by Leventhal and Watts (1966), who created fear of smoking by showing a film in which a chest X-ray led to the discovery of cancer and to surgical removal of a lung. The authors suggested that decreased X-ray taking in this condition was more likely caused by fear of the consequences of an X-ray than by defensive reactions to the fear-arousing material on cancer. The present findings do not invalidate their conclusion, but they limit the range of situations to which such an explanation might apply. Subjects spparently do not respond to small variations in the effectiveness or unpleasantness of a recommended course of action. Unless a compelling deterrent exists, people who anticipate danger prefer to do something rather than nothing.

This last statement is qualified by differences in the behavior of high- and low-self-esteem subjects. Low-self-esteem subjects showed high compliance with the recommendations in both high- and low-fear conditions, while high-self-esteem subjects showed high compliance only in the high-fear condition. In addition, personality measures of self-esteem and coping were significantly correlated ($r = .49$, $p < .01$) in the present study, as they were in the study reported by Dabbs (1964). One might conclude that high-self-esteem subjects are more active and aggressive in dealing with their environment and have developed more skill in meeting dangers with appropriate protective actions. Thus, they may recognize inoculation to be more appropriate when the danger of tetanus is greater, while low-self-esteem subjects may accept the position of the communication that inoculation is appropriate regardless of the magnitude of danger. An alternative possibility is that some differences characteristically associated with self-esteem simply disappear when there is an urgent need to combat danger (as there was with the high-fear communication).

In the conditions where inoculation was depicted as ineffective or painful, anger was negatively associated with intentions to be inoculated. It is possible that increased anger in these conditions would have lowered compliance. But since anger and compliance did not covary as the fear treatment was increased, there appears to be no causal relationship between them. It seems more likely that anger does not decrease compliance, but that under certain conditions it provides the justification for noncompliance. Under other conditions—as when recommended behaviors are highly effective, not painful, and not a reasonable target for anger—noncompliance may have to be justified in some other manner.

All the present findings could have been influenced by several factors which were not varied. For example, all subjects received the manipulations of perceived danger, effectiveness, and pain in the same order. Learning about the dangers of tetanus first may have caused some subjects to ignore differences in the effectiveness and painfulness of inoculation. In addition, all subjects were asked after reading the communications whether or not they intended to be inoculated. One might expect that stating an intention to be inoculated would commit the subject to following this course of behavior later on. However, the findings of Leventhal et al. (1965) and Leventhal et al. (1966) indicate that specific instructions must be present before intentions will be translated into behavior. Since all subjects in the present study did receive specific instructions, the high correlations between intentions and behavior may be due to this factor.

Finally, while fear and persuasions were associated, the evidence of any causal relationship between them is tenuous. As in the most studies of fear arousal and attitude change, the communications used were complex ones. They discussed the damage tetanus can cause and the likelihood of contracting it. They differed in fearfulness, interest, and feelings of nausea evoked. They also differed in length, type of language used, and amount of information about tetanus. With such confounding it is impossible to attribute increases in attitude change solely to increases in fear. This is an inherent difficulty when fear is manipulated by the use of differing descriptions of danger. Another approach should be developed by anyone wishing to manipulate fear independently of other aspects of a persuasive communication.

References

BERKOWITZ, L., AND COTTINGHAM, D. R. The interest value and relevance of fear-arousing communications. *Journal of Abnormal and Social Psychology*, 1960, **60**, 37–43.

DABBS, J. M., JR. Self-esteem, communicator characteristics, and attitude change. *Journal of Abnormal and Social Psychology*, 1964, **69**, 173–181.

GILSON, C., AND ABELSON, R. P. The subjective use of inductive evidence. *Journal of Personality and Social Psychology,* 1965, **2,** 301–310.

HAEFNER, D. P. Arousing fear in dental health education. *Journal of Public Health Dentistry,* 1965, **25,** 140–146.

INSKO, C. A., ARKOFF, A., AND INSKO, V. M. Effects of high and low fear-arousing communications upon opinions toward smoking. *Journal of Experimental Social Psychology,* 1965, **1,** 256–266.

JANIS, I. L., AND FESHBACH, S. Effects of fear-arousing communications. *Journal of Abnormal and Social Psychology,* 1953, **48,** 78–92.

JANIS, I. L., AND LEVENTHAL, H. Human reactions to stress. In E. Borgatta and W. Lambert (eds.), *Handbook of personality theory and research.* New York: Rand McNally, 1967.

JANIS, I. L., AND TERWILLIGER, R. An experimental study of psychological resistances to fear-arousing communications. *Journal of Abnormal and Social Psychology,* 1962, **65,** 403–410.

LEVENTHAL, H. Fear communications in the acceptance of preventive health practices. *Bulletin of the New York Academy of Medicine,* 1965, **41,** 1144–1168.

LEVENTHAL, H., JONES, S., AND TREMBLY, G. Sex differences in attitude and behavior change under conditions of fear and specific instructions. *Journal of Experimental Social Psychology,* 1966, **2,** 387–399.

LEVENTHAL, H., AND NILES, P. A field experiment on fear arousal with data on the validity of questionnaire measures. *Journal of Personality,* 1964, **32,** 459–479.

LEVENTHAL, H., AND NILES, P. Persistence of influence for varying durations of threat stimuli. *Psychological Reports,* 1965, **16,** 223–233.

LEVENTHAL, H., AND SINGER, R. P. Affect arousal and positioning of recommendations in persuasive communications. *Journal of Personality and Social Psychology,* 1966, **4,** 137–146.

LEVENTHAL, H., SINGER, R. P., AND JONES, S. Effects of fear and specificity of recommendations upon attitudes and behavior. *Journal of Personality and Social Psychology,* 1965, **2,** 20–29.

LEVENTHAL, H., AND WATTS, J. C. Sources of resistance to fear-arousing communications on smoking and lung cancer. *Journal of Personality,* 1966, **34,** 155–175.

NILES, P. The relationships of susceptibility and anxiety to acceptance of fear-arousing communications. Unpublished doctoral dissertation, Yale University, 1964.

SINGER, R. P. The effects of fear-arousing communications on attitude change and behavior. Unpublished doctoral dissertation, University of Connecticut, 1965. *Dissertation Abstracts,* 1966, **26,** 5574.

TAYLOR, J. A. A personality scale of manifest anxiety. *Journal of Abnormal and Social Psychology,* 1953, **48,** 285–290.

WINER, B. J. *Statistical principles in experimental design.* New York: McGraw-Hill, 1962.

8

Attribution Versus Persuasion as a Means for Modifying Behavior

Richard L. Miller, Philip Brickman, and Diana Bolen

The present research compared the relative effectiveness of an attribution strategy with a persuasion strategy in changing behavior. Study 1 attempted to teach fifth graders not to litter and to clean up after others. An attribution group was repeatedly told that they were neat and tidy people, a persuasion group was repeatedly told that they should be neat and tidy, and a control group received no treatment. Attribution proved considerably more effective in modifying behavior. Study 2 tried to discover whether similar effects would hold for a more central aspect of school performance, math achievement and self-esteem, and whether an attribution of ability would be as effective as an attribution of motivation. Repeatedly attributing to second graders either the ability or the motivation to do well in math proved more effective than comparable persuasion or no-treatment control groups, although a group receiving straight reinforcement for math problem-solving behavior also did well. It is suggested that persuasion often suffers because it involves a negative attribution (a person should be what he is not), while attribution generally gains because it disguises persuasive intent.

Despite the volume of research on attitude change and persuasion, there is surprisingly little evidence that persuasion can be effective, particularly if a criterion of persistence of change over time is applied (Festinger, 1964; Greenwald, 1965b;

Reprinted with permission from the authors and *The Journal of Personality and Social Psychology*, Vol. 31, No. 3, 1975. Copyright 1975 by the American Psychological Association.

This research was carried out as a master's thesis by the first author under the supervision of the second author and was partially supported by National Science Foundation Grant GS-28178. We are grateful to Donald Campbell and Thomas Cook for serving on the thesis committee and to Lawrence Becker and Glenn Takata for help with the data analysis. Thanks are also due to Susan Bussey and Charlotte Yeh, who served as experimenters, and to Bobbe Miller, who did much of the initial coding and typed the early drafts of this article.

This work would have been impossible without the enthusiasm, cooperation, and enlightened support of Irene Timko, principal of Anderson Elementary School. Great thanks are also due to the teachers who participated in this project. In Study 1, thanks go to Catherine Lebenyi, Wendy Weiss, and Ira Mizell for willingly helping to initiate a new behavioral modification technique. In Study 2, where an incredible number of experimental manipulations were administered by the teachers, special thanks go to Margaret Paffrath, Ronnie Briskman, June Butalla, and Susan Levie. Thanks are also due to the eighth-grade assistants who administered our tests.

Rokeach, 1968; Zimbardo and Ebbesen, 1969; Cook, 1969). The failure of persuasive efforts to produce lasting change may be taken as evidence that subjects
have not integrated the new information into their own belief systems (Kelman,
1958) or taken it as the basis for making an attribution about themselves (Kelley,
1967). We might expect that a persuasive communication specifically designed to
manipulate the attributions a person made about himself would be more effective
in producing and maintaining change. This research was designed to test the relative importance of attribution manipulations to persuasive attempts by comparing
a normal persuasion treatment and an attribution treatment.

The persuasion conditions of the present research were designed to be maximally
effective through their use of a variety of techniques which have been found to be
helpful, at least on occasion, in past research. Past research has shown that an
optimal persuasive manipulation should involve a high-credibility source (Hovland
and Weiss, 1951) delivering a repeated message (Staats and Staats, 1958) with an
explicitly stated conclusion (Hovland and Mandell, 1952) which is supported by
arguments pointing out the benefits of change (Greenwald, 1965a) and overlearned
by the audience (Cook and Wadsworth, 1972). Face-to-face communication by the
source (Jecker, Maccoby, Breitrose, and Rose, 1964), reinstatement of the source
at the time of attitude assessment (Kelman and Hovland, 1953), and active role
playing or participation by the audience in the message (Janis and King, 1954) are
also helpful.

The attribution techniques were also designed to be maximally effective through
their use of all three factors specified by Kelley (1967) as conducive to making a
stable attribution: consistency of the evidence over time, consistency of the evidence over modalities, and consistency or consensus across sources.

STUDY 1: LITTERING BEHAVIOR

Study 1 attempted to modify children's littering behavior. Behavior was monitored
before and after treatment and again after a 2-week period of nontreatment. It was
hypothesized that both the attribution and the persuasion conditions would result
in initial posttreatment behavioral change but that the attribution condition would
show greater persistence as a result of altering the basic self-concept of the subjects in a direction inconsistent with littering.

METHOD

Participants

The research took place in three fifth-grade classrooms in an inner-city Chicago
public school. Two fifth-grade classrooms were randomly assigned to the experi-

mental conditions, while a third was designated a control group. Three female experimenters, all undergraduate psychology majors at Northwestern University, were randomly assigned to a different classroom for each test.

Experimental Manipulations

There were a total of 8 days of attribution and persuasion treatments dealing with littering, with discussion intended to average about 45 minutes per day.

Attribution condition.   On Day 1, the teacher commended the class for being ecology-minded and not throwing candy wrappers on the auditorium floor during that day's school assembly. Also on Day 1, the teacher passed on a comment ostensibly made by the janitor that their class was one of the cleanest in the building. On Day 2, after a visiting class had left the classroom, the teacher commented that paper had been left on the floor but pointed out that "our class is clean and would not do that." The students at this point disagreed pointedly and remarked that they would and did indeed litter. On Day 3, one student picked up some paper discarded on the floor by another and after disposing it in the wastebasket was commended by the teacher for her ecology consciousness. On Day 4, Row 1 was pointed out as being the exceptionally neat row in the room by the teacher. Also on Day 4, the principal visited the class and commented briefly on how orderly it appeared. After the principal left the room, the students castigated the teacher for her desk being the only messy one in the room. On Day 5, a large poster of a Peanuts character saying "We are Andersen's Litter-Conscious Class" was pinned to the class bulletin board. Also on Day 5, the teacher gave a lesson on ecology and talked about what we "the class" are doing to help. On Day 6, the principal sent the following letter to the class: "As I talked to your teacher, I could not help but notice how very clean and orderly your room appeared. A young lady near the teacher's desk was seen picking up around her desk. It is quite evident that each of you are very careful in your section." On Day 7, the teacher talked about why "our class" was so much neater. In the interchange the students made a number of positive self-attributions concerning littering. On Day 8, the janitors washed the floor and ostensibly left a note on the blackboard saying that it was easy to clean.

Persuasion condition.   On Day 1 during a field trip, the children were told about ecology, the dangers of pollution, and the contribution of littering to pollution. They were then asked to role play being a trash collector and to pick up litter as they came across it. On Day 2, inside the school lunchroom the teacher talked about garbage left by students and gave reasons why it should be thrown away: it looked terrible, drew flies, and was a danger to health. On Day 3, the teacher gave a lecture on ecology, pollution, and litter and discussed with the class how the situation could be improved. Also on Day 3, the teacher passed on a comment ostensibly from the school janitor that they needed help from the students in keeping the floors clean, implying here as elsewhere that nonlittering would lead to approval and commendation by various adult authorities. On Day 4, the teacher told the students that everyone should be neat, mentioning aesthetics among other reasons for neatness. Also on Day 4, the principal visited the class and commented briefly about the need for clean and tidy classrooms. On Day 5, the teacher told the students that they should not throw candy wrappers on the floor or the playground but should dispose of them in trash cans. Also on Day 5, a large poster of a Peanuts character saying "Don't be a litterbug" with "Be neat" and "Don't Litter" bordering it was pinned to the class bulletin board. On Day 6, the principal sent the following letter to the class: "As I talked to your teacher, I could not help but notice that your room was in need of some cleaning. It is very important that we be neat and orderly in the upkeep of our school and classrooms. I hope each of you in your section will be very careful about litter." On Day 7, the teacher appointed several children in each row to watch and see if people were neat outside the building as well as in the classroom. On Day 8, a note was left on the board ostensibly from the janitors to remind the children to pick up papers off the floor.

Measurement of Littering

Pretest. To discover any existing differences among the three classrooms with respect to their tendency to litter, a specially marked reading assignment which had previously been turned in to the teachers was returned to the students 5 minutes before the end of the school day. The students were then instructed to throw the assignment away after the bell rang for dismissal. After school the experimenters counted the number of assignments thrown in the wastebasket versus left on the floor or on the shelf under the students' seats. Less than 20% of the students in each class disposed of their assignments in the wastebaskets. The precise percentages were 20% for the control group ($n = 31$), 16% for the persuasion group ($n = 26$), and 15% for the attribution group ($n = 27$).

Posttest. A two-part behavioral test was designed to tap the two aspects of the ecology-littering problem, nonlittering and cleaning up the litter of others.

On the morning of the tenth day, minutes before the first recess, each teacher introduced the experimenter for her classroom as a marketing representative of a local candy manufacturing firm and left the experimenter in charge of the room. The experimenter explained that she was testing the tastiness of a new brand of candy and passed out one piece of candy to each student. The candy was wrapped in colored cellophane with a different color used for each classroom. Following the taste test, the class was dismissed for recess. During recess the experimenters counted the number of candy wrappers in the wastebaskets, on the floor, and in the desk seats. The experimenters then relittered the classroom entrance area with seven specially marked candy wrappers. After recess the experimenters checked the hallway and playground for dicarded candy wrappers. During the lunch break, which came 1 hour after recess, the experimenters reentered the classroom and determined the disposition of the specially marked candy wrappers.

The second posttest followed the first by a period of 2 weeks. During this time no mention of ecology or littering was permitted in any of the classrooms. The second test was very similar to the first except that this time, the experimenters did not interact with the students. Ten minutes before the afternoon recess the teacher passed out toy puzzles as Christmas presents from the Parent Teachers Association. The students were asked to try to work the puzzles before recess. Each puzzle was wrapped in a color-coded container with a different color assigned to each class. During recess the experimenters entered the classroom and determined the disposition of the containers left there. They then relittered the entrance way. After recess the experimenters searched the hallway and playground for other containers. After school the experimenters reentered the classroom and determined the disposition of the relittered containers.

RESULTS AND DISCUSSION

Littering Behavior

Figure 8.1 charts the percentage of items in each group which were discarded in the wastebasket on each test. A chi-square test was used to compare frequency of littering in the three groups on the immediate and delayed posttests. Although the measures directly reflect items of litter rather than individual subjects, it was observed by the experimenters that subjects independently discarded their own candy wrappers in the wastebaskets. The three groups were significantly different at both the immediate posttest, χ^2 (2) = 18.14, $p < .001$, and at the delayed posttest, χ^2 (2) = 20.99, $p < .001$. The attribution group was significantly superior to the persuasion group on both the immediate posttest, χ^2 (1) = 7.19, $p < .01$, and the delayed posttest, χ^2 (1) = 16.15, $p < .01$. Although the persuasion group appeared to show an immediate increase in litter-conscious behavior, it was not significantly different from the control group even on the immediate posttest, χ^2 (1) = 2.57, *ns*.

Cleanup Behavior

All seven items were picked up by members of the attribution group on both the immediate and the delayed posttest. Persuasion group members picked up four items on the immediate posttest and two on the delayed posttest, while control group members picked up two items on the immediate posttest and three on the

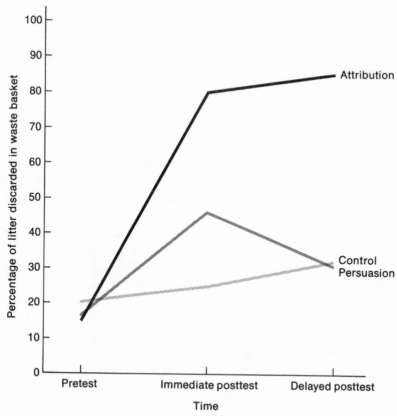

FIGURE 8.1
Nonlittering behavior of the attribution, persuasion, and control groups over time (Study 1).

delayed test. Since the total number of wrappers left was only seven and more than one wrapper may sometimes have been picked up by a single individual, a chi-square test is not fully appropriate. Nonetheless, such a test would be significant at the .05 level at each posttest, which supports the appearance of differences favoring the attribution group.

After this study the attribution teacher was advised that the nonlittering behavior could perhaps be maintained if the students were occasionally reminded of the attribution "You are neat." Three months later the teacher reported that her class was still significantly neater than it had been prior to treatment.

The results for both the littering test and the cleanup test support the hypothesis that attribution is a more effective technique than persuasion for inducing stable behavioral change. We would like to show, however, that this effect holds for other kinds of behavior. Furthermore, it would be desirable to overcome a weakness in Study 1 that arose from the fact that treatments were nested within classrooms. It is possible, if relatively unlikely, that the differences emerging over time were due to teacher differences rather than treatment differences. Study 2 avoids this by including all treatment and control conditions in each classroom.

STUDY 2: MATH ACHIEVEMENT

The results of Study 1 are certainly encouraging for an attributional approach to modifying behavior. However, while ecology consciousness and nonlittering are of some social importance, they are not the primary focus of the schools, which is, at least theoretically, to teach skills. Will attribution and persuasion techniques show the same pattern of effectiveness in generating a skilled behavior, like math achievement, as they have in generating a socially desirable but unskilled behavior, like disposing of trash in wastebaskets? Furthermore, littering behavior would certainly be only a weakly valenced aspect of self, while most of the skills taught in schools would be highly valenced and of considerable import for a student's self-concept. Will attribution be an aspect? The first purpose of Study 2 was to answer just these questions. It might also be noted that attributions in Study 2, in addition to being more central to self-concept, are specifically directed to particular individuals rather than addressed to a group as a whole.

The second purpose of Study 2 was to test the relative effectiveness of attributions of motivation versus attributions of ability in changing behavior. Both perceptions of ability and motivation are essential to the belief that a person will attain a given goal (Heider, 1958). Study 1, however, would seem to have involved primarily the attribution of motivation, since the children presumably began with a common belief that they had the ability to be neat. In the case of a skilled behavior like arithmetic, however, it would seem more likely that motivation and belief in one's motivation to do well are more common than ability and belief in one's ability to do well (Katz, 1964), so that attributions of ability would be of greater value than attributions of motivation. Nonetheless, enhancing people's perceptions of their motivation for a task may also benefit their performance. Study 2 attempts to separate ability and motivation as the bases of attribution and the targets of persuasive appeal.

So far we have only considered cognitive strategies for modifying attitudes and behavior. Staats (1965) has shown that even young children will engage in complex learning tasks if they are simply given appropriate reinforcement. According to Bandura (1969), a successful reinforcement strategy for behavior modification requires a valued reinforcer which is contingent upon the desired behavior and a reliable procedure for eliciting the desired behavior. In the present study both verbal praise and extrinsic rewards were used as reinforcers for efforts at mathematical achievement, and a number of overlapping procedures were used to elicit these efforts.

To compare the relative efficacy of the attribution and reinforcement techniques with standard attitude change approaches, a persuasive manipulation was devised similar to the one used in Study 1. The only changes were that audience participation and role playing were deleted, since neither was appropriate to the treatment conditions, while public labeling was added. It appears from the study of deviance (Becker, 1963) that public labeling of a person can lead that person to redefine himself along the lines of the label. While it was felt in Study 1 that children had to be convinced of the benefits of nonlittering in the persuasion condition, the advantages of math achievement as a means for obtaining rewards in school seemed too obvious to need pointing out.

The present study attempted to modify children's math-related self-esteem and their math scores on skill tests. The six conditions were attribution ability, attribution motivation, persuasion ability, persuasion motivation, reinforcement control, and a no-message control. It was hypothesized that all three basic techniques (attribution, persuasion, and reinforcement) should have an initial positive effect on the self-esteem and math behavior of the subjects but that attribution should have the most enduring effect over time.

METHOD

Participants

The research took place in four second-grade classrooms of the same inner-city Chicago public school involved in Study 1. Second-grade students were picked, since it was felt on the basis of Rosenthal and Jacobson (1968) that their school-related self-concepts would be more malleable than at a later age. In all, 96 students took part in the study. All five experimental conditions and one control condition were present in each of the four classrooms. From each class list of approximately 30 students, 24 were randomly assigned to the six possible conditions. Thus 4 students in each classroom appeared in each condition.

Overview of Procedure

All subjects first received a mathematics pretest and a self-esteem pretest. Subsequent treatments consisted of 8 days of attribution, persuasion, or reinforcement. Immediately following the treatment, math and self-esteem posttests were administered. Delayed posttests were given after a 2-week period of no treatment. The control group received the pretests and the immediate and delayed posttests but no treatment. Student absences for both treatments and tests were made up on the day the student returned to school.

Experimental Manipulations

Five treatment techniques were used with all groups: verbal comments, written comments, letters from the teacher, letters from the principal, and medals. The above order is followed in discussing these techniques for each experimental condition. It should also be noted that in the attribution and persuasion conditions, students were initially called to the principal's office in groups of eight, where they received a treatment-related message. The principal discontinued these treatments after the third day of the experiment, however, on the grounds that they were too time-consuming for her and that she found the false attribution treatments too difficult to deliver.

All treatment techniques were prepackaged. Before the experiment began, teachers' treatment packages were prepared which listed the treatment techniques and their recipients for each day. The order in which each subject received the treatments was randomized for classroom "A" and repeated in each of the other classrooms.

Attribution ability.   The general focus of this treatment was attributing to the students skill and knowledge in mathematics. Three different verbal comments were made by the teacher to each student on different days: "You are doing very well in arithmetic," "You are a very good arithmetic student," and "You seem to know your arithmetic assignments well." Three different written notes were tied to assignments on different days and handed back to the students: "You're doing very well," "excellent work," and "very good work." The letters from the teachers and principal underscored the students' excellent work in math and were sent home on days when a verbal or written note was not scheduled. The letter from the teacher included the phrases "very good student," "does all his assignments well," and "excellent arithmetic ability." The letter from the principal used the phrases "excellent ability," "knows his assignments," and "very good student." The medals awarded to the attribution ability students featured the words "good student—math."

Attribution motivation.   The general focus of this treatment was attributing hard work and consistent trying to the student. Three appropriate verbal comments were made by the teacher to each child privately, and three written notes were appended to a test or assignment. The verbal comments were the following: "You really work hard in arithmetic," "You're working harder in arithmetic," and "You're trying more in arithmetic." The written comments were as follows: "You're working harder, good!" "You're trying more, keep at it!" and "Keep trying harder!" The letters from the teacher and principal underscored the child's application in math. The teacher's letter used the phrases "working hard," "trying," and "applying himself." The principal's letter used the phrases "working harder," "applying himself," and "trying harder." The medals awarded to the attribution motivation students featured the words "hard worker—math."

Persuasion ability. The general focus of this treatment was to persuade the student that he should be good in arithmetic and doing well in that subject. Three verbal comments and three written comments summarizing that message were made in the same manner as that for the attribution messages. The three verbal comments were "You should be good at arithmetic," "You should be a good arithmetic student," and "You should be doing well in arithmetic." The three written comments were "should be better," "should be good at arithmetic," and "should be getting better grades." The letters from the teacher and principal used ostensible "aptitude test scores" to inform parents that their child should be making good grades in math. The teacher's letter included the injunctions "should be doing well," "should be getting high grades," and "should be becoming a good arithmetic student." The principal's letter used the phrases "should get good grades," "should do very well," and "should be a good arithmetic student." The medals awarded to these students contained the words "do better—math."

Persuasion motivation. The general focus of this treatment was to persuade the student that he should be working harder and spending more time on math. The three verbal and the three written comments asked the child to try more at math and were made in the same way as they were in the other treatment conditions. The three verbal comments were "You should spend more time on arithmetic," "You should work harder at arithmetic," and "You should try more on arithmetic." The three written notes were "Try harder," "Work more on arithmetic," and "Work harder." The letters from the teachers and the principal informed parents that their child should spend more time on math and to pass that idea along to him. The teachers' letter included the injunctions "should be trying harder," "should spend more time on arithmetic," and "should be applying himself." The principal's letter used the phrases "spend more time" and "try harder." The medals awarded the persuasion motivation students contained the words "work harder—math."

Reinforcement. The reinforcement condition also followed the same format as the attribution and persuasion conditions except that it added two additional methods and deleted the principal's comments. Three verbal comments indicating pride in the student's good work were made by the teacher. They were as follows: "I'm proud of your work," "I'm pleased with your progress in arithmetic," and "very good." Three written comments of simple praise were appended to the student's math work. They were "excellent," "very good," and "I'm very happy with your work." The letters from the teachers and principal indicated pride and satisfaction in the child's work to the parent. The teacher's letter used the phrases "proud of his work," "good grades," and "happy with his work." The principal's letter used the phrases "excellent progress," "doing very well," and "proud of him." The medals awarded those students contained the words "math award." On Days 2, 5, and 7, students were praised verbally if they chose to work on an extra math problem rather than a reading exercise. On two other days the students received silver stars by solving a problem from a math assignment.

Control. The control condition received no treatment whatsoever but took part in all tests of mathematical ability and self-esteem.

MEASUREMENTS

Self-Esteem

The self-esteem pretest was an adaptation of the questionnaire originally developed by Rogers and Dymond (1954) that includes items which measure self-esteem with regard to peers and parents, school interests, and personal interests. Four new items were added which dealt specifically with math-related self-esteem. All items were declarative sentences generally of the "I am ———" form, from which the children were asked to say whether it was "like me" or "unlike me." The self-esteem pretest and posttest were both administered individually to each student by one of a dozen eighth-grade assistants, who asked each question privately to the second graders and personally recorded the subject's answer. The self-esteem posttest was given in the morning on the day after treatment ended.

Mathematics

The pretest and all subsequent math tests were 20-item tests consisting of 25% review questions, 50% current material, and 25% new material. The math pretest was administered by the teachers as a regular arithmetic quiz. The immediate posttest was administered in the afternoon on the day after treatment ceased. The final posttest was administered after a 2-week period in which no treatment took place.

RESULTS

Self-Esteem

The mean math self-esteem scores for all six conditions on the pretest and the posttest are presented in Figure 8.2. Analysis of variance of the pretest scores indicated that there were no significant differences among the five treatment groups and the control group in math self-esteem before treatment began, $F(5, 90) = .79$, *ns*.

The major analysis of the math self-esteem scores was a repeated measures analysis of variance using the six experimental conditions and the two times of measurement as factors. Results indicated both a main effect of time, $F(1, 90) = 10.84$,

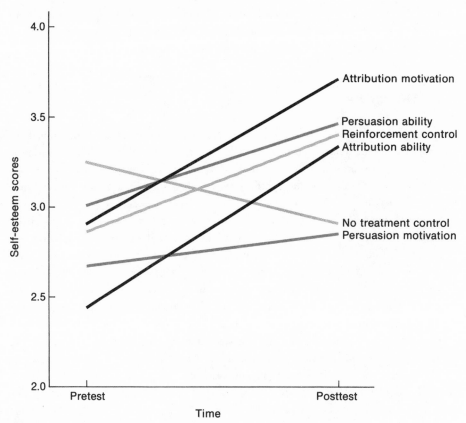

FIGURE 8.2
Self-esteem scores for the attribution, persuasion, and control groups over time (Study 2).

$p < .01$, and a significant Time $\times$ Treatment interaction, $F (5, 90) = 2.41$, $p < .05$. As can be seen from Figure 8.2, all treatment groups show an increase in self-esteem, while the control group shows a decrease, which is probably due to the lack of any treatment for control students in the face of a number of public treatments visible to them in their classroom.

A subsidiary analysis of variance was performed using only the attribution and the persuasion groups, with mode of treatment (attribution vs. persuasion), basis of treatment (ability manipulation vs. motivation manipulation), and time as the three independent variables. Results again indicated a main effect of time, $F (1, 60) = 19.02$, $p < .001$, and a significant Time $\times$ Mode of Treatment interaction, $F (1, 60) = 4.76$, $p < .05$. The math self-esteem of the attribution groups increased more sharply from pretest (2.66) to posttest (3.50) then did the math self-esteem of the persuasion groups (2.84 at pretest, 3.13 at posttest). Only the attribution ability and attribution motivation groups were significantly different from the no-treatment control group in their change from pretest to posttest, $F (1, 30) = 7.58$, $p < .01$ for attribution ability, $F (1, 30) = 7.66$, $p < .01$ for attribution motivation.

This difference approached but fell short of significance for the persuasion ability group, $F (1, 30) = 3.12$, $p < .11$. Similar analyses showed no significant effects on general school-related and non-school-related self-esteem.

Mathematics

The mean total math scores for all six conditions on the pretest, the immediate posttest, and the delayed posttest are presented in Figure 8.3. Analysis of variance of the pretest scores indicated that there were no significant differences among the five treatment groups and the control group in their total math scores before treatment began, $F (5, 90) = .54$, *ns*.

The major analysis of the math test scores was a repeated measures analysis of variance using the six experimental conditions and the three times of measurement as factors. The results show both a main effect of time, $F (2, 180) = 36.69$, $p < .001$, and a significant Time × Treatment interaction, $F (10, 180) = 5.57$, $p < .001$. From Figure 8.3 it can be seen that both attribution conditions show marked increases on the immediate posttest followed by a slight tendency to enlarge that increase after the 2-week delay. Both persuasion conditions appear to show an increase on the immediate posttest but fail to maintain that increase over the 2-week delay. The reinforcement condition shows a pattern similar to that of the attribution conditions but with a lesser degree of improvement.

To assess the extent to which attribution and persuasion differed in their initial effectiveness versus the extent to which they differed in their ability to maintain their effectiveness over time, separate analyses were made of changes from the pretest to the immediate posttest and from the immediate posttest to the delayed posttest. Mode of treatment (attribution vs. persuasion) and basis of treatment (ability vs. motivation) were used as factors along with time of test. For change from pretest to the immediate posttest, a significant Mode × Time interaction, $F (1, 60) = 11.97$, $p < .001$, indicated that attribution was significantly more effective than persuasion in inducing change. Only the attribution ability and attribution motivation groups were significanlty different from the no-treatment control group in their change from pretest to immediate posttest, $F (1, 30) = 14.75$, $p < .01$ for attribution ability; $F (1, 30) = 11.42$, $p < .01$ for attribution motivation. This difference approached but fell short of significance for the persuasion ability group, $F (1, 30) = 2.92$, $p < .10$.

For change from the immediate posttest to the delayed posttest, another significant Mode × Time interaction, $F (1, 60) = 13.67$, $p < .001$, indicated that the attribution treatments were also superior to the persuasion treatment in maintaining what change they produced. The attribution treatments show an increase of .50 from the immediate to the delayed posttest, while the persuasion treatments show a decrease of .63.

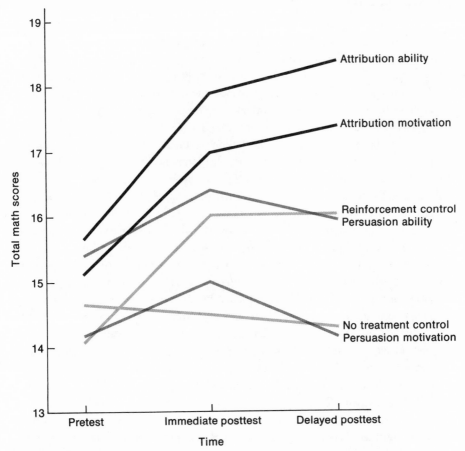

FIGURE 8.3
Total math scores of the attribution, persuasion, and control groups over time (Study 2).

Finally, to make a preliminary test as to whether attribution and persuasion were similar in their effects on high- and low-ability students, subjects in each condition were divided at the cell median according to their math pretest performance. An analysis of variance on posttest math scores was conducted using pretest ability (high vs. low), treatment (attribution vs. persuasion, ignoring basis of treatment), and posttest time (immediate vs. delayed) as factors. Initially more able students continued to outperform initially less able students, $F(1, 60) = 46.01$, $p < .001$, on the posttests. Attribution was more effective than persuasion for both groups, $F(1, 60) = 19.39$, $p < .001$, but the attribution–persuasion difference was significantly larger for the low-ability students than for the high-ability students, interaction $F(1, 60) = 5.66$, $p < .05$. Average posttest scores for the high-ability students were 18.8 for attribution and 17.8 for persuasion, while for low-ability students they were 16.5 for attribution for 13.0 for persuasion.

DISCUSSION

In both studies the attribution treatments caused a significant change which persisted over time. These treatments were strong enough to overcome counterarguing by subjects in Study 1 and a history of at best modest success among many
subjects in Study 2. Both attributions based upon subjects' ability to do something
and those based upon subjects' motivation to do it appeared effective. The effects
of persuasion were, in general, insignificant and dissipated over time.

The fact that the superiority of attribution treatments over persuasion treatments
was demonstrated in two different field experiments using different behaviors
(nonlittering and math problem solving), two different subject populations (fifth
graders and second graders), and two different designs (a between-classroom design and a within-classroom design) gives us some confidence that these effects
are real and generalizable. Neither study is without weakness. The nesting of
treatments within classrooms in Study 1 leaves teacher or group differences as a
possible, if unlikely, alternative explanation for the treatment effects that emerged
over time. The public nature of the treatments in each classroom in Study 2 means
that treatment effects may have been aided by implicit comparisons students were
making between their own condition and other treatments, a process which must
lie behind the unexpected drop in math self-esteem shown by the no-treatment control
group. (This last result must also be counted among the ethical costs of a within-
classroom design.) The weaknesses of the two studies, however, are quite different, while their effects are quite similar, which suggests that the results are not
due to idiosyncrasies of design.

The present research provides a general framework into which previous work
concerning the effects of teacher expectancies on pupil performance (Beez, 1968;
Rosenthal and Jacobson, 1968; Seaver, 1973) can be fit. The means by which teacher
expectancies are communicated is at best a dependent variable in previous studies
(Meichenbaum, Bowers, and Ross, 1969) and often mysteriously unobserved. The
present study made the communication of expectancies, in the form of attributions, its central manipulation. The fact that this programmed communication of
expectancies worked as it did provides some support for the essential validity of
the often elusive teacher expectancy effect.

Effects of Attribution

That attributions based upon ability and attributions based upon motivation did
not differ in their effectiveness implies that direct linkage of skill-specific attributions to the self-system is more important than the basis on which this linkage is
made. The message "You are a particular kind of person" is more important than
the specification of "why." It should be noted, however, that the present research
contrasted only two kinds of internal attributions, ability and motivation. Attribu-

tions made to a person on the basis of external factors ("You are neat because I am watching you") would presumably be less effective in producing lasting change.

Attribution can, of course, involve elements of persuasion.. As we have seen, the statement "You are a neat person" may be a most effective means of persuading someone to be neat. Nonetheless, such attributional statements need not involve persuasive intent but may instead be simple statements of fact. Indeed, their guise as truth statements may be thought of as their most effective advantage. Not only does this enable them to work directly on a person's self-concept, as noted, but it may also enable them to slip by the defenses a person ordinarily employs against persuasive attempts effective because it is less easily recognized as persuasion, and hence less likely to arouse resistance, counterarguing, or reactance.

In Study 1 the attribution treatment did elicit counterarguing by the students, which suggests a possible reactance (Brehm, 1972) or boomerang effect (McGuire, 1969), but there was no evidence of such an effect by the end of the experiment. It is possible that reactance, like other attitude-change forces, dissipates over time as the issue is worked through and the treatment is maintained. It is also possible that the elementary school students who were subjects in the present studies are less likely to perceive the manipulative intent of an attribution treatment and less likely to show reactance than a comparable adult group.

Attribution and Persuasion

In accounting for the relative ineffectiveness of persuasion, we may note first of all that persuasive communications urging a person to do something do not necessarily tap the internal self-concept of their target. Worse yet, to the extent that they do implicate self-concept, they may involve the negative attribution that the person is not the kind of individual who engages in the recommended behavior. An appeal to be neat or an appeal to work hard can involve the implicit attribution that the person is not currently the sort of individual who is neat or works hard. If convincing people that they are neat or hard working is the key to making them neat or hard working, a naive persuasive attempt can cancel out its own message. At best it attributes to its target the potential for becoming the sort of person recommended, but this is clearly much weaker than the attribution that the person already embodies the desired behavior.

In this study, moreover, the persuasive messages implied something negative about the subjects, while the attribution treatments implied something positive. Although blame as well as praise has been shown to be effective in eliciting improved performance (see Kennedy and Willcutt, 1964, for a review of the mixed results in this area), one of the best designed studies (Hurlock, 1925) found that the improvement elicited by blame dissipated over time, while that elicited by praise persisted. It may be speculated that persuasion and punishment both remain effective in motivating behavior only so long as the actors feel that they can accomplish what is being called for. As they accept the implicit negative attributions of a persuasive message, the effectiveness of the message diminishes. This may explain why persuasion was relatively more effective for high-ability students than for low-ability students. The high-ability students may have been better able to respond to the appeal to do better without becoming discouraged by the implicit attribution (which they could to some extent discount) that they were not currently demonstrating appropriate accomplishment.

The implicit attributions of the persuasion treatments in this study were negative because the behaviors they were calling for (which the subjects were presumably not emitting) were positive. Likewise, the attribution treatments were positive because the behaviors they attributed to the subjects were positive. For practical and ethical reasons, all of the treatments in the present study were aimed at producing positive or socially desirable behaviors. It is our idea that the implicit or explicit labeling itself, and not merely the rewardingness or punishingness of the labels, made the present attribution treatments effective and the persuasion treatments ineffective. This can only be known for sure, however, by a study that aims at producing undesirable behavior, in which case attributions of the behavior would have negative implications, which persuasion to the behavior would have implicit positive implications for the present self-concept. If people were responding to the attributions in the present experiment only because they were rewarding, we would expect that attributions of an undesirable behavior would have reverse effects and, indeed, that persuasion under these circumstances might even be more effective. The sociological literature on deviance (e.g., Schur, 1971), however, suggests that negative attributions (e.g., labeling as delinquent) can indeed produce or support the attributed behavior, as does a recent experimental study of labeling "charitable" or "uncharitable" behaviors (Kraut, 1973). We may suspect, then, that the positive implications of the attributions in the present study were not the sole key to their effectiveness and that if a suitable experiment could be designed in which the target behavior were socially undesirable, attribution would continue to be more effective than persuasion in generating that behavior.

Attribution and Reinforcement

A straightforward reward contingency program seems to modify behavior because it makes that behavior worthwhile to the subject. However, the separation between reinforcement and attribution seems somewhat confounded. Symbolic reinforcement, as used in the present study, has some attributional aspects. Simple praise is often interpreted as a "You are X" statement. Furthermore, attribution can contain elements of reinforcement especially when socially desirable behaviors are the focus of the attributional process. Thus to some extent a reinforcement procedure that produces enduring change may require elements of attribution, while a successful attribution treatment may involve elements of reinforcement.

The remaining feature distinguishing attribution from reinforcement would seem to be the noncontingent nature of the attribution. In this regard it is interesting to note that Kazdin (1973) has recently found that under circumstances where the desired behavior was emitted at a fairly high base rate and subjects believed that reinforcement was contingent, noncontingent reinforcements were as effective as contingent ones in modifying behavior. Attribution treatments, however, may have the very important further advantage over simple reinforcement of serving to elicit behavior (like modeling; see Bandura, 1969) as well as to maintain it.

Practical and Ethical Implications

The present study supports the idea that an effort to improve the child's academic self-concept will help improve academic performance, if only because the improved self-image will make actual success less inconsistent, less unexpected, and less likely to be discounted or rejected (Brickman, 1972). A distinguishing feature of the present attribution treatments is that they focused on raising self-esteem in a specific area of skill rather than raising global or general self-esteem. The failure of more general ''cultural enrichment'' or general ''self-concept enhancement'' programs may be due in part to the fact that these global manipulations have only vague and diffuse implications for particular areas of academic performance (but cf. deCharms, 1972).

Nonetheless, there are a number of reasons to be cautious about considering these results as the basis for a solution to any social problem. It is unlikely that long-standing individual differences in accomplishment will be overcome by short-term manipulations of motivation, incentive, or self-regard. A gain of three problems solved after a week of treatment is not very substantial in terms of life chances, and there is no reason to assume that 10 weeks of treatment will necessarily result in a gain of 30 problems. Second, the attribution treatments in Study 2 were difficult to administer and, unlike the treatments in Study 1, did not produce any immediate and visible indications of success to sustain teacher enthusiasm. While the teachers involved in Study 1 were quite positive toward the study, those involved in Study 2 had decidedly mixed feelings about the value of the time and energy involved. More seriously, the false attributions came increasingly to be felt by at least one teacher and by the principal as an intolerable risk to their credibility. As indicated, the principal terminated her office meetings with the students after the third day of treatment mainly on these grounds. While future research could tailor attributions not to be too discrepant from individual pretest baselines, the practical and ethical difficulties involved in maintaining such attributions will not thereby be eliminated. All of these matters warn against an uncritical application of the present results to matters of educational importance.

References

BANDURA, A. *Principles of behavior modification.* New York: Holt, Rinehart & Winston, 1969.

BECKER, H. S. *Outsiders: Studies in the sociology of deviance.* New York: Free Press, 1963.

BEEZ, W. V. Influence of biased psychological reports on teacher behavior and pupil performance. *Proceedings of the 76th Annual Convention of the American Psychological Association,* 1968, **3**, 605–606. (Summary)

BREHM, J. *Responses to loss of freedom: A theory of psychological reactance.* Morristown, N.J.: General Learning Press, 1972.

BRICKMAN, P. Rational and non-rational elements in reactions to disconfirmation of performance expectancies. *Journal of Experimental Social Psychology,* 1972, **8**, 112–123.

COOK, T. D. *The persistence of induced attitude change: A critical review of pertinent theories.* Unpublished manuscript, Northwestern University, 1969.

COOK, T. D., AND WADSWORTH, A. Attitude change and the paired-associate learning of mimal cognitive elements. *Journal of Personality,* 1972, **40**, 50–61.

DECHARMS, R. Personal causation training in the schools. *Journal of Applied Social Psychology,* 1972, **2**, 95–113.

FESTINGER, L. Behavioral support for opinion change. *Public Opinion Quarterly,* 1964, **28**, 404–417.

GREENWALD, A. G. Behavior change following a persuasive communication. *Journal of Personality,* 1965a, **33**, 370–391.

GREENWALD, A. G. Effects of prior commitment on behavior change after a persuasive communication. *Public Opinion Quarterly,* 1965b, **29**, 595–601.

HEIDER, F. *The psychology of interpersonal relations.* New York: Wiley, 1958.

HOVLAND, C. I., AND MANDELL, W. An experimental comparison of conclusion drawing by the communicator and the audience. *Journal of Abnormal and Social Psychology,* 1952, **47**, 581–588.

HOVLAND, C. I., AND WEISS, W. The influence of source credibility on communication effectiveness. *Public Opinion Quarterly,* 1951, **15**, 635–650.

HURLOCK, E. B. An evaluation of certain incentives used in school work. *Journal of Educational Psychology,* 1925, **16**, 145–159.

JANIS, I. L., AND KING, B. T. The influence of role playing on opinion change. *Journal of Abnormal and Social Psychology,* 1954, **49**, 211–218.

JECKER, J., MACCOBY, N., BREITROSE, H. S., AND ROSE, E. D. Teacher accuracy in assessing cognitive visual feedback from students. *Journal of Applied Psychology,* 1964, **48**, 393–397.

KATZ, I. Review of evidence relating to effects of desegregation on the intellectual performance of Negroes. *American Psychologist,* 1964, **19**, 381–399.

KAZDIN, A. E. The role of instructions and reinforcement in behavior changes in token reinforcement programs. *Journal of Educational Psychology,* 1973, **64**, 63–71.

KELLEY, H. H. Attribution theory in social psychology. *Nebraska Symposium on Motivation,* 1967, **14**, 192–238.

KELMAN, H. C. Compliance, identification, and internalization: Three processes of opinion change. *Journal of Conflict Resolution,* 1958, **2**, 51–60.

KELMAN, H. C., AND HOVLAND, C. I. "Reinstatement" of the communicator in delayed measurement of opinion change. *Journal of Abnormal and Social Psychology,* 1953, **48**, 327–335.

KENNEDY, W. A., AND WILLCUTT, H. C. Praise and blame as incentives. *Psychological Bulletin,* 1964, **62**, 323–332.

KRAUT, R. E. Effects of social labeling on giving to charity. *Journal of Experimental Social Psychology,* 1973, **9**, 551–562.

MCGUIRE, W. J. The nature of attitudes and attitude change. In G. Lindzey and E. Aronson (eds.), *Handbook of social psychology* (2nd ed.). Reading, Mass.: Addison-Wesley, 1969.

MEICHENBAUM, D. H., BOWERS, K. S. AND ROSS, R. R. A behavioral analysis of teacher expectancy effect. *Journal of Personality and Social Psychology,* 1969, **13**, 306–316.

Rogers, C. and Dymond, R. *Psychotherapy and personality change.* Chicago: University of Chicago Press, 1954.

Rokeach, M. *Beliefs, attitudes, and values.* San Francisco: Jossey-Bass, 1968.

Rosenthal, R., and Jacobson, L. *Pygmalion in the classroom: Teacher expectation and pupils' intellectual development.* New York: Holt, Rinehart & Winston, 1968.

Schur, E. M. *Labeling deviant behavior.* New York: Harper & Row, 1971.

Seaver, W. B. Effects of naturally induced teacher expectancies. *Journal of Personality and Social Psychology,* 1973, **28,** 333–342.

Staats, A. W. A case in and a strategy for the extension of learning principles to problems of human behavior. In L. Krasner and L. P. Ullman (eds.), *Research in behavior modification,* New York: Holt, Rinehart & Winston, 1965.

Staats, A. W., and Staats, C. K. Attitudes established by classical conditioning. *Journal of Abnormal and Social Psychology,* 1958, **57,** 37–40.

Zimbardo, P., and Ebbesen, E. B. *Influencing attitudes and changing behavior.* Reading, Mass.: Addison-Wesley, 1969.

9

Television Criminology: Network Illusions of Criminal Justice Realities

Craig Haney and John Manzolati

Twenty million Americans watch as a burly police lieutenant converses with a district attorney about a case that has the city police force baffled.

"Somebody's gotta stop that nut before he kills again."

"Right. But we got rules around here. You guys can't just rough up witnesses because you think they're holding out. Wait until the Commissioner finds out about this."

"Look, somebody's got to protect people from animals like that. All the laws in the world won't stop one man with a gun. It's going to take me or someone like me. And you know what? I'll do it any way I can."

And so it goes. In the last fifteen years the amount of time on television devoted to crime drama has increased by over 50 percent. So, too, has the amount of time most Americans spend watching television. By the time a child graduates from high school, he or she will have spent an average of over fifteen thousand hours watching television and only eleven thousand hours in school (Kaye, 1974).

Television is primarily an educational medium—its important effects are informational. In sometimes subtle ways, it teaches us and our children about our world and the way it works. The importance of these lessons is enhanced when we have no alternative source of information about whatever television shows us. It is in this context that the significance of "television criminology"—what the television communicates to us about crime and criminal justice—must be taken.

Crime is not only enormously pervasive on television, but it is pervasive *in the absence of virtually any other source of information about crime*. Even when we are victimized by criminals, we learn very little about the crime itself—except that our stereo is gone or that our purse has been taken. We learn nothing about *why* the crime was committed and usually not even *who* committed it. If a suspect is apprehended, our criminal justice system—for perhaps understandable reasons— functions to keep victim and perpetrator apart. A friend of ours in California was robbed recently. The police actually threatened *him* with arrest if he did not cease in his attempts to talk with the suspect in his case. Our friend wanted to find out why he had been selected as the target of this crime and to learn more about the motives of the person who had victimized him, but he was prevented from doing so. Where do most people turn when they wish to find out this and more general information about crime? To television, of course.

Recognizing the importance of television in "informing" people about crime, we and a group of our students began content analyzing television crime drama in 1974.[1] Since that time we have analyzed several hundred hours of randomly selected crime drama on television. We have reached some conclusions about the accuracy of the information that is presented and about some of the consequences of prolonged exposure to these images of crime.

TELEVISION CRIMINALS: THE MAD AND THE BAD

Television has become perhaps our single most important source of information about crime and criminal justice.[2] In a society where there are few opportunities to learn directly about the causes of crime or the realities of law enforcement, television tells us what crimes to fear most, who commits crime and for what reason, and how law enforcement officials can, do, and should respond to it. In many important ways, television drama seriously misrepresents the realities of the criminal justice system. These misrepresentations influence our expectations about crime and law enforcement.

Gerbner and Gross (1976) have found that television actually inflates people's estimates of the likelihood that they will be victimized by crime, and also that it

---

[1] We are especially grateful to Valerie Kam, Tina Lum, Guy Kawasaki, and Michael Rubin, who served as research assistants in various phases of this project.

[2] News media report primarily on the occurrence and frequency of crime, not on its reasons or context. Since crime is "news" when it first occurs, the identity of the perpetrators as well as their motives is typically unknown. By the time such facts are uncovered—except for the most sensational cases—the story is "stale" and consequently goes unreported.

increases the amount of fear they experience about crime. But this finding should be interpreted in light of studies which show that the more people think that crime is on the increase and the more they fear victimization themselves, the more likely they are to favor harsh punishment for criminals. Thus, television may convince people that their world is more threatened by criminals than in fact it is, and in so doing lead them to advocate more punitive solutions to crime than otherwise they would have.

But television conveys a misleading impression of not only the *amount* but also the *kind* of crime that threatens our society. In our content analyses, we found that in contrast to countless hours spent depicting "street crime" on television, white collar crime was vastly underrepresented. Corporate crime (like consumer fraud) and "ecological crime" (like industrial pollution) were *nowhere* to be seen. Conversely, violent crime, especially murder, predominated on the air even though these crimes occur with relative infrequency in real life. This may help to explain why Americans are so much more concerned about street crime than about white collar or corporate crime, even though the latter may inflict far greater economic and social harm.

More importantly, perhaps, we found that television gave a consistent but extremely inaccurate picture of *why* crimes are committed. Crime on television is almost never justified or justifiable. It is rarely, if ever, engaged in for reasons the audience could identify as "good"—ones for which they could feel some sympathy. Often, when such reasons seem to exist at the outset of the program, the criminals turn out to be *mistaken* in their beliefs upon which their criminal actions were based. So, for example, we see family members who had attacked several policemen in avenging the death of their slain brother learn that the cops did not kill him after all. Or we watch a husband who had murdered a blackmailer in an attempt to protect his young wife's honor discover finally that she is both ungrateful and truly dishonorable.

But the anger and frustration that leads to crime in real life is frequently valid. How such emotions are expressed may be another matter, but the feelings themselves are genuine and not often based on erroneous or mistaken beliefs. By always depicting them as otherwise, television helps teach people that they don't have good reasons for their strong feelings. It also diverts our attention away from the sources of these feelings.

Much television crime is presented in the absence of any context or meaning at all. Criminals typically sweep in from nowhere with no backround, no actual or "normal" relationships with anybody—in short, with few human qualities or contacts—and commit their dastardly deed. In fact, in nearly 40 percent of the programs we scored, crime was committed in the first three minutes of the show, before any time *could* be devoted to depicting the criminal as a person in a real human context or situation. This is in stark juxtaposition to the victims and the police, who are usually elaborately portrayed in a network of human, personal, and caring relationships.

Not surprisingly, then, crime on television is almost never the product of situational causes or force of circumstance. Television criminals are depicted as free and autonomous beings, committing crime because they more or less freely choose to. Thus, the causes of crime on television are almost exclusively *personal*, relating directly to the traits or dispositions of the individual lawbreaker and to little else. It is individual pathology and never pathological social conditions that cause television crime and violence.

For example, in less than 5 percent of the shows that we analyzed was *unemployment*—a situational determinant—even remotely implicated as an explanation for the crime committed. This should be interpreted in light of a recent Government Accounting Office study demonstrating that nearly 90 percent of the increase in the number of prisoners admitted to federal penitentiaries was directly attributable to the recent rises in unemployment.

Since criminals on television choose to do things we obviously would not do, they must be different from us. But this inference begs the more relevant question: How would we behave in circumstances like those experienced by most people who commit crime? Since television never shows us these circumstances, it fails to promote this kind of empathetic awareness. Moreover, it diverts our attention away from these "criminogenic" situations and their consequences.

Not only does television blame crime almost exclusively on the traits or characteristics of its perpetrators, but it is highly selective in *which* traits or characteristics it employs as explanations. Specifically, we found that only *two* motives accounted for over 70 percent of the crimes depicted on the screen: Television criminals engage in crime either because they are "crazy" or because they are pathologically greedy.

The crazy or insane criminal is either explicitly labelled as such (e.g., the police sergeant tells the young rookie that the murderer they are looking for "had a mental breakdown after he got back from 'Nam'") or depicted in such a way as to make the inference of mental disturbance inescapable (e.g., the culprit is shown behaving bizarrely with strangers in a context unrelated to the crime). In many shows a psychiatrist appears on a semi-regular basis to do an on-the-spot psychoanalysis of the elusive and irrational criminal.

Pathological greed, on the other hand, is generally communicated indirectly. These criminals are shown living in relatively luxurious surroundings—a big, fancy house, expensive car, and so on—yet they continue to engage in material or economic crime. They have much, but are driven by insatiable greed to criminal acts that will get them still more.

Neither of these caricatures accurately reflects criminal justice realities. Studies of people arrested show that they are *not* significantly more likely to suffer from mental disturbances than comparable groups of nonarrested persons. And to the extent that slightly higher rates of psychological disorder do exist inside prisons, they are readily attributable to the extreme and stressful conditions under which prisoners are confined. Conversely, former mental patients are no more likely to engage in crimes or violence than any other comparable group of "normals." And the typical prisoner is not, nor has he ever been economically well-off. In 1975, for example, the average income of federal prisoners the year before their arrest was $4,000.

Real criminals, then, are usually sane, and they are usually poor. People who are down and out, who are poor, who are leading marginal existences, who are angry—to be sure, who are confused—perhaps, but justifiably so, these people rarely end up on television crime drama. They appear only in our prisons. By distorting the actual picture of crime, television implies solutions to the problem of criminal behavior that may be ill-founded and erroneous, not to say unjust.

Now we are not suggesting that good reasons do not sometimes exist for the distortions that occur in the televised image of crime. The practical realities of having to cram crime drama plots neatly into fifty minute packages may mean that certain kinds of crimes are easier to depict than others. It may be difficult in these short intervals to develop the sensitive and complex crime stories of a Charles Dickens or Victor Hugo.

But neither are we excluding the possibility that additional mechanisms are operating which are less benign. Television is a commercial medium, paid for by sponsors who in turn depend on viewers for their revenue. Corporate and industrial crime may be difficult to portray on television when the bills are being paid by companies that may be engaging in it. In addition, the viewing audience may be less willing to consume the vast array of gadgets and psuedo-products that are paraded before them during commercials if they are reminded regularly of the connection between poverty and crime. Finally, writers concerned with the marketability of their scripts may opt for a "least common denominator" approach to their craft, producing stories that appeal to the most and offend the least number of people. With respect to crime drama this means playing to the popular stereotypes—that criminals are mad and bad while the police are pure and good. The whites hats versus the black.

But whatever the reasons—good or bad—for these inaccuracies and misportrayals, they still occur. And their potential real world consequences are difficult to ignore or dismiss.

GETTING THE PICTURE: THE REAL CONSEQUENCES OF AN UNREAL WORLD

The accuracy of television criminology is of concern not simply because there are few alternative sources from which we can obtain this information. We actually derive "information" about some other topics almost exclusively from television—from what it was like to be a teenager during the 1950s to what the newest Hollywood starlet thinks of her last movie director. What is crucial about television criminology, however, is that not only do we get this information from television exclusively, but we are called upon to use and act on it in a way that has real consequences for people and society.

As citizens we cast votes for particular candidates, in part because of the positions they take on the issue of crime. We evaluate those candidates and the programs and solutions they propose in terms of our own beliefs about crime, the theories *we* have about what causes crime. As citizens, also, we approve or disapprove, applaud or condemn the actions of judges, police, and district attorneys, again because of what we *believe* about crime and its solutions. And the theories and beliefs we employ in making these political decisions appear to be derived primarily from television.

Sometimes we are called upon to make decisions with more immediate consequences than even these, as jurors who must decide upon the guilt or innocence of defendants charged with crimes. Jurors enter the courtroom with a set of preconceptions, expectations, or "prior probabilities," not necessarily about this specific case, but about crime in general. Based on what they believe about "typical" crime, the prosecution and defense theories in the case seem more or less plausible to them. This framework affects how they assess and evaluate the evidence and arguments that are presented to them.

Prosecution and defense attorneys may be similarly affected. Not only do they form their own stereotypes and expectations, but, in addition, they are sensitive to and influenced by what they believe the expectations of the judges and jurors to be. Since 90 percent of all criminal cases never go to trial, attorneys use their "knowledge" of popular stereotypes to predict how jurors *would* behave in this case, so that they can decide whether and how to plea bargain in it.

If people are operating with theories of crime that are derived from television, and these theories are *wrong,* then real world decisions will be made and actual consequences follow from these unreal perspectives. An idea that begins for whatever reasons in the mind of a television crime writer may end up influencing millions of potential voters and jurors. Of course, before such speculation about the way real world decisions are actually influenced by the content of television becomes plausible, we need to determine whether persons are really influenced in their beliefs at all by television criminology. However consistently distorted television crime shows might be, if people disregard or ignore their lessons, these distortions become unimportant.

To get some idea of whether or not people actually internalize television criminology, we administered a questionnaire to several hundred persons, asking them about a variety of criminal justice issues. We also asked them about how much television they watched. We found that people do indeed appear to internalize the belief system that television crime drama provides for them. People who watched a lot of television ("heavy viewers" watching for four or more hours each day) had theories about crime that reflected almost perfectly those that television presented to them.[3] They were, for example, significantly more likely to employ personal or "dispositional" explanations for crime than persons who watched less television ("light viewers," who viewed two or fewer hours per day). Heavy viewers were more likely than light viewers to discount the role of unemployment in crime. They were more likely to associate crime with "abnormality," although they were no more willing than light viewers to absolve criminals of responsibility because of "mental disturbance." They believed crime was irrational and rarely if ever justified by the circumstances.

Interestingly, there was one area where heavy viewers did not adopt the stereotype that television provided for them—with respect to the demographic (especially racial) characteristic of the "typical" criminal. Even though young, black males rarely appear on television as criminals, heavy television viewers were no less likely to identify them as the group most frequently involved in crime. What such viewers apparently do is internalize the television *explanation* for crime, while substituting their own version of its perpetrators. Television images are racially transformed by people to be more consistent with their own personal and stereotypic villains.

CRIMESTOPPERS NOTEBOOK: THE NETWORK COPS

Of course, for every villain in a television crime drama, there is a hero—the policemen who tracks him down. But the image of the police, too, sacrifices realism at the hands of television crime drama writers. Television police work is almost exclusively investigative—the police are constantly shown in dedicated and dogged pursuit of vicious criminals. In real life, however, it is nothing of the sort. Most police work is routine, and only a very small percentage of it can be classified as being in any way investigative.

In fact, a major portion of police budgets is devoted to prosecuting victimless crimes, which are rarely shown on television at all. In 1965, for example, "drunkenness" accounted for almost *one-third* of the total arrests made. When "drunkenness-related" offenses (for example, driving under the influence, disorderly conduct) are included, they constitute nearly 50 percent of all criminal arrests. Yet they too are nowhere to be found on network crime drama.

More significantly, though, television cops *never make a mistake.* In the several hundred shows we scored, we were unable to find a single instance in which the

---

[3] A methodological problem plagues this and all similar studies that divide people into categories of "heavy" and "light" viewers. Since the subjects in these studies are not randomly assigned to conditions, the causes of differences between them cannot be unambiguously identified.

Our "heavy viewers" may have had attitudes consistent with television criminology *before* they began watching television. The pervasiveness of the media stereotypes and the relative absence of alternative sources of information about crime in this culture suggest television as a causal agent. But without elaborate cross-lag correlational designs that are beyond the scope of this study, we cannot be absolutely sure.

wrong man was in custody at the end of the show. (From their position of relative omniscience, of course, viewers typically know when the right man has been caught.) Moreover, television cops are tremendously efficient. In over 70 percent of the shows we scored, the *very first suspect* in the case eventually proved to be the culprit. In about 90 percent of the cases, the first person to be arrested by the television police turned out to be the actual perpetrator.

One consequence of this is the creation of an *illusion of certainty* in police work. Because they actually saw the crime committed, most television viewers are in a position police and jurors never are—they *know* who committed the crime. In real life, police work and prosecution is fraught with uncertainty. This false image of sureness and certainty in criminal justice may actually create in the minds of most viewers a *presumption of guilt* rather than innocence. We found that "heavy viewers" were significantly more likely than "light viewers" to believe that defendants "must be guilty of something, otherwise they wouldn't be brought to trial." It is not too farfetched to speculate that juries composed of heavy television watchers begin with a presumption of guilt and are thus more likely to convict. Indeed, as increasing numbers of Americans watch more and more television crime drama, the phenomenon will become more widespread and juries may become generally more conviction-prone. And this conviction-proneness will be based upon the false reality of television criminology.

THE TELEVISED CONSTITUTION: YOU HAVE THE RIGHT TO . . . WHAT?

Television crime drama provides another kind of "education" for its viewers. Americans are remarkably unknowledgeable about the law and topics as basic as their constitutional rights. Television fills this void by providing a source of information about what police can and cannot legally do, and what rights citizens possess in their contacts with the law. Like much of the other information conveyed about criminal justice, however, what television depicts here is quite inaccurate.

We examined the extent to which the police regularly violate the law and the Constitution in television drama and found that such violations actually occur with alarming frequency—averaging between two and three per hour-long program and ranging from constitutional infractions to outright police brutality. But beyond the simple fact of violation, few television shows actually label the acts as such. In consequence, the viewing public may be lulled into believing that real life police can do more or less *whatever they want*.

We became aware of the magnitude of this problem when we began training our students to code the behavior of the television police. One evening we spent several hours together, taking in a concentrated dose of prime time crime drama. At the end of the session we asked the students to read back to us the infractions they had observed. When they failed to identify many of the most blatant police violations (assaulting witnesses, breaking into suspects' homes, stealing "evidence," and so on), we were surprised and asked them about it. "They're not crimes," they told us, "if the police do it." This belief is as erroneous as it is prevalent. and television has played a large part in its creation.

On television, the police are never reprimanded or punished for their violations, whether they involve infringements of constitutional rights or outright violations of the criminal law. Moreover, in most shows these infractions are depicted as necessary and essential—the *only* way a hard working law enforcement agency can stop vicious, heartless criminals. Illegal searches and seizures on television always turn up the vital missing piece of evidence, and a witness brutalized by a television detective always provides the crucial lead just in time to prevent the next heinous crime from occurring. Indeed, we found that heavy viewers were more likely to endorse the statement, "police must often break the law to stop crime," but were themselves *less* able to identify such violations than were light viewers.

A recent episode of a popular crime drama dealt—in its own peculiar way— with precisely this issue. A young and obviously misguided liberal reporter writes a highly critical article about the two star detectives' unnecessary violence. Near the end of the show, however, she is grabbed by the psychopathic killer that the detectives have been tracking down. Her composure (and her liberal sympathies) disappear immediately as she screams hysterically to the detectives, "Shoot him, shoot him!" But the two detectives, who earlier were shown indiscriminately bullying witnesses and assaulting even the most remote suspects, calmly put down their guns. Even as the crazed killer points his pistol menacingly at them, they advance slowly, compassionately, and unarmed towards him. Finally, soothed by their insightfully chosen words, he surrenders in a burst of tears.

When the naive reporter, still shaken by her brush with the harsh reality of police work, demands, "Why didn't you shoot him?," one of the detectives replies, routinely but with disdain, "Because it wasn't necessary, lady." The message? That only the police, and *never* an outsider, can know what is appropriate in these situations. However unjustified the earlier violence might have appeared to the untrained eye, it was absolutely essential to cracking the case. When violence is not necessary—even when the well-intentioned but uninitiated would use it—the police never do. Would that it were so.

Thus, in the rare case when police violations are labelled as such, they are portrayed as absolutely necessary to effective police work. Moreover, we found that when constitutional rights *are* acknowledged or mentioned on television, this is done so disparagingly. In over 70 percent of the instances in which television cops mention judges, the courts, or the Constitution, the reference is a *negative* one— in terms of their interference with, ignorance of, or obstruction to police work. In less than 10 percent of these cases are these institutions mentioned in a positive way, as making a contribution to solving the problem of crime. The message is a simple one: The law and the Constitution *stand in the way* of effective solutions to our crime problem.

Actually, for the television police, the laws and the Constitution are either impediments to crime control or they are expressions of misplaced concern—civilized protections offered to an uncivilized group of criminals that neither deserves nor understands them. Thus, one interesting ending of television crime shows involves the ironic and sarcastic "reading of rights." In the final scene, after his diabolical scheme has been unravelled by the police, the heartless criminal faces his captors. He has lied, stolen, raped, and murdered, and he stands snivelling before those who have relentlessly pursued him. He begs for mercy. (Keep in mind, since the wrong man is almost never arrested on television, the innocent are never read their rights. We *know* the guy did it. On television the Constitution only protects the guilty.) In a show of remarkable self-restraint—as viewers we want to strangle the guy—the detective stares pointedly at the pitiful figure before him, turns to the police sergeant nearby, and says disdainfully, "Read him his rights."

Finally, in addition to miseducating us about the realities of crime and misinforming us of our rights, television crime drama may create in people certain unrealistic expectations about the fairness and efficiency of the criminal justice system. On television only the "bad guys" are roughed up and harassed, while the truly innocent and uninvolved are treated justly and with dignity. And the system functions with admirable efficiency to fit everything neatly into those fifty minute time slots!

But there is much evidence that people's expectations about the legal system are disconfirmed by their contact with it (Sarat, 1977). Increased contact with the law leaves people frustrated and critical—the more they know about it the less they like it. Perhaps television has given them unrealistic expectations about the criminal justice system, expectations that are impossible for the system to meet. Perhaps, also, people who are not "fortunate" enough to have this personal contact with the system are loathe to criticize it. Television has taught them how well it works. The causes of rising crime rates and increased disrespect for the law must lie elsewhere. Or so the network cops would have us believe.

THE MEDIUM AND ITS MESSAGE

In a recent issue of *TV Guide,* Edward Linn wrote that "When it comes to fictionalizing crime, the TV networks have achieved heights of realism that bring cries of complaint from parent-teacher organizations and smiles of gratitude from their stockholders" (August 20, 1977). We hope to have made clear the extent to which our research casts serious doubts upon the "realism" of television crime drama. But note that most of the present viewing audience is part of a generation completely "educated" about crime by television. For them, "realism" may have come to mean a similarity to popularly held images and stereotypes that television itself has created.

The complaints of parents and teachers, however, center not so much around the lack of "realism" as the absolute amount of violence in television crime drama.

In this sense, their emphasis may be misplaced. While there is some evidence that violence on television may affect violent behavior—especially in children—it seems to us that this is not the main impact that television watching has on us. There is a sense in which television may have turned us *all* into a society of observers, a culture of people who watch rather than do, who look at others acting on the screen rather than participating ourselves. With respect to crime and criminal justice, television is an especially effective tranquilizer. We watch a detective "solve" the problems for us while we sit comfortably in our living rooms, thinking little and doing less about the real causes of crime.

The television medium is visual and verbal, and its effects are primarily cognitive rather than behavioral. It captures our attention by virtually forcing our eyes to concentrate and focus on a stream of constantly changing visual images. It teaches important lessons to people who are *not aware they are learning*. As viewers we suspend our critical and evaluative capacities, and in this passivity we are most vulnerable. Television teaches us about worlds we have never experienced, and it often misinforms. What we see is often not real, and the lessons we learn from it are frequently wrong.

The PTA is justified in its anger about the relationship of television to crime. Television crime drama undoubtedly contributes to the amount of crime and violence in our society. But its contribution is primarily one of confusion rather than incitement. It confuses us about the causes of crime and thus leads us to base our "solutions" on erroneous assumptions. It confuses us about our rights vis-à-vis the police and so permits them to violate the law and the Constitution. And it confuses us about the equity and fairness of the criminal justice process and so renders us less capable of honestly evaluating this system.

Perhaps we are always more comfortable dealing with symptoms rather than causes. We count the number of violent acts on television and hope that by doing something so simple as reducing that number we will reduce the amount of crime and violence in society. Such logic is itself part of a television stereotype: Violence and crime are not increasing in this society simply because television has made our children "sick" or violence prone. There will be less crime and violence in our society only when we rid ourselves of the conditions that cause it— situations that anger, frustrate, provoke, humiliate, and demean people.

And if we must be concerned about provocation on television, then we would do well to look *not* at the content or violence levels of the stories but rather at the commercials that support them. Here are messages intentionally designed to overstimulate acquisitive drives and competitive urges. They teach the lesson that the measure of a person is to be found only in the number of things that he has. Television crime drama, then, may give us the worst of both worlds: commercials that provoke and programs that confuse. If we are to solve problems like crime and violence, then we must look accurately at their causes. Television stands as a buffer between us and the truth.

References

BLOCK, R. Support for civil liberties and support for the police. *American Behavioral Scientist,* 1970, **13,** 781.

BLOCK, R. Fear of crime and fear of the police. *Social Problems,* 1971, **19,** 91.

GERBNER, G., AND GROSS, L. The scary world of TV's heavy viewer. *Psychology Today,* April 1976, 41–46, 89.

LINN, E. *TV Guide,* August 20, 1977.

SARAT, A. Studying American legal culture: An assessment of survey evidence. *Law & Society Review,* 1977, **11** (3), 427–488.

THOMAS, C., AND FOSTER, S. A sociological perspective on public support for capital punishment. *American Journal of Orthopsychiatry,* 1975, **45,** 641.

10

The Impact of Mass Violence
on U.S. Homicides

David P. Phillips

The impact of mass media violence on aggression has almost always been studied in the laboratory; this paper examines the effect of mass media violence in the real world. The paper presents the first systematic evidence indicating that a type of mass media violence triggers a brief, sharp increase in U.S. homicides. Immediately after heavyweight championship prize fights, 1973–1978, U.S. homicides increased by 12.46 percent. The increase is greatest after heavily publicized prize fights. The findings persist after one corrects for secular trends, seasonal, and other extraneous variables. Four alternative explanations for the findings are tested. The evidence suggests that heavyweight prize fights stimulate fatal, aggressive behavior in some Americans.

Since 1950 more than 2500 studies have attempted to discover whether mass media violence triggers additional aggressive behavior (Comstock et al., 1978; Murray and Kippax, 1979; Roberts and Bachen, 1981; National Institutes of Mental Health, 1982). With few exceptions (reviewed in Phillips, 1982b), researchers have studied aggression *in the laboratory,* and there is consensus that media violence can trigger additional aggression in the laboratory setting. However, policy makers, unlike researchers, have been primarily concerned with violence *outside* the laboratory, particularly with serious, fatal violence like homicide. Studies of media effects on homicide have been extremely rare and there is no systematic

Reprinted with permission of the author and *American Sociological Review,* Vol. 48, 1983.

Direct all correspondence to: David P. Phillips, Department of Sociology, University of California, La Jolla, CA 92093.

This paper benefited very substantially from technical discussions with H. White and C. Granger (Economics Department) and N. Beck (Political Science Department). I am grateful to these colleagues for leading me beside the still waters of time-series regression analysis. I am also grateful to M. Cole and M. Schudson (Communications Department), V. Konecni (Psychology Department), and B. Berger, A. Cicourel, F. Davis, C. Mukerji, and C. Nathanson (Sociology Department), all at the University of California at San Diego. Finally, I would like to thank two anonymous referees for helpful comments.

evidence to date indicating that mass media violence elicits additional murders.[1] As Andison has noted (1980:564)), we do not know whether "there are deaths and violence occurring in society today because of what is being shown on the TV screen."

This paper presents what may be the first systematic evidence suggesting that some homicides are indeed triggered by a type of mass media violence. The current study builds on earlier research (Phillips, 1974, 1977, 1978, 1979, 1980, 1982a) which showed that: (1) U.S. suicides increase after publicized suicide stories. This finding has been replicated with American (Bollen and Phillips, 1982) and Dutch (Ganzeboom and de Haan, 1982) data. (2) The more publicity given to the suicide story, the more suicides rise thereafter. (3) The rise occurs mainly in the geographic area where the suicide story is publicized. (4) California (Phillips, 1979), Dutch (Ganzeboom and de Haan, 1982), and Detroit (Bollen and Phillips, 1981) auto fatalities all increase just after publicized suicide stories. (5) The more publicity given to the stories, the greater the increase, and (6) the increase occurs mainly in the area where the story is publicized. (7) Single-car crash fatalities increase more than other types, and (8) the driver in these crashes is significantly similar to the person described in the suicide story, while the passengers are not. These results are statistically significant and persist after correction for day-of-the-week and seasonal fluctuations, holiday weekends, and linear trends. After testing alternative explanations, Phillips concluded that suicide stories appear to elicit additional suicides, some of which are disguised as auto accidents.

It would be interesting to discover whether *homicide* stories elicit additional homicides. But it is difficult to conduct such a study because, unlike suicide stories, homicide stories occur so often that it is very difficult to separate the effect of one story from the effect of the others. However, some other types of violent stories occur much less often, and it is possible to discover whether these types of stories trigger a rise in U.S. homicides.

MASS MEDIA VIOLENCE AND U.S. HOMICIDES

In reviewing the literature on media effects, Comstock (1977) concluded that violent stories with the following characteristics were most likely to elicit aggression: When the violence in the story is presented as (1) rewarded, (2) exciting, (3) real, and (4) justified; when the perpetrator of the violence is (5) not criticized for his behavior and is presented as (6) intending to injure his victim.[2]

---

[1]Some anecdotal data link a particular murder with subsequent murders or murder attempts (e.g., the Tylenol "copycat" crimes). But I know of only one *systematic* study of the topic (Berkowitz and Macaulay, 1971). This study found no increase in homicides after three publicized murder stories.

[2]Comstock also notes that a story is more likely to be imitated if the aggressor in the story is like the person exposed to the story, and if the victim in the story is like the imitator's victim. These points will be taken up later in this paper.

One type of story that meets all of these criteria is the heavyweight prize fight, which is almost universally presented as highly rewarded, exciting, real, and justified. Furthermore, the participants are not critized for their aggressive behavior and are presented as trying to injure each other.

In a well-known series of studies, Berkowitz and various associates (1963, 1966, 1967, 1973) examined the impact of a filmed prize fight in the laboratory. They found that angered laboratory subjects behaved more aggressively after seeing a filmed prize fight scene. In contrast, angered laboratory subjects exposed to a track meet film displayed a significantly lower level of aggression.

In sum, the heavyweight prize match is a promising research site because (1) it meets Comstock's criteria for stories most likely to elicit aggression, and (2) it is known to elicit aggression in the laboratory.

DATA SOURCES

An exhaustive list of championship heavyweight prize fights and their dates was obtained from *The Ring Book Boxing Encyclopedia,* which is the standard reference on the topic. The period 1973–1978 has been chosen for analysis because, for this period, daily counts of all U.S. homicides are publicly available from the National Center for Health Statistics.[3]

METHOD OF ANALYSIS

A standard time-series regression analysis is used.[4] Homicides are known to fluctuate significantly by day of the week, by month, and by year (Conklin, 1981). In addition, as we will see, homicides rise markedly on public holidays. All these "seasonal" effects must be corrected for before one can assess the effect of prize fights on homicides.

A 0–1 dummy variable was constructed for all days that were Mondays, another dummy variable was coded for Tuesdays, and in general a different dummy vari-

---

[3]Data for 1973–1977 consist of computerized death certificates generated by the National Center for Health Statistics and made available by the Inter-University Consortium for Political Science Research. As of this writing, 1978 computerized death certificates are not yet publicly available. Consequently, for 1978, a published table (National Center for Health Statistics, 1978: Table I-30) has been used instead. A 50 percent sample of 1972 deaths is also available but will not be analyzed, because its inclusion with the complete, 100 percent sample data for 1973–1978 would violate the assumption of homoscedasticity required in the analysis that follows. It is theoretically possible to correct for this type of heteroscedasticity and then include the 1972 data in the analysis. But it was judged unnecessary to do so, because the data set is already very large even without the 1972 information. In all, there are 2192 data points for the daily data, 1973–1978.

[4]For the application of this approach to daily mortality data, see Bollen and Phillips (1981, 1982). For general introductions to time-series regression techniques, see Ostrom (1978), Rao and Miller (1971) and Johnston (1972).

able was assigned to each day of the week, with Sunday being the omitted variable. Similarly, a 0-1 variable was coded for each month of the year (with January being the omitted variable), and for each year (with 1978 being the omitted variable). In addition, a dummy variable was assigned to each of the public holidays (New Year's Day, Memorial Day, Independence Day, Labor Day, Thanksgiving, and Christmas). Finally, a dummy variable, PFIGHT(X), was used to indicate the presence of a championship prize fight. The regression coefficient of PFIGHT(X) gives the effect of a prize fight on homicides X days later (i.e., the effect of a prize fight lagged X days). Initially, the effect of the prize fight is examined for the ten-day period following it; later, a longer period is studied.

RESULTS

Table 10.1 gives the size and statistical significance of each coefficient.[5] This table shows that, after the average championship prize fight, homicides increase mark-

TABLE 10.1
U.S. HOMICIDES regressed on heavyweight prize fight, controlling for daily, monthly, yearly, and holiday effects, 1973–1978

Regressand HOMICIDES	R^2 $\bar{R}^2$	D.F. N
	.671 .665	2148 2190

Regressor	Regression coefficient	t-statistic
Intercept	55.34*	30.16
HOMICIDE(1)	.12*	5.64
PFIGHT(−1)	1.97	.94
PFIGHT(0)	1.95	.93
PFIGHT(1)	−.26	−.13
PFIGHT(2)	1.32	.63
PFIGHT(3)	7.47***	3.54

[5]These statistical significances are biased if there is serial correlation among the regression residuals. The conventional test for serial correlation, the Durbin-Watson test, is appropriate when a lagged dependent variable is included in the regression model (Nerlove and Wallis, 1966), as is the case in Table 10.1. A common alternative test, using Durbin's h statistic, cannot be used here for reasons described in Bollen and Phillips (1982: fn. 7). Consequently, another test for serial correlation proposed by Durbin (1970) was used instead. This test reveals no evidence of first-order autocorrelation. Autocorrelation of higher orders was sought by the methods described in Bollen and Phillips (1982), with no evidence of serial correlation being uncovered. One other feature of Table 1 should be mentioned briefly. The table shows that homicides increase markedly on all U.S. public holidays except Memorial Day. To my knowledge, this finding has not been previously demonstrated with U.S. daily homicide data.

TABLE 10.1 (*continued*)
U.S. HOMICIDES regressed on heavyweight prize fight, controlling for daily, monthly, yearly, and holiday effects, 1973–1978

Regressand HOMICIDES	R^2 .671	$\bar{R}^2$ .665	D.F. 2148	N 2190

Regressor	Regression coefficient	t-statistic
PFIGHT(4)	4.15†	1.97
PFIGHT(5)	−.60	−.29
PFIGHT(6)	3.28	1.57
PFIGHT(7)	.35	.17
PFIGHT(8)	.99	.47
PFIGHT(9)	3.10	1.48
PFIGHT(10)	2.28	1.09
Monday	−16.46*	−21.74
Tuesday	−16.71*	−17.97
Wednesday	−18.42*	−19.13
Thursday	−15.81*	−15.88
Friday	−8.02*	−8.41
Saturday	14.54*	16.95
February	1.88**	1.99
March	1.13	1.23
April	.43	.46
May	−.69	−.73
June	1.61	1.74
July	4.16*	4.46
August	4.46*	4.83
September	3.91*	4.16
October	2.79*	3.02
November	3.04*	3.25
December	5.86*	6.30
1973	−1.11	−1.70
1974	1.71*	2.62
1975	1.28	1.96
1976	−3.01*	−4.60
1977	−1.73*	−2.62
New Year's Day	41.08*	10.29
Memorial Day	−1.05	−.28
Independence Day	21.61*	5.89
Labor Day	16.92*	4.56
Thanksgiving	18.34*	4.98
Christmas	10.25*	2.79

Note: The variable HOMICIDE(1) indicates homicides lagged one day. Two-tailed t-tests are used for all seasonal variables; one-tailed t-tests for prize fight variables.
*Significant at .01 or better.
**Significant at .05 or better.
***Significant at .0002.
†Significant at .025.

edly on the third day (by 7.47) and on the fourth day (by 4.15), for a total increase of 11.62.[6] The rise in homicides after the prize fight is statistically significant.[7]

Table 10.1 shows that the third day displays by far the largest peak in homicides. It is interesting to note that this "third-day peak" appears not only in the present study but also, repeatedly, in several earlier investigations: California auto fatalities peak on the third day after publicized suicide stories (Phillips, 1979), as do Detroit auto fatalities (Bollen and Phillips, 1981) and U.S. noncommercial airplane crashes (Phillips, 1978, 1980). At present we do not know the precise psychosocial mechanisms producing the third day lag, but this phenomenon has now been replicated so often in different data sets that it seems to be a relatively stable effect which will repay future investigation.

The observed peak in homicides after a prize fight cannot be ascribed to day-of-the-week, monthly, yearly, or holiday effects, because all of these "seasonal" variables were corrected for in the regression analysis. In addition, one cannot plausibly ascribe the homicide peak to random fluctuations, because the peak is statistically significant.

SOME ALTERNATIVE EXPLANATIONS FOR THE PEAK IN HOMICIDES

Two different explanations can be tested with the data in Table 10.2. For each fight, this table indicates: (1) The number of homicides observed three days after the prize fight. (2) The number of homicides expected on the third day, under the null

---

[6]The coefficient of .12 for HOMICIDE(1) indicates that there is a small, lagged endogenous effect. This implies, for example, that each of the lagged prize fight dummies has its impact distributed over more than one day. Therefore, the effect of prize fights on homicides one day later (for example) does not take place one day later, but is realized over a longer period of time. The small coefficient for the endogenous variable (.12) means that the long-run effects of PFIGHT(X) and other variables decay very rapidly and aren't much more than their immediate ones, but the distributed effects do exist. Thus, the pattern of lags is more complicated than is immediately apparent from Table 10.1. In sum, because of small, lagged endogenous effects, the impact of one PFIGHT(X) variable overlaps to a small extent with the impact of another. However, the presence of the lagged endogenous variable does not affect the validity of the statistical tests of the hypotheses (see also footnote 7).

[7]In Table 10.1 we are examining the series of eleven coefficients, PFIGHT(0), PFIGHT(1), . . . PFIGHT(10). Under the null hypothesis, none of these eleven prize fight coefficients is likely to be very large. On the other hand, under the alternative hypothesis that prize fights trigger homicides, one or more of these eleven coefficients is likely to be large and positive. If one or more of the PFIGHT(X) coefficients is sufficiently large, we can reject the null hypothesis in favor of the alternative. One way to discover whether H_0 can be rejected is to proceed as follows. Because the covariance matrix indicates that the estimates of the coefficients for PFIGHT(X) are uncorrelated, and because of the asymptotic normality of the coefficient estimates, it follows that these coefficient estimates are in fact independent. This in turn implies that the t-statistics for each of these coefficients are independent. Under these circumstances, one can use the binomial test to evaluate the probability of finding that r or more of the PFIGHT(X) coefficients are statistically significant at a given level. Table 10.1 indicates that there are two PFIGHT(X) coefficients — PFIGHT(3), PFIGHT(4) — which are statistically significant at .025 or better. For $n = 11$, $P = 0.25$, $r \geq 2$, the binomial test indicates that the probability of finding two or more significance levels of .025 in 11 independent trials is .0296. Thus, we can reject the null hypothesis on the joint evidence provided by the eleven PFIGHT(X) coefficients.

TABLE 10.2
Fluctuation of U.S. homicides three days after each heavyweight prize fight, 1973–1978

Name of fight	Observed no. of homicides	Expected no. of homicides	Observed minus expected	Fight held outside U.S.?	On network evening news?
Foreman/Frazier	55	42.10	12.90	yes	yes
Foreman/Roman	46	49.43	−3.43	yes	no
Foreman/Norton	55	54.33	.67	yes	no
Ali/Foreman	102	82.01	19.99	yes	yes
Ali/Wepner	44	46.78	−2.78	no	yes
Ali/Lyle	54	47.03	6.97	no	yes
Ali/Bugner	106	82.93	23.07	yes	no
Ali/Frazier	108	81.69	26.31	yes	yes
Ali/Coopman	54	45.02	8.98	yes	no
Ali/Young	41	43.62	−2.62	no	no
Ali/Dunn	50	41.47	8.53	yes	yes
Ali/Norton	64	52.57	11.43	no	yes
Ali/Evangelista	36	42.11	−6.11	no	no
Ali/Shavers	66	66.86	−.86	no	no
Spinks/Ali	89	78.96	10.04	no	yes
Holmes/Norton[a]	53	48.97	4.03	no	no
Ali/Spinks	59	52.25	6.75	no	yes
Holmes/Evangelista[a]	52	50.24	1.76	no	no

*Sponsored by World Boxing Council; all other fights sponsored by the World Boxing Association.

hypothesis that prize fights have no effect on homicides.[8] (3) The difference between the observed and expected number of homicides. (A positive difference indicates that homicides are higher than expected just after the prize fight.) (4) Whether the fight was held outside the United States. (5) Whether the fight was discussed on the network evening news.

"Personal Experience" hypothesis. Perhaps the prize fight affects only those actually attending the fight, not those experiencing it through the mass media. If this is so, one cannot claim that *mass media* violence is triggering a rise in homicides.

If one must personally experience the prize fight in order to be affected by it, then prize fights occurring outside the United States should trigger few if any U.S. homicides. In contrast, prize fights held inside the United States should elicit much larger rises in homicides. The evidence in Table 10.2 contradicts these predictions. After the average "foreign" fight, U.S. homicides rise by 12.128, while a much

---

[8]Under the null hypothesis, PFIGHT(X) has no impact on the number of homicides; thus, for Table 10.2 the expected number of homicides under H_0 is calculated by omitting PFIGHT(X) from the regression variables and rerunning the regression equation.

smaller rise, 2.862, occurs after the average U.S. fight.[9] Thus, the "personal experience" hypothesis does not seem plausible.

"Modeling" hypothesis—first test. A different hypothesis can also be tested with the data in Table 10.2. Prize fights may trigger some homicides through some type of modeling of aggression. If this is so, then prize fights receiving much publicity should have a greater effect than prize fights receiving less publicity.

One way to test this hypothesis is to see whether prize fights discussed on the network evening news are followed by relatively large increases in homicides, while relatively small increases occur after the remaining, less-publicized prize fights.[10] The evidence in Table 10.2 is consistent with this "modeling" explanation. Homicides rise by 11.127 after the average "publicized" fight, and by only 2.833 after the average unpublicized one. The difference between these two figures is statistically significant at .0286 (two-sample *t*-test, one-tailed).[11]

It is perhaps worth noting that the most touted of all the prize fights in this period, the so-called "Thrilla in Manilla" between Ali and Frazier, displays the largest third-day peak in homicides.

"Modeling" hypothesis—second test. The modeling hypothesis can also be tested in another way. The laboratory literature on the modeling of mass media aggression (see footnote 2) repeatedly suggests that (1) a person is more likely to imitate an aggressor on the screen if he is similar to that aggressor; (2) a person is more likely to aggress against a target victim if his target is similar to the victim on the screen.[12] In sum, the laboratory literature suggests that there is modeling of both the *aggressor and* of the aggressor's *victim.*

If *aggressor* modeling exists after a prize fight, then after a young, black male wins a boxing match, murders by young, black males should increase (but murders by young, *white* males should not). Conversely, after a young, white male wins a boxing match, the opposite findings should occur. Unfortunately, *aggressor* modeling cannot be studied with the death certificates examined in this paper, because these certificates do not reveal the identity of the murderer, only of the victim.

However, it is possible to use these death certificates to discover whether *victim* modeling exists after a heavyweight prize fight. If such modeling occurs, then, just

---

[9]At present, we do not know why U.S. homicides rise so much more after foreign than domestic fights. Perhaps a detailed study of the characteristics of these fights would help to resolve this question.

[10]A thorough analysis of this topic is desirable but would be extremely laborious. Future studies might attempt to measue the additional publicity derived from advertisements in all the media, not only at the time of the fight, but also in the weeks and months preceding it. In addition, one might wish to measure closed circuit television receipts, corrected for inflation.

[11]The formula used for this particular *t*-test does not require that the two compared populations have equal variances. For a description of this test, see Brownlee (1965:299–303). One might prefer to substitute the Mann-Whitney for the *t*-test. When this is done, $P = .0211$.

[12]Berkowitz and associates (1963, 1966, 1967, 1973) have shown this in a series of ingenious studies particularly relevant to this paper. They showed that laboratory subjects were most likely to inflict shocks on a target if that target had the same name as the losing boxer on the screen.

after a prize fight, homicide victims should be unusually similar to the losing boxer. Specifically, after a young, *white* male is beaten in a prize fight, the homicide deaths of young, white male victims should increase; no such increase should appear for young, *black* male victims. Conversely, after a young, *black* male is beaten in a prize fight, the homicide deaths of young, *black* male victims should increase, while the homicide deaths of young, white males should not.

These predictions can be tested with the information in Tables 10.3 and 10.4,

TABLE 10.3
Impact of "White Loser" (WL) and "Black Loser" (BL) prize fights on the homicides of young, white male victims, U.S., 1973–1977[a]

Regressand HOMICIDES	R^2 $\bar{R}^2$ .378 .360	D.F. N 1772 1825
Regressor	Regression coefficient	t-statistic
Intercept	10.43*	23.78
HOMICIDE(1)	.01	.23
WL(−1)	.70	.46
WL(0)	3.86**	2.54
WL(1)	.30	.20
WL(2)	3.14**	2.07
WL(3)	.48	.31
WL(4)	.57	.37
WL(5)	−.29	−.19
WL(6)	.93	.61
WL(7)	.35	.23
WL(8)	2.97†	1.96
WL(9)	.53	.35
WL(10)	.58	.39
BL(−1)	1.36	1.27
BL(0)	.04	.03
BL(1)	−1.44	−1.34
BL(2)	−.59	−.55
BL(3)	1.20	1.11
BL(4)	.69	.64
BL(5)	−1.06	−.99
BL(6)	1.61	1.50
BL(7)	−.14	−.13
BL(8)	.32	.30
BL(9)	.28	.26
BL(10)	−.53	−.49

[a]As in Table 10.1 the effect of prize fight variables is calculated, controlling for seasonal variables. For reasons of clarity, the coefficients for these seasonal variables have not been displayed in Table 10.2 since the prime purpose of this table is to contrast the impact of "White Loser" and "Black Loser" prize fights. One-tailed t-tests are used for the prize fight variables; two-tailed t-tests for all other variables.
*Significant at less than .001.
**Significant at .006.
***Significant at .019.
†Significant at .0251.

TABLE 10.4

Impact of "White Loser" (WL) and "Black Loser" (BL) prize fights on the homicides of young, black male victims, U.S., 1973–1977

Regressand HOMICIDES	R^2 $\bar{R}^2$ .452 .436	D.F. N 1772 1825
Regressor	Regression coefficient	t-statistic
Intercept	10.59*	20.79
HOMICIDE(I)	.04	1.63
WL(−1)	.83	.48
WL(0)	−1.30	−.75
WL(1)	−1.60	−.93
WL(2)	.19	.11
WL(3)	2.82	1.59
WL(4)	−.82	−.47
WL(5)	−1.19	−.69
WL(6)	−1.66	−.96
WL(7)	2.80	1.62
WL(8)	−.78	−.45
WL(9)	1.62	.94
WL(10)	.59	.34
BL(−1)	−.25	−.21
BL(0)	1.19	.98
BL(1)	−.60	−.49
BL(2)	.18	.15
BL(3)	.67	.54
BL(4)	2.68**	2.19
BL(5)	2.28***	1.86
BL(6)	−.22	−.18
BL(7)	.04	.03
BL(8)	−.76	−.62
BL(9)	1.50	1.23
BL(10)	.30	.25

*Significant at less than .001.
**Significant at .014.
***Significant at .032.
See also Footnotes to Table 10.3.

which distinguish between the impact of "black loser" prize fights (in which a black is beaten) and "white loser" prize fights (in which a white is beaten).[13] The detailed mortality data necessary to generate these tables can be found only in the computerized death certificates cited in footnote 3. These are available only for 1973–

---

[13]In the period under study (1973–1977) nearly all the losing boxers were 20–34.9; consequently I have defined "young males" as men in this age range. Nearly all the losing boxers were white (Wepner, Bugner, Coopman, Dunn) or black (Frazier, Norton, Foreman, Lyle, Young, Shavers). However, two of the losing boxers were Hispanic Americans (one Uruguayan and one Puerto Rican). There is no separate classification for Hispanic Americans on the computerized death certificate, and it is unclear whether one can treat these fighters as either white or black. Consequently, these two fights have been excluded from the analysis that follows.

1977. Thus, it should be stressed that the period to be examined in the remainder of this paper is 1973–1977, not 1973–1978, as in Tables 10.1 and 10.2.

Table 10.3 examines the impact of "white loser" and "black loser" prize fights on the homicides of young, *white* male victims. The evidence supports the hypothesis of victim modeling. *White* loser prize fights are followed by significant increases in young, white male homicide deaths; in contrast, *black* loser prize fights do not seem to trigger young, white male homicide deaths.[14]

White homicides increase significantly on the day of the prize fight (by 3.86 per fight), two days thereafter (by 3.14 per fight), and eight days after the fight (by 2.97 per fight). Thus young, white male homicides rise by a total of 9.97 (= 3.86 + 3.14 + 2.97) per white loser prize fight. Interestingly, the typical white loser prize fight has a larger total impact (9.97) than almost any other variable in the table. Of the 27 "seasonal" variables examined, only one (New Year's Day) has a larger impact on young, white male homicides.[15] This suggests that the impact of a white loser prize fight is not only statistically significant, but practically significant as well. At present, it is not known why this type of prize fight seems to exert so large an effect.

Table 10.4 examines the impact of "white loser" and "black loser" prize fights on the homicides of young, *black* male victims. Once again, the evidence supports the hypothesis of victim modeling. Black loser prize fights are followed by significant increases in young, black male homicide deaths. In contrast, white loser prize fights do not trigger significant increases in black male homicides.

Black homicides rise significantly on the fourth and fifth days after black loser fights by a total of 4.96 (= 2.68 + 2.28) per fight.[16] The total impact of the black loser prize fight exceeds the impact of almost all seasonal variables. Only New Year's Day and Thanksgiving trigger larger increases in homicides (the coefficients for these holidays being 8.88 and 8.00, respectively). Evidently, a black loser prize fight has a significant, substantive effect on young, black male homicides.

---

[14]The analysis described in footnote 7 (and applied to Table 10.1) can be reapplied to the results in Table 10.3. Examining the coefficients, WL(0), ... WL(10), we see that three are statistically significant at .0251 or better. The covariance analysis indicates that the estimates of the coefficients WL(X) are uncorrelated. Because of this and the asymptotic normality of the coefficient estimates, we can treat as independent the t-statistics for WL(0), ... WL(10). Using the binomial test, with $n = 11$, $P = .0251$, $r \geqslant 3$, one finds that the probability of finding three or more coefficients significant at .0251 in eleven independent trials is .0022. Hence, the homicides of young, white males increase significantly just after "White Loser" prize fights.

[15]The coefficient for this holiday is 15.75. Although the effect of any given prize fight is large compared with the effect of seasonal variables, the cumulative effect of all prize fights combined is not large compared with the cumulative effect of all seasonal variables combined. This is because there are relatively few prize fights.

[16]The analysis of Table 10.4 is parallel to that of Table 10.3. Once again, statistical theory and the covariance analysis justify treating as independent the eleven t-statistics for BL(0), ... BL(10). We observe two BL(X) coefficients significant at .032 or better. The probability of finding two or more BL(X) coefficients significant at this level in eleven independent trials is .0465. Hence, young, black male homicide deaths increase significantly just after "Black Loser" prize fights.

Precipitation hypothesis.   The above evidence is consistent with the notion that prize fights sometimes serve as aggressive models and trigger some U.S. homicides. But perhaps the prize fight merely precipitates a murder that would have occurred anyway, even in the absence of the prize fight.

If a prize fight merely "moves up" a murder so that it occurs a little sooner than it otherwise would have, then the peak in homicides after a prize fight should be followed by a *dip* in homicides soon after. An examination of the three-week period following the prize fight reveals no significant dip in homicides. None of the negative coefficients for PFIGHT(1), PFIGHT(2), . . . , PFIGHT(21) is significant, even at the .10 level. Hence, the precipitation hypothesis seems to be implausible.

Gambling hypothesis.   Perhaps the prize fight provokes no aggressive modeling whatsoever. It merely triggers an increase in gambling, which in turn provokes anger, fighting, and murder. If this explanation is correct, then homicides should rise not only after prize fights but also after other occasions that provoke a great deal of gambling. In the United States, the Super Bowl probably provokes more gambling than any other single event. Yet homicides do *not* rise significantly after these occasions.

One can construct a variable, SUPERBOWL(X), to assess the impact of the Super Bowl on homicides X days later, and one can include this variable in the regression model specified in Table 10.1. The coefficients for SUPERBOWL(X) are listed in Table 10.5. There is some weak evidence that homicides actually *decrease* on the day of the Super Bowl and one day later, and then rise above the expected rate on the third day. Even if one considers these coefficients to be statistically significant (which they are not), it is evident that the Super Bowl is associated with a net drop in homicides rather than a rise. This is not what one would expect if the gambling hypothesis were correct. This hypothesis is also rendered implausible by some of the other evidence presented above: If the gambling hypothesis were true, then it is difficult to see why the traits of the homicide victims should be similar to the traits of the losing boxer.[17]

In sum, we have now assessed four possible explanations for the rise in homicides after a heavyweight prize fight. At present, the best available explanation is that the prize fight provokes some imitative, aggressive behavior, which results in an increase in homicides. The size of this increase will be considered in the next section.

---

[17]The evidence presented does not support the notion that the gambling hypothesis is a necessary and sufficient explanation for the rise in homicides after a prize fight. But it remains possible that gambling *in combination with* aggressive modeling is helping to provoke the increase in homicides. One way to test this hypothesis is to examine police case histories of murders occurring three and four days after a prize fight. These case histories would have to be compared with case histories taken from control periods.

TABLE 10.5
Impact of the Superbowl on U.S. homicides, controlling
for the effect of seasonal and prize fight variables, U.S.,
1973–1977

Regressor	Regression coefficient	t-statistic
SUPERBOWL(−1)	2.78	.68
SUPERBOWL(0)	−5.03	−1.22
SUPERBOWL(1)	−6.36	−1.55
SUPERBOWL(2)	2.26	.55
SUPERBOWL(3)	6.00	1.46
SUPERBOWL(4)	−1.41	−.34
SUPERBOWL(5)	1.10	.27
SUPERBOWL(6)	−1.67	−.41
SUPERBOWL(7)	−2.24	−.54
SUPERBOWL(8)	3.61	.87
SUPERBOWL(9)	−4.82	−1.17
SUPERBOWL(10)	2.16	.52

Note: The coefficients for the other regressor variables
(i.e., prize fight and seasonal variables) are not displayed.

Size of the increase in homicides after prize fights.   Column 3 of Table 10.3 gives the amount by which homicides increase on the third day after each prize fight. The sum of the numbers in this column is 125.64, indicating that U.S. homicides rose by this amount on the third day after championship heavyweight prize fights, 1973–1978. The sum of the numbers in column 2 gives the total number of homicides expected on the third day—1008.36. Dividing 125.64 by 1008.36 gives the percentage increase in homicides on the third day—12.46 percent. Thus, whether one considers the percentage increase or the absolute increase, it appears that homicides rise by a nontrivial amount on the third day after a championship heavyweight prize fight.

The rise in homicides on the fourth day is smaller but still not negligible. Employing calculations similar to those in Table 10.2, one can determine that homicides increase by 67.97 on the fourth day. The percentage increase is 6.58 percent (= 67.97/1033.03). For the third and fourth days combined, homicides increase by a little less than 200 (193.61 = 125.64 + 67.97). The percentage increase for the two-day period is 9.48 percent (= 193.61 / [1008.36 + 1033.03]).

This paper has presented evidence which suggests that heavyweight prize fights provoke a brief, sharp increase in homicides. Some implications of this evidence will be briefly considered in the final section of this paper.

SUMMARY

Many researchers have claimed that one cannot generalize with confidence from the impact of mass media violence *in the laboratory* to the impact of mass media violence *in the real world.*[18] These critics point out that laboratory experiments have been set in highly artificial contexts. Typically, the sorts of aggression studied in a laboratory (like hitting plastic dolls or inflicting electric shocks) have not been representative of serious, real-life violence, such as murder or rape. In almost all studies, the laboratory subjects have been nursery school children or college students and thus not representative of the U.S. television audience. Typically, the laboratory subject is presented with a brief, violent excerpt of a television program. In contrast, the "real-life" viewer may watch several hours of television at a sitting, and the violence may be interspersed with humor, commercials, and trips to the bathroom. In contrast to the laboratory subject, who watches television alone, the real-life viewer may well be surrounded by family or friends. Their comments may distract from the television or shape the perception of its many messages. For these reasons, it is inappropriate to generalize from the laboratory to the real world.

The above argument appears to be seriously challenged by the evidence provided in this paper. The data presented in this paper indicate that mass media violence *does* provoke aggression in the real world as well as in the laboratory. In contrast to laboratory studies, the present investigation assesses the effect of mass media violence in a natural context. Unlike laboratory studies, the present study examines a type of violence which is of serious concern to policy makers. Finally, the present investigation does not focus exclusively on a mass media audience consisting of college students and nursery school children. The laboratory study, with its great potential for rigor, has always been capable of establishing the internal validity of findings. The present study has helped to establish that these findings have external validity as well.

---

[18]Comstock (1975:30–40) provides a valuable summary of the debate on this topic. In addition, see Phillips ("Behavioral impact," 1982), who also indicates why it is difficult to generalize from the few field experiments that exist.

References

ANDISON, E. S. Television violence and viewer aggression: A cumulation of study results 1956–1976. In G. C. Wilhoit, and H. de Bock (eds.), *Mass Communication Review Yearbook.* Vol. I. Beverly Hills, Cal.: Sage, 1980.

BERKOWITZ, L., AND RAWLINGS E. Effects of film violence on inhibitions against subsequent aggression. *Journal of Abnormal and Social Psychology,* 1963, **66,** 405–12.

BERKOWITZ, L., AND GEEN R. Film violence and the cue properties of available targets. *Journal of Personality and Social Psychology,* 1966, **3,** 525–30.

BERKOWITZ, L., AND GEEN, R. Stimulus qualities of the target of aggression: A further study. *Journal of Personality and Social Psychology,* 1967, **5,** 364–68.

BERKOWITZ, L., AND ALIOTO J. T. The meaning of an observed event as a determinant of its aggressive consequences. *Journal of Personality and Social Psychology,* 1973, **28,** 206–17.

BERKOWITZ, L., AND MACAULAY J. The contagion of criminal violence. *Sociometry,* 1971, **34,** 238–60.

BOLLEN, K. A., AND PHILLIPS D. P. Suicidal motor vehicle fatalities in Detroit: A replication. *American Journal of Sociology,* 1981, **87,** 404–12.

BOLLEN, K. A., AND PHILLIPS D. P., 1981, Imitative suicides: A national study of the effects of television news stories. *American Sociological Review,* 1982, **47,** 802–809.

BORNLEE, K. A. *Statistical theory and methodology in science and engineering.* New York: Wiley, 1965.

COMSTOCK, G. *Television and human behavior: The key studies.* Santa Monica, Cal.: Rand, 1975.

COMSTOCK, G. Types of portrayal and aggressive behavior. *Journal of Communication,* 1977, **26,** 189–98.

COMSTOCK, G., CHAFFEE, S., KATZMAN, N., McCOMBS, M., AND ROBERTS, D. *Television in America.* Beverly Hills, Cal.: Sage, 1978.

CONKLIN, J. *Criminology.* New York: Macmillan, 1981.

DURBIN, J. Testing for serial correlation in least-squares regression when some of the regressors are lagged dependent variables. *Econometrica,* 1970, **38,** 410–21.

GANZEBOOM, H. B. G., AND DE HAAN D. Gepubliceerde zelfmoorden en verhoging van sterfte door zelfmoord en ongelukken in Nederland 1972–1980. *Mens en Maatschappij,* 1982, **57,** 55–69.

JOHNSTON, J. *Econometric methods.* New York: McGraw-Hill, 1972.

MURRAY, J., AND KIPPAX, S. From the early window to the late night show: international trends in the study of television's impact on children and adults. *Advances in Experimental Social Psychology, 1979,* **12,** 253–320.

NATIONAL CENTER FOR HEALTH STATISTICS. *Vital statistics of the United States.* Washington, D.C.: U.S. Government Printing Office, 1978.

NATIONAL INSTITUTES OF MENTAL HEALTH. *Television and behavior: Ten years of scientific progress and implications for the eighties. vol. 1: summary report.* Washington, D.C.: U.S. Government Printing Office, 1982.

NERLOVE, M., AND WALLIS, K. F. Use of the Durbin-Watson statistic in inappropriate situations. *Econometrica,* 1966, **34,** 235–38.

OSTROM, C. W. *Time series analysis: Regression techniques.* Beverly Hills, Cal.: Sage, 1978.

PHILLIPS, D. P. The influence of suggestion on suicide: Substantive and theoretical implications of the Werther effect. *American Sociological Review,* 1974, **39,** 340–54.

PHILLIPS, D. P. Motor vehicle fatalities increase just after publicized suicide stories. *Science,* 1977, **196,** 1464–65.

PHILLIPS, D. P. Airplane accident fatalities increase just after stories about murder and suicide. *Science,* 1978, **201,** 148–50.

PHILLIPS, D. P. Suicide, motor vehicle fatalities, and the mass media: Evidence toward a theory of suggestion." *American Journal of Sociology,* 1979, **84,** 1150–74.

PHILLIPS, D. P. Airplane accidents, murder, and the mass media: Towards a theory of imitation and suggestion. *Social Forces,* 1980, **58,** 1001–24.

PHILLIPS, D. P. The impact of fictional television stories on U.S. adult fatalities: New evidence on the effect of the mass media on violence. *American Journal of Sociology,* 1982, **87,** 1340–59.

PHILLIPS, D. P. The behavioral impact of violence in the mass media: A review of the evidence from laboratory and nonlaboratory investigations. *Sociology and Social Research,* 1982, **66,** 387–98.

RAO, P., AND MILLER, R. L. *Applied econometrics.* Belmont, Cal.: Wadsworth, 1971.

THE RING BOOK SHOP *The Ring boxing encylcopedia and record book.* New York: The Ring Book Shop, 1980.

ROBERTS, D. F., AND BACHEN, C. M. Mass communication effects. In M. R. Rosenzweig and L. W. Parker (eds.), *Annual Review of Psychology.* Palo Alto, Cal.: Annual Reviews, 1981.

IV

SELF-JUSTIFICATION

11

Persuasion via Self-Justification: Large Commitments for Small Rewards

Elliot Aronson

HISTORICAL BACKGROUND

In the 1950s and early 1960s, there was a great deal of excitement among social psychological researchers in this country. Partly this was the result of Kurt Lewin's legacy of enthusiasm and optimism about the potential benefits which a new scientific social psychology could bring to society. Partly it flowed from the more general zeitgeist of those decades, the development of computers and spaceships, the seemingly limitless possibilities of technology, and the unprecedented linkage of humanitarian and scientific concerns exemplified by scientists like Lewin. A generation of social psychologists grew up with the conviction that through laboratory experiments it would be possible to discover scientific principles of human behavior which, in turn, would lead to social change. It was in this era that Leon Festinger invented and developed his theory of cognitive dissonance, and in my opinion, social psychology has not been the same since. The first dissonance experiments introduced ideas and approaches which had a strong effect on the mainstream of the field. They served as a pivotal point, allowing the integration of a cognitively oriented approach to behavior into an experimental framework.

It is too early to write a historical or retrospective account of dissonance theory. For one thing, the theory itself is still expanding and growing and is not ready for either a eulogy or an obituary. In addition, I think that it is one of the occupational hazards of being an experimental social psychologist to be oriented toward the present and future and to become impatient, after a time, with soul searching and backward looks. Yet the early dissonance experiments provided certain hopes and expectations—hopes and expectations which are sometimes forgotten and at times

Reprinted with permission of the author and *Retrospection on Social Psychology*, Leon Festinger, ed. Copyright © 1980 by Oxford University Press.

This article was written while I was on a grant from NIMH. My research assistant, Erica Goode, provided a great deal of hard work, much encouragement, and even some ideas. I am grateful.

need to be recalled. And undeniably there are occasions when it is useful to take stock of what we have learned over a period of years. This, in fact, is the general purpose of the present volume. Thus, rather than write still *another* descriptive analysis of the theory of cognitive dissonance—something which I have done in the recent and not so recent past (Aronson, 1968, 1969, 1978) and which others have done far more completely than I (Wicklund and Brehm, 1976)—I would like, in this chapter, to discuss the role dissonance theory has played in the field and the directions in which it has expanded in the last several years. As with most things I write, it is both personal and idiosyncratic.

When I first became interested in social psychology as an undergraduate in the early 1950s, there was already a sizeable experimental literature on persuasion. If, at that time, I had been able to assemble a panel of expert social psychologists and asked them what we know for sure about how to persuade people, my guess is that they would have included the following techniques:

1. Offer tangible rewards for compliance and clear punishments for espousal of opposing views.

2. Present an audience with a reasonable communication attributing it to a highly credible communicator (e.g., Hovland and Weiss, 1951).

3. Present the individual with the illusion that everyone else in sight agrees with one another and disagrees with him (e.g., Asch, 1951).

In the early 1950s the overwhelming trend in American psychology was, "Let's find the external reward." If a person does something, there must be a reason, and that reason had to be the gaining of an identifiable reward such as food, money, or praise, or the removing of a noxious state of affairs such as pain, fear, or anxiety. If food will induce a hungry rat to press the lever of a Skinner box or turn left in a Y-maze, surely similar rewards can induce a person to hold a given opinion (see Miller and Dollard, 1941). Let us look at Solomon Asch's well-known experiment in which a unanimous majority apparently disagrees with the individual subject on a simple, unambiguous perceptual judgment. Why do so many people conform to this kind of group pressure? Perhaps it makes them anxious to be alone against a unanimous majority; they fear being considered crazy, being held in low esteem, and so on. It's comforting to be in agreement with others. Why do people tend to believe a credible source? Perhaps it increases the probability of being right—and being right reduces anxiety and makes them feel good, smart, and esteemed.[1]

While the research on which these general propositions were based was reasonably clear and replicable, the effects do not seem very powerful or long-lasting. For example, in the classic experiment by Hovland and Weiss (1951) it was found

---

[1]There were, of course, some notable exceptions to this general zeitgeist—most notably, the work of Kurt Lewin and his colleagues. But this work was not in the mainstream. More will be said on this later on in this chapter.

that, while students tended to believe the distinguished physicist J. Robert Oppenheimer when he allegedly wrote that atomic submarines were feasible, the effects of his credibility faded rather rapidly. That is, one month after the communication was presented, the high credibility communicator showed a marked decrease in effectiveness. Indeed, the slippage was such that there was virtually no difference left between the effectiveness of the highly credible source and a source having low credibility. Hovland himself (1959) clearly recognized the weakness and flimsiness of the existing methods of persuasion. The brief duration of the effectiveness of the "persuasion" is probably even more pronounced in the Asch experiment. Asch, of course, realized that most of the effects produced in his procedure were the result of temporary conformity rather than an actual change in belief. That is, the typical yielding subject did not become convinced that the majority was right; rather, he went along with the majority in order to avoid unpleasantness. This process was subsequently confirmed in a more elaborate experiment by Deutsch and Gerard (1955). In one condition of this experiment, subjects were separated by partitions so that they might respond anonymously and in privacy. In this condition, there was far less conformity than in the condition that replicated the more public situation that existed in Asch's study.

ENTER DISSONANCE

Most experimental social psychologists seemed sanguine about their inability to produce important and long-lasting changes in opinions or attitudes and apparently were content with reward reinforcement theory as an explanation for conformity and persuasion phenomena. Then, in 1959, an experiment was performed which presented a strikingly different approach to persuasion. Picture the following scene:

> A young man is instructed to perform a monotonous, boring task as part of an experiment. After completing the task, he is informed by the experimenter that his formal participation as a subject is over. The experimenter then appeals to him for help. He states that his research assistant was unable to be there and asks the subject if he would help him run the experiment. Specifically, the experimenter explains that what he is investigating is the effect of people's preconceptions on their performance of a task. He goes on to explain that he wants to see if a person's performance is influenced by whether he's told either good things about the task (in advance), bad things about the task (in advance), or nothing at all about the task. There is another participant about to arrive; that person is designated to be in the "favorable information condition." The experimenter asks the subject if he would tell the incoming participant that he had just completed the task (which is true) and that he found it to be an exceedingly enjoyable one (which is not true, according to the subject's own experience). The subject is offered either $1 or $20 for telling this lie and for remaining on call in case the regular assistant cannot show up in the future.

This is the scenario of the classic experiment by Festinger and Carlsmith (1959). The results of this experiment are now old hat to students of social psychology.

The subjects who said that they found the task enjoyable in order to earn $1 came to believe that it actually *was* enjoyable to a far greater extent than those who said so for $20.

The experiment was, of course, derived from Festinger's (1957) theory of cognitive dissonance. Basically, the theory states that if an individual simultaneously holds two cognitions that are *psychologically* inconsistent, he will experience discomfort. Consequently, he will strive to reduce the inconsistency (dissonance) by changing one or both cognitions to make them more consonant or by adding a third cognition which will render the original cognitions less inconsistent with one another. Thus, if I were a subject in the Festinger-Carlsmith experiment, my cognition that the task I performed was boring is dissonant with the fact that I informed another person that it was enjoyable. If I were paid $20 for making that statement, this cognition provides external justification for my action. However, if I were paid only $1, I lack much external justification for having made the statement. This produces cognitive dissonance. One way to reduce dissonance is for me to convince myself that the task was somewhat more interesting than it seemed at first. In effect, I convince myself that my statement to the other student was not a great lie, in the process persuading myself that the task actually was interesting. This change in opinion through *self-justification* is not limited to trivial judgments (like the dullness of a boring task). It has been extended to much more important opinions, such as a reassessemnt of the dangers of smoking marijuana among students at the University of Texas (Nel, Helmreich, and Aronson, 1969) and the softening of Yale students' attitudes toward the alleged anti-student brutality of the New Haven police (Cohen, 1962). And theoretically, because the attitudes dealt with are important ones and are not directly linkable to an obvious source of persuasion, one would expect the effects to last longer than those resulting from the persuasion techniques in earlier experiments.

This point deserves some elaboration. The major reason for the power of dissonance effects is that the arousal of dissonance always contains *personal involvement,* and therefore the reduction of dissonance always involves some form of *self-justification.* This self-justification is necessary because the individual has usually done something that makes him feel either stupid or immoral (see Aronson, Chase, Helmreich, and Ruhnke [1974] for an elaboration of this point). Moreover, the greater the commitment or self-involvement implied by the action and the smaller the *external* justification for the action, the greater the dissonance and, therefore, the more powerful the attitude change. Thus, in the Festinger-Carlsmith experiment, to deceive another human being would make an individual feel immoral or guilty. So, he convinces himself that he didn't really deceive anyone—that it *was* a rather interesting task. This justifies his previous action. A similar process occurs in Aronson and Mills' (1959) experiment on the initiation effect. In this experiment, college women chose to undergo an initiation in order to be admitted to a discussion group that turned out to be a boring waste of time. As predicted, those who underwent a severe initiation convinced themselves that the group was *more* worthwhile than those who underwent either a mild initiation or no initiation at all. To go through a difficult or embarrassing initiation in order to become a member of a silly and uninteresting discussion group would make the individual feel stupid; therefore, she convinces herself that the group was *not* silly and was *not* uninteresting. This justifies her decision to work hard in order to become a member of the group.

Compare this with a typical communication-persuasion study (e.g., Hovland and Weiss, 1951). Here the reason I'm changing my opinion is that someone smart and trustworthy thinks that something is true. Very little of me is invested. Accordingly, it's easy for me to forget the source, forget my opinion, etc. Moreover, if my new opinion is challenged, I can always change back. I have nothing invested in it. In the dissonance studies, the individual's self-esteem is involved. Therefore, it is expected that an individual's opinions on important matters can be changed, that these changes can be large and truly significant (not just *statistically* significant!), and that they will be permanent.

Some evidence for large and truly significant changes about important opinions comes from an experiment by Nel, Helmreich, and Aronson (1969) in which college students who believed that marijuana should not be legalized were induced to make a counterattitudinal videotape extolling the use of marijuana. In the no dissonance condition, there was little or no change in opinion. In the maximum dissonance condition (small external justification *and* being informed in advance that the tape would be shown to malleable high school students), there were huge and dramatic changes in opinion. Subjects in this condition came to believe that marijuana should be legalized. The change in opinion was enormous—73 percent of what was possible.

In that experiment, we would predict that those large changes in opinion would last a considerable amount of time, partly because in the late 1960s college students would not lightly change their opinions about an important (to them) issue such as the legalization of marijuana. Moreover, it is unlikely that opinion change based on the need for self-justification would shift sporadically. Any regression in the subjects' opinions on the legalization of marijuana would be certain to reinstate their feelings of dissonance and guilt about having misled some innocent and malleable high school students. Unfortunately, we have no direct data on the permanence of the change of opinion about marijuana. Why not? On the surface, it would seem easy to come back thirty or forty days later and reassess the subjects' opinions. But, of course, it would have been highly unethical for us *not* to have debriefed our subjects immediately after assessing their initial change in opinion—in view of the fact that we *were* dealing with an important and highly salient issue. To leave people running around for thirty days believing that atomic submarines are feasible (as Hovland and Weiss did) is relatively harmless. To leave them running around for 30 days with a laboratory-induced belief that marijuana is a good deal less harmful than they had previously thought would have been unthinkable. Thus, ethically, as the importance of the issue increases, so too does the need for immediate debriefing and reinstatement of the initial opinion. This makes it impossible to assess the permanence of the effect under these conditions.

Fortunately, there are some solid data on the permanence of dissonance-induced opinion change. These come from an experiment by Jonathan Freedman (1965) and confirm our speculations. Freedman performed a conceptual replication of Aronson and Carlsmith's "toy" study (1963) in which children were threatened with either severe or mild punishment if they played with an attractive toy. In the original experiment, Aronson and Carlsmith predicted and found that the children complying in the face of mild threats subsequently convinced themselves that the forbidden toy was *less* attractive than did those children who were confronted with severe threats. The derivation should be clear to the reader. If I am a child eager to play with a specific attractive toy and you (an adult authority figure) tell me that you will break my arm if I play with it, I will not play with the toy and I will experience no dissonance. The cognition that you will break my arm is more than ample justification for my abstinence. But suppose you issued a mild threat which was just barely sufficient to induce short-term compliance but was not a superabundant justification. Here, my not playing with the toy will be dissonant with the attractiveness of the toy. I will seek additional justification for my abstinence—by convincing myself, for example, that the toy isn't all that attractive anyway.

Freedman conceptually replicated these results. The experimenter admonished children (either with mild or severe threats) not to play with an extraordinarily attractive toy (a battery-powered robot). As in the Aronson and Carlsmith experiment, all of the children resisted the temptation to play with it. Several weeks later, a different person visited the school and, in a totally unrelated activity, happened to be administering a test to the students in the same room where the toy study had been conducted. The toys were casually scattered about the room. The visitor administered a paper and pencil test to each child. She then asked him to stick around while she graded it—and suggested very casually that, while he was waiting, he might want to amuse himself by playing with those toys that "someone must have left here." The results were striking. Those children who had been admonished previously with a severe threat tended to play with the forbidden toy. After all, the person who had threatened them was no longer around; why *not* play with an attractive toy? On the other hand, the children who had previously refrained from playing with the robot in the face of a mild threat had had little external justification for this restraint. Accordingly, at the time of the initial threat, they had a much stronger need for *self*-justification. In other words, they needed to convince themselves that they weren't playing with the toy because they didn't want to—and that they didn't want to because they didn't like the toy. As predicted, this cognitive activity had a relatively permanent impact. As much as nine weeks after the mild threat, the overwhelming majority of the children in the mild threat condition refrained from playing with the toy that virtually all children (who were not subjected to dissonance arousal) would certainly consider to be by far the most attractive and exciting toy in the room. The power of self-justification *does* produce important and long-lasting changes in attitude.

While the results of these early experiments on cognitive dissonance (especially Festinger and Carlsmith [1959] and Aronson and Mills [1959]) may seem boring and obvious to the reader in the 1980s, they startled the social psychological community at the time of their publication. For many working social psychologists, these results generated a great deal of enthusiasm and excitement; for others, skepticism and anger. As a community, we have yet to recover from the impact of this research—fortunately! The reasons for the *skepticism* and *anger* are complex and will dribble out from time to time throughout this essay. One reason will be mentioned here. The findings departed from the general orientation accepted either tacitly or explicitly by most social psychologists in the 1950s: High reward (not *low* reward, and certainly not high punishment) has *always* been accompanied by greater learning, greater conformity, greater performance, greater satisfaction, greater persuasion, and so on.

The reason for the *enthusiasm* was that for some social psychologists, these results represented a striking and convincing act of liberation from the dominance of a general reward-reinforcement theory. The results of these early experiments demonstrated dramatically that, at least under certain conditions, reward theory is inadequate. Moreover, the early dissonance experiments sounded a clarion call to cognitively oriented social psychologists, proclaiming in the most striking manner that *human beings think;* they do not always behave in a mechanistic manner. When in situations which allow them to think, human beings engage in all kinds of cognitive gymnastics aimed at justifying their own behavior. Perhaps most important, this theory inspired an enormous number and variety of hypotheses which were specific to the theory and could be tested in the laboratory.

In the past twenty-five years, the wide array of research that dissonance theory has inspired has been truly astonishing. As Sears and Abeles pointed out in 1969, in the early years of the theory "no behavioral phenomenon was safe from the ravages of the imperialistic dissonance hordes." Dissonance research runs the gamut from decision making to the socialization of children; from interpersonal attraction to the antecedents of hunger and thirst; from color preference to selective informational exposure; from the proselytizing behavior of religious zealots to the behavior of gamblers at a race track—and much, much more. And because it led into domains that had never before been tested in the laboratory, the theory inspired a range and depth of methodological ingenuity unparalleled in the history of social psychology. This proved to be a curse as well as a blessing—providing still another reason for the annoyance and skepticism of its critics, as Jones observed in his perceptive Foreword to Wicklund and Brehm's (1976) *Perspectives on Cognitive Dissonance:*

> The methodological style of the dissonance proponents also fed the flames of controversy. Those who fashioned the dissonance literature were, by and large, uncommonly bright and inventive. But their inventiveness was usually coupled with the use of deceptive scenarios and a procedural complexity that made precise replication difficult. Those bred in the tradition of cumulative research in experimental psychology thought they saw serious vulnerability in this complexity and in the apparent disregard of standardized dependent variable measurement.
>
> Instead, and this was difficult to realize at any given moment, there was devel-

oping a body of literature that was conceptually cumulative and therefore more impressive than permutations on a measuring device or a statistical inference rule. No one can read this book without appreciating the development of a solid research consensus on a number of important and once controversial issues. No one can any more claim that dissonance effects are artifacts or that dissonance is at most a contrived laboratory experience. Undoubtedly many readers of this volume will have reservations about specific interpretations and prefer their own alternatives. Nonetheless, the cumulative reach of dissonance research is remarkable. We may now have reached a less flamboyant stage of tidying of loose ends and charting out the relations between dissonance theory and other psychological conceptions, but whatever the future holds, the dissonance research "movement" has been the most important development in social psychology to date. (p. x)

PRECURSORS TO DISSONANCE THEORY

Obviously, Leon Festinger was not the first psychologist to notice that human beings try to justify their actions in complex cognitive ways. Nor was he the first to theorize about this fact. In the 1930s, for example, both Kurt Lewin and Prescott Lecky were interested in the new physics of Einstein and Planck. Both were familiar with European psychological schools, and both were developing theories which emphasized the cognitive, active nature of the human organism. Lecky, who had studied with the psychoanalyst Alfred Adler, working at Columbia University within an experimental framework, took issue with the idea of reinforcement as the sole mechanism through which humans learn. Reinforcement was unable to account for a great deal of human behavior that Lecky considered to be important. He developed a theory of self-consistency, which portrayed human beings as active, organized, unified organisms seeking to predict and having the tendency to create (1945). "Prediction and control," Lecky maintained, "is the problem of the subject as well as the experimenter." Human learning, according to Lecky's view, involves conflict between the organization of the individual's expectations or values and new information or stimuli. In resolving this conflict, the person tries to maintain consistency. In some cases, the conflicting or inconsistent stimulus may be distorted or ignored, particularly if it is threatening to the self-concept. In other cases, the individual actively tries to integrate the new information.

> The point is that all of an individual's values are organized into a single system the preservation of whose integrity is essential. The nucleus of the system, around which the rest of the system revolves, is the individual's valuation of himself. The individual sees the world from his own viewpoint, with himself as the center. Any value entering the system which is inconsistent with the individual's valuation of himself cannot be assimilated; it meets with resistance and is likely . . . to be rejected. (pp. 152–153)

Lecky's thinking is close to Festinger's. In Lecky's formulation there is a willingness, and even a need, to view humans as active, organizing beings.

Kurt Lewin, too, particularly in the group decision studies carried out as part of his channel theory, foreshadowed dissonance theory. Lewin astutely observed that the process by which group members' attitudes are changed in a more or less permanent way could be described in three steps: 1. in which the old attitude is "unfrozen"; 2. in which the person is "moved" to the new attitude level; and 3. in which the new attitude is "frozen." This "unfreezing"—"moving"—"refreezing" conception became central in the development of dissonance theory. Festinger used similar terminology to describe the process of postdecision dissonance reduction. In fact, Lewin's explanation of why a group's voluntary decision is more effective in changing attitudes than a lecture or command to the group is carried all but intact to the later dissonance formulation and has been borne out by subsequent research. This group decision, Lewin maintained, provided a link between motivation and action. The relatively small behavioral commitment of the group decision served to "freeze" the decision and link it to later, more concrete behavioral changes. The individual who has committed himself as a member of the group will show a tendency to "stick to his decision."

The link between behavioral commitment and change is, of course, a very important aspect of Festinger's work, as we have already seen. The specific carryover of Lewin's studies may be seen clearly in Freedman and Fraser's (1966) experiment on the "foot-in-the-door-technique," in which it was demonstrated that once a person has made a small behavioral commitment, he becomes more willing to make larger commitments in the future. The results of the "foot-in-the-door" experiment are clearly and easily derivable from the theory of cognitive dissonance: Commitment produces a need for self-justification, which increases the probability of future action. The astute reader will be able to make the same prediction from Lewin's more general theoretical approach.

Trained in the Gestalt tradition, Lewin was dissatisfied with a purely mechanistic view of causality and convinced that to adopt a Newtonian approach to the psychology of human behavior was inadequate. The causes of behavior could not be traced solely to sources internal or external to the person. Rather, human behavior was the result of a complex interaction of the person and his environment: the life space. Humans, according to Lewin, were not completely passive, driven organisms but took an active role in constructing and deciding. In short, they were able to think, and this ability had enormous consequences for persuasion and social action. Tolman (1948), in a tribute to Lewin, saw Lewin's ideas of systematic rather than historical causation as intimately bound up with his commitment to social action, to the notion that human behavior could be changed. Festinger himself, in his contribution to Marrow's biography of Lewin, saw change as central to Lewin's work:

> This theme, that in order to gain insight into a process one must create a change and then observe its variable effects and new dynamics runs through all Lewin's work. To Lewin life was not static; it was changing, dynamic, fluid. Lewin's unfreezing-stabilizing-refreezing concept of change continues to be highly relevant to-

day. His understanding of the importance of change was part of his philosophical approach to science and a basic ingredient of his "metatheory." As such, it helped change much of social psychology from art into science. (Marrow, 1969, p. 235)

Fritz Heider's extension of Lewin's work also shares a similar cognitively oriented approach to behavior. Heider's conception of causal attribution, of course, provided the basis for what was later developed by Kelley (1973) and Jones and Davis (1965) into attribution theory. In addition, Heider's balance theory works on a common premise with dissonance theory and other consistency theories. Consistency of relationship is, in many cases, of itself desirable for the individual; and lack of consistency or its loss, an unbalanced relationship, causes the individual to seek it, restoring balance.

WHAT'S THE HULLABALOO ABOUT?

If dissonance theory is firmly rooted in earlier conceptualizations, why do I credit dissonance theory with a larger direct impact on social psychology than the earlier theories of Lewin or Lecky? Why did dissonance theory provide such enthusiasm and anger? Some of the answers to these questions are implicit in the statement by Jones (presented earlier). While Lecky's ideas were clearly consistent with dissonance theory, there was little or no original research generated. On the other hand, Lewin's general theory was not only rich with complex cognitive and dynamic subtleties, Lewin and his students produced a good deal of interesting research. Yet, this research did not have the impact on the field that dissonance research did, partly because as research it did not directly engage or challenge existing mainstream theories. That is, while Lewin's theory was both cognitive and dynamic, for the most part the data themselves could (at least superficially) be fit under the rubric of reward theory. For example, if one looks at the classic study of leadership (Lewin, Lippitt, and White, 1939), the results are not troublesome to a reinforcement theorist. After all, if you expose youngsters to a stiff, dogmatic, punitive, autocratic leader, this would be less rewarding than a more casual, attractive leader who allows them to make their own decisions. It falls easily within the general reward zeitgeist to assume that the latter would produce more satisfaction and greater productivity among the children. Similarly, while the theorizing underlying the group decision research (Lewin, 1947) is complex and cognitive, the actual data are not so far removed from the data of the Asch experiment that the conformity cannot be explained in terms of anxiety reduction.

What Festinger and his band of irreverent students brought to bear on the scene was not a brand new conceptualization. Rather, it was a theory stated in terms that opened the door wide for a diverse set of hypotheses and, more important, stated those hypotheses in a way that existing theories *had* to pay attention to: Either reward theory made no prediction at all or the opposite prediction.

I recall vividly the atmosphere in which Judson Mills and I, as graduate students at Stanford University in 1957, presented the hypothesis and procedure of the initiation experiment to a group of our peers: "Undergoing an embarrassing initiation in order to become a member of a group will increase the attractiveness of that group." The dominant response was that our hypothesis was silly—that it went against common sense. Clearly a stimulus (the group discussion) associated with unpleasantness (the embarrassing initiation) would be liked less through association. "Common sense" is, of course, defined as the climate created by the global acceptance of existing theories. When the results supported our "nonsensical" hypothesis, the first response of serious critics (e.g., Chapanis and Chapanis, 1964) was to conclude with great confidence that our data were an artifact—that what we called an embarrassing initiation was actually a pleasant, even an exciting, event. As the reader knows, the experiment has been successfully replicated several times, using such clearly unpleasant initiation procedures as severe electric shock (e.g., Gerard and Mathewson, 1966).

Of course, a prediction that is nonsensical in one decade can become "common sense" several years later—and even old hat and boring in a third decade. In one sense, it is the duty of a theory to become boring, and fortunately, this fate seems to have befallen dissonance theory. But in the early days of the theory, a great deal was made of the "nonobvious" predictions which were generated. The battle over who can be most nonobvious was part of the growing pains of dissonance theory in the 1950s and 1960s. The claim of nonobviousness was uttered both as a criticism of the theory and as a source of pride and pleasure among the people working most closely with the theory. Kiesler, Collins, and Miller (1969) have pointed out that "nonobvious" is a misnomer in the sense that a new theory *should* be able to handle predictions which other theories cannot, and that dissonance predictions are no more nonobvious than predictions in other experiments where competing hypotheses have been ruled out. But "nonobvious" is a misnomer in another sense, as I have mentioned elsewhere (1968). Dissonance predictions are nonobvious only when seen against the background of a reward-reinforcement framework. They are quite consistent, however, when taken in the context of other theoretical traditions such as Lewin's and Lecky's.

While dissonance theory was in its brash infancy, Sears and Abeles (1969) were not unfair in characterizing us as "the imperialistic dissonance hordes." However, it eventually became clear that we were not opposed to peaceful coexistence. Most specifically, we spelled out both theoretically and empirically under which conditions reward-incentive theory would make its clearest predictions and under which conditions dissonance theory would make its predictions. Most generally, people will strive for rewards and may even change their opinions in the face of a reward under conditions where they are not *personally involved* (Aronson, 1969; Carlsmith, Collins, and Helmreich, 1966; Linder, Cooper, and Jones, 1967). This state of affairs prompted Sears and Abeles (1969) to exult over the fact that the "youthful brashness of dissonance theory is [being] replaced by well-fed middle-aged generosity."

Increasingly, in recent years, dissonance experiments have been aimed less at breaking new ground and certainly have not attempted to outrage the establishment by violating either traditional notions of common sense or traditional notions of the conduct of scientific inquiry. As the theory has continued to amble through middle age, it has moved more and more in the direction of defining the parameters of a variety of social psychological situations, clarifying the conditions under which dissonance phenomena do or do not exist. Here I refer to the admirable research on forced compliance by Collins and Hoyt (1972); on the necessity of arousal for attitude change (by Cooper, Zanna, and Taves, 1977); on the conditions under which dissonance theory and self-perception theory apply (Fazio, Zanna, and Cooper, 1977); on personality differences in selective exposure to information (Olson and Zanna, 1977); and on dissonance and attribution (Zanna and Cooper, 1974). I have recently reviewed this research (1978) and see no good reason to repeat the review here. Rather, I would prefer to spend my last few pages on a longish summary of what we now know for sure about persuasion that we didn't know some three decades ago.

A LONGISH SUMMARY

The difficulties which researchers were having in achieving long-lasting effects for persuasion techniques, based on a reward-reinforcement incentive theory, were part of what made the first dissonance experiments appealing. Dissonance theory provided a theory and a method for affecting deep-seated, rather than superficial, attitudes. "Who says what to whom with what effect," while seemingly effective in inducing short-term opinion change, did not change important attitudes—and didn't change even unimportant ones for very long. Although a credible communicator under the best of circumstances (the appropriate technique and audience) could probably sell you one brand of toothpaste or breakfast cereal over another, it was not clear that he could convince you to be less prejudiced against blacks, induce you to support the development of solar energy, or modify your opinion about the legalization of marijuana.

Dissonance theory was different in this respect. In linking the cognitive and motivational components of attitudes to behavior, it offered the possibility of advancing the field in two directions: 1. providing insight into the mechanisms behind deeper set attitudes and 2. pointing the way toward methods of effecting long-term changes in attitudes of a nonsuperficial nature. Festinger achieved this by suggesting that, at least in regard to important issues, people would not simply respond to the communication package with the greatest incentive value. Rather, people are involved in justifying their attitudes and behavior and in maintaining consistency. And again, this insistence on the link between motivation and behavior was carried over from the Lewinian tradition. The implications of this approach were completely different from those of a strict reinforcement theory, suggesting that the best way to change attitudes was to get people to persuade themselves. And as Lewin implied, techniques which made use of behavioral commitment had the possibility of leading to attitude changes of a more permanent nature. The potential of such an approach added to the excitement and enthusiasm surrounding the beginnings of dissonance theory.

Research concerned with clarifying and defining the parameters of the theory has expanded and continues to expand in many directions. And because of the research generated by the early controversies, this later research stands on a solid foundation—one made up of basics which we really do understand now and did not understand thirty years ago. Before we go on to discuss some of this research, it may be useful to list some of these by now well-known basics:

1. Choice, or volition, plays an important role in the operation of dissonance. In situations where an individual has voluntarily agreed to perform a discrepant behavior, dissonance is most likely to be aroused.

2. If a person's commitment to perform a discrepant behavior, such as arguing against his own beliefs or lying to another person, is high, he is more likely to experience dissonance and engage in subsequent dissonance reduction.

3. In situations in which a person feels responsible for the consequences of a discrepant behavior and is able to foresee its negative consequences, dissonance is aroused.

4. In some cases, even when the negative outcome is *not* foreseen, people will assign responsibility irrationally to themselves and will experience dissonance. This is particularly likely in situations where the behavior has had negative consequences for others besides the individual.

5. What all of these situations have in common is that the self or the self-concept is engaged.

The fact that dissonance predicts most clearly in situations in which the self is engaged and the person feels responsible for the consequences of his action may help to clarify the ability of the theory to address important rather than trivial attitudes. Dissonance predicts more clearly the firmer the expectancies are involved and the less they are subject to individual, idiosyncratic views of what are consonant and what are dissonant cognitions. Almost by definition, our expectancies about our own behavior are firmer than our expectancies about another person's behavior. The fact that most people share some elements of a self-concept (that they are moral and intelligent people, for example) means that predictions which involve the disconfirmation of these elements will be less equivocal and more generally valid than predictions involving other types of expectations.

This line of reasoning may be taken a step further. Our expectations about ourselves are not only firmer, they are more important to us than our expectations about general subjects. Intuitively, the fact that I have misled someone when I consider myself to be a moral person will be more important to me (providing for individual differences) than the fact that you (whom I thought was honest) have misled someone. When cognitions about ourselves are threatened or contradicted, we are likely to devote more energy to restoring consonance than if the inconsistency involves things about which we care little.

As we have seen, the research on dissonance and responsibility confirms this view. In order for dissonance to be aroused in most situations, the person must feel a connection between his own behavior and its consequences. Thus, as mentioned previously, in the experiment by Nel, Helmreich, and Aronson (1969), subjects who knew in advance that their pro-marijuana videotape would be used to change the attitudes of a highly persuasible audience—a group of young people who had no previous opinion on the issue—experienced more dissonance and exhibited more attitude change than those who believed the tape would be played to an audience confirmed in its beliefs. Moreover, subsequent work (Aronson, Chase, Helmreich, and Ruhnke, 1974) indicates that this connection between action and consequences need not necessarily be a rational one. If the results of a behavior are severe, particularly when they are harmful to others, people tend to take responsibility for the consequences *even when they could not reasonably have been expected to foresee them.* In the case of positive consequences, responsibility seems to be assumed only when those consequences had been foreseen (Brehm and Jones, 1970; Lepper, Zanna, and Abelson, 1970).

Let us return for a moment to the self-concept. The notion of responsibility is tied to the self-concept in a number of ways. Wicklund and Brehm (1976) point to the dual components of choice and foreseeability in the concept of responsibility. Both of these describe the connection between the self and the consequences of action. For the consequences of a behavior to arouse dissonance, we must not only see them as connected to us but as connected to us in a way which contradicts an element of our self-concept. The fact that we have harmed others arouses dissonance because most of us carry the expectancy that we are not harmdoers, even unintentionally.

The self-concept, for our purposes, may be seen as composed of a number of statements—such as "I am an honest person," "I am an intelligent person," and so on. Each of these statements, by necessity, must be further defined for the individual, and the subdefinitions must consist of the attitudes, values, and expectancies with which the statement is associated. Thus, for the subjects in the Nel, Helmreich, and Aronson experiment, the general heading "I am not a harmdoer" was further defined by the fact that they regarded marijuana as a harmful substance and therefore their advocacy of smoking it as a harmful thing to do. This further definition, of course, is implicit in the experimental procedure, being determined by the initial attitude measurement.

The suggestion here is that it may be possible to define "deep-seated" attitudes as attitudes which are associated with the self-concept as extended definitions. A peripheral opinion, on the other hand (and it may be peripheral because it was recently acquired, because it falls in a region in which the person claims no expertise, or for numerous other reasons), may be seen as one which has no such association. Whether I buy one brand of soap or another may be reasonably seen as having little association with issues of my intelligence, honesty, and so on.

If this argument holds, it is more easily seen why behavioral commitment in a dissonance-arousing situation may lead to more permanent attitude change. Self-justification, the process of dissonance reduction, involves the reorganization either of the headings themselves or of the subdefinitions. Thus, again in the Nel, Helmreich, and Aronson study, "harmful" is redefined to exclude the advocacy of smoking marijuana. It is likely that in a situation of behavioral commitment where the attitude must be changed to achieve consonance since the behavior cannot, such reorganization will result in relatively long-lasting change. The change of a peripheral opinion, since it requires no substantial reorganization, carries no guarantee of lasting over time.

Although the above is purely speculative, I believe that research along these lines—responsibility, foreseeability, and the self-concept—sustain dissonance as a vital theory. When combined with the other directions of dissonance research, there is indication that continued excitement and enthusiasm are justified.

Twenty-five years of dissonance research have not only illuminated our understanding of social influence and persuasion but, in addition, the cumulated findings have changed our thinking about how people get persuaded on the deepest possible level. Accordingly, researchers of social psychology have shifted their primary focus from an interest in variations in either the nature of a communication or in attributes of the communicator and have placed it on variables involving the person's own behavior—which, in turn, affect his active perception of himself in the situation. Dissonance studies have directed attention to the link between a person's cognition, his affect, and his behavior, to the process by which the subject actively attempts to bring these elements in line with one another. Choice, foreseeability, commitment—all are forces which actively engage the subject, linking him to the situation of which he is a part and inducing him to reconstruct his cognitive environment. What we now know for sure that we didn't know when I was an undergraduate is that when an individual is personally involved in a situation wherein he might consider himself to be stupid or immoral, he engages in self-justifying behavior which involves some form of self-persuasion. That self-persuasion can affect important attitudes and opinions (e.g., as in the experiments by Nel, Helmreich and Aronson [1969] and Cohen [1962]); cause enormous shifts in those opinions (up to 73 percent of the possible range, as in the Nel, Helmreich, and Aronson study); and persist over relatively long periods of time (e.g., in the Freedman [1965] toy study).

Moreover, this process occurs in a wide variety of situations which, before they were linked by dissonance theory, would have appeared to be disparate and unrelated. In addition, what the work on dissonance theory brought into focus more clearly than any other body of work is the fact that the social psychological laboratory, with all of its contrivances and complex scenarios, can produce clear, powerful effects which are conceptually replicable in both the laboratory and the real world.

Having begun this chapter with a rather dramatic description of social psychology in the 1950s and 1960s, I feel some obligation to end it on an equally dramatic note. It seems reassuring to me that over twenty-five years of research one is able to see both continuity and progress. It seems that this should be reassuring to others as well. That both of these qualities have been tied to careful experimental and theoretical work indicates that there is validity in a method which at times feels unproductive or disorganized. In short, there is no reason to think that the work of this decade will be any less productive than that of past decades—or any less exciting.

References

ARONSON, E. Dissonance theory: Progress and problems. In R. P. Abelson, E. Aronson, W. J. McGuire, T. M. Newcomb, M. J. Rosenberg, and P. H. Tannenbaum (Eds.), *Theories of cognitive consistency: A sourcebook.* Chicago: Rand-McNally, 1968, pp. 5–27.

ARONSON, E. The theory of cognitive dissonance: A current perspective. In L. Berkowitz (Ed.), *Advances in experimental social psychology,* (Vol. 4). New York: Academic, 1969, pp. 1–34.

ARONSON, E. The theory of cognitive dissonance: A current perspective. In L. Berkowitz (Ed.), *Cognitive theories in social psychology.* New York: Academic, 1978, pp. 215–220.

ARONSON, E. AND CARLSMITH, J. M. Effect of severity of threat on the valuation of forbidden behavior. *Journal of Abnormal and Social Psychology,* 1963, *66,* pp. 584–588.

ARONSON, E., CHASE, T., HELMREICH, R., AND RUHNKE, R. A two-factor theory of dissonance reduction: The effect of feeling stupid or feeling "awful" on opinion change. *International Journal of Communication Research,* 1974, *3,* 340–352.

ARONSON, E. AND MILLS, J. The effects of severity of initiation on liking for a group. *Journal of Abnormal and Social Psychology,* 1959, *59,* 177–181.

ASCH, S. E. Effects of group pressure upon the modification and distortion of judgments. In H. Guetzkow (Ed.), *Groups, leadership and men.* Pittsburgh: Carnegie Press, 1951, pp. 177–190.

BREHM, J. W. AND JONES, R. A. The effect on dissonance of surprise consequences. *Journal of Experimental Social Psychology,* 1970, *6,* 420–431.

CARLSMITH, J. M., COLLINS, B. E., AND HELMREICH, R. L. Studies in forced compliance: I. The effect of pressure for compliance on attitude change produced by face-to-face role playing and anonymous essay writing. *Journal of Personality and Social Psychology,* 1966, *4,* 1–13.

CHAPANIS, N. P. AND CHAPANIS, A. C. Cognitive dissonance: Five years later. *Psychological Bulletin,* 1964, *61,* 1–22.

COHEN, A. R. An experiment on small rewards for discrepant compliance and attitude change. In J. W. Brehm and A. R. Cohen (Eds.), *Explorations in cognitive dissonance.* New York: Wiley, 1962, pp. 73–78.

COLLINS, B. E. AND HOYT, M. F. Personal responsibility-for-consequences: An integration and extension of the "forced compliance" literature. *Journal of Experimental Social Psychology,* 1972, *8,* 558–593.

COOPER, J., ZANNA, M. P., AND TAVES, P. A. Arousal as a necessary condition for attitude change following induced compliance. Unpublished paper, 1977.

DEUTSCH, M. AND GERARD, H. A study of normative and informational social influences on individual judgment. *Journal of Abnormal and Social Psychology*, 1955, *51*, 629–636.

FAZIO, R. H., ZANNA, M. P., AND COOPER, J. Dissonance and self-perception: An integrative view of each theory's proper domain of application. *Journal of Experimental Social Psychology*, 1977, *13*, 464–479.

FESTINGER, L. *A theory of cognitive dissonance*. Stanford, Calif.: Stanford University Press, 1957.

FESTINGER, L. AND CARLSMITH, J. M. Cognitive consequences of forced compliance. *Journal of Abnormal and Social Psychology*, 1959, *58*, 203–210.

FREEDMAN, J. L. Long-term behavioral effects of cognitive dissonance. *Journal of Experimental Social Psychology*, 1965, *1*, 145–155.

FREEDMAN, J. L. AND FRASER, S. C. Compliance without pressure: The foot-in-the-door technique. *Journal of Personality and Social Psychology*, 1966, *4*, 195–202.

GERARD, H. B. AND MATHEWSON, G. C. The effects of severity of initiation on liking for a group: A replication. *Journal of Experimental Social Psychology*, 1966, *2*, 278–287.

HOVLAND, C. I. Reconciling conflicting results derived from experimental and survey studies of attitude change. *American Psychologist*, 1959, *14*, 8–17.

HOVLAND, C. I. AND WEISS, W. The influence of source credibility on communication effectiveness. *Public Opinion Quarterly*, 1951, *15*, 635–650.

JONES, E. E. AND DAVIS, K. E. From acts to dispositions: The attribution process in person perception. In L. Berkowitz (Ed.), *Advances in experimental social psychology*. New York: Academic, 1965.

KELLEY, H. H. The processes of causal attribution. *American Psychologist*, 1973, *28*, 107–128.

KIESLER, C. A., COLLINS, B. E., AND MILLER, N. *Attitude change: A critical analysis of theoretical approaches*. New York: Wiley, 1969.

LECKY, P. *Self-consistency: A theory of personality* (2nd ed.). Hamden, Conn.: Shoe String Press, 1961 (rpt. of 1951 edition).

LEPPER, M. R., ZANNA, M. P., AND ABELSON, R. P. Cognitive irreversibility in a dissonance reduction situation. *Journal of Personality and Social Psychology*, 1970, *16*, 191–198.

LEWIN, K. Group decision and social change. In T. H. Newcomb and E. L. Hartley (Eds.), *Readings in social psychology*. New York: Holt, 1947.

LEWIN, K., LIPPITT, R., AND WHITE, R. Patterns of aggressive behavior in experimentally created "social climates." *Journal of Social Psychology*, 1939, *10*, 271–299.

LINDER, D. E., COOPER, J., AND JONES, E. E. Decision freedom as a determinant of the role of incentive magnitude in attitude change. *Journal of Personality and Social Psychology*, 1967, *6*, 245–254.

MARROW, A. J. *The practical theorist: The life and work of Kurt Lewin*. New York: Basic Books, 1969.

MILLER, N. E. AND DOLLARD, J. *Social learning and imitation*. New Haven: Yale University Press, 1941.

NEL, E., HELMREICH, R., AND ARONSON, E. Opinion change in the advocate as a function of the persuasibility of his audience: A clarification of the meaning of dissonance. *Journal of Personality and Social Psychology*, 1969, *12*, 117–124.

OLSON, J. M. AND ZANNA, M. P. A new look at selective exposure. Paper presented at the Annual Meeting of the American Psychological Association, San Francisco, August, 1977.

SEARS, D. O. AND ABELES, R. Attitudes and opinions. In P. H. Mussen and M. Rosenzweig (Eds.), *Annual Review of Psychology*, 1969, *20*, 253–288.

TOLMAN, E. C. AND LEWIN, K. Obituary. *Psychological Review*, 1948, *55*, No. 1, 1–4.

WICKLUND, R. A. AND BREHM, J. W. *Perspectives on Cognitive Dissonance*. New York: Wiley, 1976.

ZANNA, M. P. AND COOPER, J. Dissonance and the pill: An attribution approach to studying the arousal properties of dissonance. *Journal of Personality and Social Psychology*, 1974, *29*, 703–709.

12

Compliance Without Pressure:
The Foot-in-the-Door Technique

Jonathan L. Freedman and Scott C. Fraser

Two experiments were conducted to test the proposition that once someone has agreed to a small request he is more likely to comply with a larger request. The first study demonstrated this effect when the same person made both requests. The second study extended this to the situation in which different people made the two requests. Several experimental groups were run in an effort to explain these results, and possible explanations are discussed.

How can a person be induced to do something he would rather not do? This question is relevant to practically every phase of social life, from stopping at a traffic light to stopping smoking, from buying Brand X to buying savings bonds, from supporting the March of Dimes to supporting the Civil Rights Act.

One common way of attacking the problem is to exert as much pressure as possible on the reluctant individual in an effort to force him to comply. This technique has been the focus of a considerable amount of experimental research. Work on attitude change, conformity, imitation, and obedience has all tended to stress the importance of the degree of external pressure. The prestige of the communicator (Kelman and Hovland, 1953), degree of discrepancy of the communication (Hovland and Pritzker, 1957), size of the group disagreeing with the subject (Asch, 1951), perceived power of the model (Bandura, Ross, and Ross, 1963), etc., are the kinds of variables that have been studied. This impressive body of work, added to the research on rewards and punishments in learning, has produced convincing evidence that greater external pressure generally leads to greater compliance with

Reprinted with permission from the authors and *The Journal of Personality and Social Psychology*, Vol. 4, No. 2, 1966. Copyright 1966 by the American Psychological Association.

The authors are grateful to Evelyn Bless for assisting in the running of the second experiment reported here. These studies were supported in part by Grant GS-196 from the National Science Foundation. The first study was conducted while the junior author was supported by an NSF undergraduate summer fellowship.

the wishes of the experimenter. The one exception appears to be situations involving the arousal of cognitive dissonance in which, once discrepant behavior has been elicited from the subject, the greater the pressure that was used to elicit the behavior, the less subsequent change occurs (Festinger and Carlsmith, 1959). But even in this situation one critical element is the amount of external pressure exerted.

Clearly, then, under most circumstances the more pressure that can be applied, the more likely it is that the individual will comply. There are, however, many times when for ethical, moral, or practical reasons it is difficult to apply much pressure when the goal is to produce compliance with a minimum of apparent pressure, as in the forced-compliance studies involving dissonance arousal. And even when a great deal of pressure is possible, it is still important to maximize the compliance it produces. Thus, factors other than external pressure are often quite critical in determining degree of compliance. What are these factors?

Although rigorous research on the problem is rather sparse, the fields of advertising, propaganda, politics, etc., are by no means devoid of techniques designed to produce compliance in the absence of external pressure (or to maximize the effectiveness of the pressure that is used, which is really the same problem). One assumption about compliance that has often been made either explicitly or implicitly is that once a person has been induced to comply with a small request he is more likely to comply with a larger demand. This is the principle that is commonly referred to as the foot-in-the-door or gradation technique and is reflected in the saying that if you "give them an inch, they'll take a mile." It was, for example, supposed to be one of the basic techniques upon which the Korean brainwashing tactics were based (Schein, Schneier, and Barker, 1961), and, in a somewhat different sense, one basis for Nazi propaganda during 1940 (Bruner, 1941). It also appears to be implicit in many advertising campaigns which attempt to induce the consumer to do anything relating to the product involved, even sending back a card saying he does not want the product.

The most relevant piece of experimental evidence comes from a study of conformity done by Deutsch and Gerard (1955). Some subjects were faced with incorrect group judgments first in a series in which the stimuli were not present during the actual judging and then in a series in which they were present, while the order of the memory and visual series was reversed for other subjects. For both groups the memory series produced more conformity, and when the memory series came first there was more total conformity to the group judgments. It seems likely that this order effect occurred because, as the authors suggest, once conformity is elicited at all it is more likely to occur in the future. Although this kind of conformity is probably somewhat different from compliance as described above, this finding certainly lends some support to the foot-in-the-door idea. The present research attempted to provide a rigorous, more direct test of this notion as it applies to compliance and to provide data relevant to several alternative ways of explaining the effect.

EXPERIMENT I

The basic paradigm was to ask some subjects (Performance condition) to comply first with a small request and then three days later with a larger, related request. Other subjects (One-Contact condition) were asked to comply only with the large request. The hypothesis was that more subjects in the Performance condition than in the One-Contact condition would comply with the larger request.

Two additional conditions were included in an attempt to specify the essential difference between these two major conditions. The Performance subjects were asked to perform a small favor, and, if they agreed, they did it. The question arises whether the act of agreeing itself is critical or whether actually carrying it out was necessary. To assess this a third group of subjects (Agree-Only) was asked the first request, but, even if they agreed, they did not carry it out. Thus, they were identical to the Performance group except that they were not given the opportunity of performing the request.

Another difference between the two main conditions was that at the time of the larger request the subjects in the Performance condition were more familiar with the experimenter than were the other subjects. The Performance subjects had been contacted twice, heard his voice more, discovered that the questions were not dangerous, and so on. It is possible that this increased familiarity would serve to decrease the fear and suspicion of a strange voice on the phone and might accordingly increase the likelihood of the subjects agreeing to the larger request. To control for this a fourth condition was run (Familiarization) which attempted to give the subjects as much familiarity with the experimenter as in the Performance and Agree-Only conditions with the only difference being that no request was made.

The major prediction was that more subjects in the Performance condition would agree to the large request than in any of the other conditions, and that the One-Contact condition would produce the least compliance. Since the importance of agreement and familiarity was essentially unknown, the expectation was that the Agree-Only and Familiarization conditions would produce intermediate amounts of compliance.

Method

The prediction stated above was tested in a field experiment in which housewives were asked to allow a survey team of five or six men to come into their homes for two hours to classify the household products they used. This large request was made under four different conditions: after an initial contact in which the subject had been asked to answer a few questions about the kinds of soaps she used, and the questions were actually asked (Performance condition); after an identical contact in which the questions were not actually asked (Agree-Only condition); after an initial contact in which no request was made (Familiarization condition); or after no initial contact (One-Contact condition). The dependent measure was simply whether or not the subject agreed to the large request.

Procedure.   The subjects were 156 Palo Alto, California, housewives, 36 in each condition, who were selected at random from the telephone directory. An additional 12 subjects distributed about equally among the three two-contact conditions could not be reached for the second contact and are not included in the data analysis. Subjects were assigned randomly to the various conditions, except that the Familiarization condition was added to the design after the other three conditions had been completed. All contacts were by telephone by the same experimenter who identified himself as the same person each time. Calls were made only in the morning. For the three groups that were contacted twice, the first call was made on either Monday or Tuesday and the second always three days later. All large requests were made on either Thursday or Friday.

At the first contact, the experimenter introduced himself by name and said that he was from the California Consumers' Group. In the Performance condition he then proceeded:

> We are calling you this morning to ask if you would answer a number of questions about what household products you use so that we could have this information for our public service publication, ''The Guide.'' Would you be willing to give us this information for our survey?

If the subject agreed, she was asked a series of eight innocuous questions dealing with household soaps (e.g., ''What brand of soap do you use in your kitchen sink?'') She was then thanked for her cooperation, and the contact terminated.

Another condition (Agree-Only) was run to assess the importance of actually carrying out the request as opposed to merely agreeing to it. The only difference between this and the Performance condition was that, if the subject agreed to answer the questions, the experimenter thanked her, but said that he was just lining up respondents for the survey and would contact her if needed.

A third condition was included to check on the importance of the subject's greater familiarity with the experimenter in the two-contact conditions. In this condition the experimenter introduced himself, described the organization he worked for and the survey it was conducting, listed the questions he was asking and then said that he was calling merely to acquaint the subject with the existence of his organization. In other words, these subjects were contacted, spent as much time on the phone with the experimenter as the Performance subjects did, heard all the questions, but neither agreed to answer them nor answered them.

In all of these two-contact conditions some subjects did not agree to the requests or even hung up before the requests were made. Every subject who answered the phone was included in the analysis of the results and was contacted for the second request regardless of her extent of cooperativeness during the first contact. In other words, no subject who could be contacted the appropriate number of times was discarded from any of the four conditions.

The large request was essentially identical for all subjects. The experimenter called, identified himself, and said either that his group was expanding its survey (in the case of the two-contact conditions) or that it was conducting a survey (in the One-Contact condition). In all four conditions he then continued:

> The survey will involve five or six men from our staff coming into your home some morning for about two hours to enumerate and classify all the household products that you have. They will have to have full freedom in your house to go through the cupboards and storage places. Then all this information will be used in the writing of the reports for our public service publication, "The Guide."

If the subject agreed to the request, she was thanked and told that at the present time the experimenter was merely collecting names of people who were willing to take part and that she would be contacted if it were decided to use her in the survey. If she did not agree, she was thanked for her time. This terminated the experiment.

Results

Apparently even the small request was not considered trivial by some of the subjects. Only about two-thirds of the subjects in the Performance and Agree-Only conditions agreed to answer the questions about household soaps. It might be noted that none of those who refused the first request later agreed to the large request, although as stated previously all subjects who were contacted for the small request are included in the data for those groups.

Our major prediction was that subjects who had agreed to and carried out a small request (Performance condition) would subsequently be more likely to comply with a larger request than would subjects who were asked only the larger request (One-Contact condition). As may be seen in Table 12.1, the results support the predic-

TABLE 12.1
Percentage of subjects complying
with large request in Experiment I

Condition	%
Performance	52.8
Agree-Only	33.3
Familiarization	27.8*
One-Contact	22.2**

Note: $N = 36$ for each group. Significance levels represent differences from the Performance condition.
*$p < .07$.
**$p < .02$.

tion. Over 50 percent of the subjects in the Performance condition agreed to the larger request, while less than 25 percent of the One-Contact condition agreed to it. Thus it appears that obtaining compliance with a small request does tend to increase subsequent compliance. The question is what aspect of the initial contact produces this effect.

One possibility is that the effect was produced merely by increased familiarity with the experimenter. The Familiarization control was included to assess the effect on compliance of two contacts with the same person. The group had as much contact with the experimenter as the Performance group, but no request was made during the first contact. As the table indicates, the Familiarization group did not differ appreciably in amount of compliance from the One-Contact group, but was different from the Performance group ($\chi^2 = 3.70$, $p < .07$). Thus, although increased familiarity may well lead to increased compliance, in the present situation the differences in amount of familiarity apparently were not great enough to produce any such increase; the effect that was obtained seems not to be due to this factor.

Another possibility is that the critical factor producing increased compliance is simply agreeing to the small request (i.e., carrying it out may not be necessary). The Agree-Only condition was identical to the Performance condition except that in the former the subjects were not asked the questions. The amount of compliance in this Agree-Only condition fell between the Performance and One-Contact conditions and was not significantly different from either of them. This leaves the effect of merely agreeing somewhat ambiguous, but it suggests that the agreement alone may produce part of the effect.

Unfortunately, it must be admitted that neither of these control conditions is an entirely adequate test of the possibility it was designed to assess. Both conditions are in some way quite peculiar and may have made a very different and extraneous impression on the subject than did the Performance condition. In one case, a housewife is asked to answer some questions and then is not asked them; in the other, some man calls to tell her about some organization she has never heard of. Now, by themselves neither of these events might produce very much suspicion. But, several days later, the same man calls and asks a very large favor. At this point it is not at all unlikely that many subjects think they are being manipulated, or in any case that something strange is going on. Any such reaction on the part of the subjects would naturally tend to reduce the amount of compliance in these conditions.

Thus, although this first study demonstrates that an initial contact in which a request is made and carried out increases compliance with a second request, the question of why and how the initial request produces this effect remains unanswered. In an attempt to begin answering this question and to extend the results of the first study, a second experiment was conducted.

There seemed to be several quite plausible ways in which the increase in compliance might have been produced. The first was simply some kind of commitment to or involvement with the particular person making the request. This might work, for example, as follows: The subject has agreed to the first request and perceives that the experimenter therefore expects him also to agree to the second request. The subject thus feels obligated and does not want to disappoint the experimenter; he also feels that he needs a good reason for saying "no"—a better reason than he would need if he had never said "yes." This is just one line of causality—the particular process by which involvement with the experimenter operates might be quite different, but the basic idea would be similar. The commitment is to the particular person. This implies that the increase in compliance due to the first contact should occur primarily when both requests are made by the same person.

Another explanation in terms of involvement centers around the particular issue with which the requests are concerned. Once the subject has taken some action in connection with an area of concern, be it surveys, political activity, or highway safety, there is probably a tendency to become somewhat more concerned with the area. The subject begins thinking about it, considering its importance and relevance to him, and so on. This tends to make him more likely to agree to take further action in the same area when he is later asked to. To the extent that this is the critical factor the initial contact should increase compliance only when both requests are related to the same issue or area of concern.

Another way of looking at the situation is that the subject needs a reason to say "no." In our society it is somewhat difficult to refuse a reasonable request, particularly when it is made by an organization that is not trying to make money. In order to refuse, many people feel that they need a reason—simply not wanting to do it is often not in itself sufficient. The person can say to the requester or simply to himself that he does not believe in giving to charities or tipping or working for political parties or answering questions or posting signs, or whatever he is asked to do. Once he has performed a particular task, however, this excuse is no longer valid for not agreeing to perform a similar task. Even if the first thing he did was trivial compared to the present request, he cannot say he never does this sort of thing, and thus one good reason for refusing is removed. This line of reasoning suggests that the similarity of the first and second requests in terms of the type of action required is an important factor. The more similar they are, the more the "matter of principle" argument is eliminated by agreeing to the first request, and the greater should be the increase in compliance.

There are probably many other mechanisms by which the initial request might produce an increase in compliance. The second experiment was designed in part to test the notions described above, but its major purpose was to demonstrate the effect unequivocally. To this latter end it eliminated one of the important problems with the first study which was that when the experimenter made the second request he was not blind as to which condition the subjects were in. In this study the second request was always made by someone other than the person who made the first request, and the second experimenter was blind as to what condition the subject was in. This eliminates the possibility that the experimenter exerted systematically different amounts of pressure in different experimental conditions. If the effect of the first study were replicated, it would also rule out the relatively uninteresting possibility that the effect is due primarily to greater familiarity or involvement with the particular person making the first request.

EXPERIMENT II

The basic paradigm was quite similar to that of the first study. Experimental subjects were asked to comply with a small request and were later asked a considerably larger request, while controls were asked only the larger request. The first request varied along two dimensions. Subjects were asked either to put up a small sign or to sign a petition, and the issue was either safe driving or keeping California beautiful. Thus, there were four first requests: a small sign for safe driving or for beauty, and a petition for the two issues. The second request for all subjects was to install in their front lawn a very large sign which said "Drive Carefully." The four experimental conditions may be defined in terms of the similarity of the small and large requests along the dimensions of issue and task. The two requests were similar in both issue and task for the small-sign, safe-driving group, similar only in issue for the safe-driving-petition group, similar only in task for the small "Keep California Beautiful" sign group, and similar in neither issue nor task for the "Keep California Beautiful" petition group.

The major expectation was that the three groups for which either the task or the issue was similar would show more compliance than the controls, and it was also felt that when both were similar there would probably be the most compliance. The fourth condition (Different Issue-Different Task) was included primarily to assess the effect simply of the initial contact which, although it was not identical to the second one on either issue or task, was in many ways quite similar (e.g., a young student asking for cooperation on a noncontroversial issue). There were no clear expectations as to how this condition would compare to the controls.

Method

The subjects were 114 women and 13 men living in Palo Alto, California. Of these, 9 women and 6 men could not be contacted for the second request and are not included in the data analysis. The remaining 112 subjects were divided about equally among the five conditions (see Table 12.2). All subjects were contacted between 1:30 and 4:30 on weekday afternoons.

Two experimenters, one male and one female, were employed, and a different one always made the second contact. Unlike the first study, the experimenters actually went to the homes of the subjects and interviewed them on a face-to-face basis. An effort was made to select subjects from blocks and neighborhoods that were as homogeneous as possible. On each block every third or fourth house was approached, and all subjects on that block were in one experimental condition. This was necessary because of the likelihood that neighbors would talk to each other about the contact. In addition, for every four subjects contacted, a fifth house was chosen as a control but was, of course, not contacted. Throughout this phase of the experiment, and in fact throughout the whole experiment, the two experimenters did not communicate to each other what conditions had been run on a given block nor what condition a particular house was in.

The small-sign, safe-driving group was told that the experimenter was from the Community Committee for Traffic Safety, that he was visiting a number of homes in an attempt to make the citizens more aware of the need to drive carefully all the time, and that he would like the subject to take a small sign and put it in a window or in the car so that it would serve as a reminder of the need to drive carefully. The sign was three inches square, said "Be a safe driver," was on thin paper without a gummed backing, and in general looked rather amateurish and unattractive. If the subject agreed, he was given the sign and thanked; if he disagreed, he was simply thanked for his time.

The three other experimental conditions were quite similar with appropriate changes. The other organization was identified as the Keep California Beautiful Committee and its sign said, appropriately enough, "Keep California Beautiful." Both signs were simply black block letters on a white background. The two petition groups were asked to sign a petition which was being sent to California's United States Senators. The petition advocated support for any legislation which would promote either safer driving or keeping California beautiful. The subject was shown a petition, typed on heavy bond paper, with at least twenty signatures already affixed. If she agreed, she signed and was thanked. If she did not agree, she was merely thanked.

The second contact was made about 2 weeks after the initial one. Each experimenter was armed with a list of houses which had been compiled by the other experimenter. This list contained all four experimental conditions and the controls, and, of course, there was no way for the second experimenter to know which condition the subject had been in. At this second contact, all subjects were asked the same thing: Would they put a large sign concerning safe driving in their front yard? The experimenter identified himself as being from the Citizens for Safe Driving, a different group from the original safe-driving group (although it is likely that most subjects who had been in the safe-driving conditions did not notice the difference). The subject was shown a picture of a very large sign reading "Drive Carefully" placed in front of an attractive house. The picture was taken so that the sign obscured much of the front of the house and completely concealed the doorway. It was rather poorly lettered. The subject was told that: "Our men will come out and install it and later come and remove it. It makes just a small hole in your lawn, but if this is unacceptable to you we have a special mount which will make no hole." She was asked to put the sign up for a week or a week and a half. If the subject agreed, she was told that more names than necessary were being gathered and if her home were to be used she would be contacted in a few weeks. The experimenter recorded the subject's response and this ended the experiment.

Results

First, it should be noted that there were no large differences among the experimental conditions in the percentages of subjects agreeing to the first request. Although somewhat more subjects agreed to post the "Keep California Beautiful" sign and somewhat fewer to sign the beauty petition, none of these differences approach significance.

The important figures are the number of subjects in each group who agreed to the large request. These are presented in Table 12.2. The figures for the four experimental groups include all subjects who were approached the first time, regardless of whether or not they agreed to the small request. As noted above, a few subjects were lost because they could not be reached for the second request, and, of course these are not included in the table.

It is immediately apparent that the first request tended to increase the degree of compliance with the second request. Whereas fewer than 20 percent of the controls agreed to put the large sign on their lawn, over 55 percent of the experimental subjects agreed, with over 45 percent being the lowest degree of compliance for any experimental condition. As expected, those conditions in which the two requests were similar in terms of either issue or task produced significantly more compliance than did the controls (X^2's range from 3.67, $p < .07$ to 15.01, $p < .001$). A somewhat unexpected result is that the fourth condition, in which the first request had relatively little in common with the second request, also produced more compliance than the controls ($X^2 = 3.40$, $p < .08$). In other words, regardless of whether or not the two requests are similar in either issue or task, simply having

TABLE 12.2
Percentage of subjects complying with large request
in Experiment II

Issue[a]	Task[a]			
	Similar	N	Different	N
Similar	76.0**	25	47.8*	23
Different	47.6*	21	47.4*	19

One-Contact 16.7 ($N = 24$)

Note: Significance levels represent differences from the One-Contact condition.
[a]Denotes relationship between first and second requests.
*$p < .08$.
**$p < .01$.

the first request tends to increase the likelihood that the subject will comply with a subsequent, larger request. And this holds even when the two requests are made by different people several weeks apart.

A second point of interest is a comparison among the four experimental conditions. As expected, the Same Issue-Same Task condition produced more compliance than any of the other two-contact conditions, but the difference is not significant (X^2's range from 2.7 to 2.9). If only those subjects who agreed to the first request are considered, the same pattern holds.

DISCUSSION

To summarize the results, the first study indicated that carrying out a small request increased the likelihood that the subject would agree to a similar larger request made by the same person. The second study showed that this effect was quite strong even when a different person made the larger request, and the two requests were quite dissimilar. How may these results be explained?

Two possibilities were outlined previously. The matter-of-principle idea which centered on the particular type of action was not supported by the data, since the similarity of the tasks did not make an appreciable difference in degree of compliance. The notion of involvement, as described previously, also had difficulty accounting for some of the findings. The basic idea was that once someone has agreed to any action, no matter how small, he tends to feel more involved than he did before. This involvement may center around the particular person making the first request or the particular issue. This is quite consistent with the results of the first study (with the exception of the two control groups which as discussed previously were rather ambiguous) and with the Similar-Issue groups in the second experiment. This idea of involvement does not, however, explain the increase in compliance found in the two groups in which the first and second request did not deal with the same issue.

It is possible that in addition to or instead of this process a more general and diffuse mechanism underlies the increase in compliance. What may occur is a change in the person's feelings about getting involved or about taking action. Once he has agreed to a request, his attitude may change. He may become, in his own eyes, the kind of person who does this sort of thing, who agrees to requests made by strangers, who takes action on things he believes in, who cooperates with good causes. The change in attitude could be toward any aspect of the situation or toward the whole business of saying "yes." The basic idea is that the change in attitude need not be toward any particular issue or person or activity, but may be toward activity or compliance in general. This would imply that an increase in compliance would not depend upon the two contacts being made by the same person, or concerning the same issue or involving the same kind of action. The similarity could be much more general, such as both concerning good causes, or requiring a similar kind of action, or being made by pleasant, attractive individuals.

It is not being suggested that this is the only mechanism operating here. The idea of involvement continues to be extremely plausible, and there are probably a number of other possibilities. Unfortunately, the present studies offer no additional data with which to support or refute any of the possible explanations of the effect. These explanations thus remain simply descriptions of mechanisms which might produce an increase in compliance after agreement with a first request. Hopefully, additional research will test these ideas more fully and perhaps also specify other manipulations which produce an increase in compliance without an increase in external pressure.

It should be pointed out that the present studies employed what is perhaps a very special type of situation. In all cases the requests were made by presumably nonprofit service organizations. The issues in the second study were deliberately noncontroversial, and it may be assumed that virtually all subjects initially sympathized with the objectives of safe driving and a beautiful California. This is in strong contrast to campaigns which are designed to sell a particular product, political candidate, or dogma. Whether the technique employed in this study would be successful in these other situations remains to be shown.

References

ASCH, S. E. Effects of group pressure upon the modification and distortion of judgments. In H. Guetzkow (ed.), *Groups, leadership and men; research in human relations.* Pittsburgh: Carnegie Press, 1951, Pp. 177–190.

BANDURA, A., ROSS, D., AND ROSS, S. A. A comparative test of the status envy, social power, and secondary reinforcement theories of identificatory learning. *Journal of Abnormal and Social Psychology,* 1963, **67,** 527–534.

BRUNER, J. The dimensions of propaganda: German short-wave broadcasts to America. *Journal of Abnormal and Social Psychology,* 1941, **36,** 311–337.

DEUTSCH, M., AND GERARD, H. B. A study of normative and informational social influences upon individual judgment. *Journal of Abnormal and Social Psychology,* 1955, **41,** 629–636.

FESTINGER, L., AND CARLSMITH, J. Cognitive consequences of forced compliance. *Journal of Abnormal and Social Psychology,* 1959, **58,** 203–210.

HOVLAND, C. I., AND PRITZKER, H. A. Extent of opinion change as a function of amount of change advocated. *Journal of Abnormal and Social Psychology,* 1957, **54,** 257–261.

KELMAN, H. C., AND HOVLAND, C. I. "Reinstatement" of the communicator in delayed measurement of opinion change. *Journal of Abnormal and Social Psychology,* 1953, **48,** 327–335.

SCHEIN, E. H., SCHNEIER, I., AND BARKER, C. H. *Coercive pressure.* New York: Norton, 1961.

13

Reducing Weight by Reducing Dissonance: The Role of Effort Justification in Inducing Weight Loss

Danny Axsom and Joel Cooper

The role of effort justification in psychotherapy was examined. It was hypothesized that the effort involved in therapy, plus the conscious decision to undergo that effort, leads to positive therapeutic changes through the reduction of cognitive dissonance. An experiment was conducted in which overweight subjects attempted to lose weight through one of two forms of ''Effort Therapy.'' These therapies were bogus in that they were based solely on the expenditure of effort on a series of cognitive tasks that were unrelated to any existing techniques or theory addressing weight loss. One of the therapies called for a high degree of effort, while the degree of effort in the second therapy was low. Decision freedom to enter into and continue with the study was also varied. It was predicted that weight loss would occur only when both effort and decision freedom were high. Results supported these predictions, although some ambiguity arose from the failure to clearly manipulate decision freedom. Over an initial three-week period, High Effort subjects lost slightly more weight than Low Effort subjects. A six-month follow-up revealed that the effects of effort on weight loss had increased and were highly significant. Internal analyses indicated a further influence of the decision freedom variable. Possible mechanisms mediating the dissonance effect were discussed, as were several alternative explanations.

Several theorists have noted that psychotherapy is potentially a fertile arena for the application of social psychological principles (Frank, 1961; Goldstein, Heller, and Sechrest, 1966; Brehm, 1976; Strong, 1978). Frank (1961), for example, has characterized therapy as a relationship between a sufferer and a socially sanctioned authority who attempts to produce certain changes in the emotions, attitudes, and behaviors of the sufferer. Clearly this implies the importance of social psychological processes dealing with attitude change and social influence. These would seem to have an important bearing on the interpersonal influence setting we call psychotherapy.

Reprinted by permission of the authors.

One approach that has been applied to the study of attitude change is the concept of effort justification derived from the theory of cognitive dissonance (Festinger, 1957). We believe that the notion of effort justification is also potentially important in understanding both the process and outcome of psychotherapy. Effort justification concerns the consequences of engaging in an effortful activity in order to obtain some goal. The fact that one has engaged in an effortful event is discrepant with the notion that one does not usually engage in such effort. And for what purpose? In the typical effort justification sequence, either the goal or the means for achieving that goal is not attractive at the outset of the sequence. Consider, for example, the classic experiment by Aronson and Mills (1959). Subjects in a High Effort condition were made to undergo an event that was difficult and embarrassing. Their goal was to join a sexual discussion group that was, in reality, dull, boring, and a general waste of time. Yet those subjects who underwent the highly effortful procedure came to indicate that the group and its members were generally interesting and enjoyable.

Why did such changes occur? The reason given by Aronson and Mills was based upon the tension state of dissonance that was created by the voluntary expenditure of effort. "Why did I undergo such embarrassment and effort?" a subject may have asked. "Because I really did like the discussion group," might be the reply. In other words, the goal was elevated in attractiveness as a way of justifying the expenditure of effort.

Cooper (forthcoming) argues that the effort justification sequence might lie at the basis of many psychotherapeutic systems. Typically, psychotherapy involves a patient volunteering for an effortful and sometimes emotionally draining process. A client may have a fear of certain objects, may find relationships with others unpleasant, or may find it noxious to behave in certain socially adaptable ways. Yet in any of the myriad of procedures generically called psychotherapy, clients often make changes in their attitudes, emotions, and behavior. The goal states— be they phobic objects, interpersonal relations, or particular behaviors—become more acceptable or attractive. At least part of this change may result from an attempt to justify the expenditure of effort, just as in Aronson and Mills's study the discussion group became more attractive.

Which systems of therapy are more effective in producing change? The answer does not appear to rest decisively with any school. Despite the fact that some systems of therapy rely upon the production of behavior, others on the arousal of anxiety and others on the discussion of emotionally traumatic events, none seems to be clearly superior in producing effective changes (e.g., Sloane et al., 1975). Frank (1961) has aptly noted that the commonalities in psychotherapeutic systems may be more germaine than their differences. And one of the common factors that would seem to underlie virtually all effective therapy systems is the expenditure of effort. Psychoanalytic schools rely upon emotional catharses, desensitization approaches evoke degrees of anxiety, and virtually all therapies rely upon the use of a considerable amount of time and often expense. To the degree that participation in the therapies is voluntary, they are conceptually similar to Aronson and Mills's study of the effort justification sequence.

Cooper (forthcoming) conducted a pair of experiments to test the role of effort justification in psychotherapy. He reasoned that if successful therapy outcomes were due to effort justification, the particular form of effort should not matter, as long as that effort is seen as related to the goal in question. In one experiment, college students who were afraid of snakes were recruited. They participated in one of two experimental therapies. One of the therapeutic procedures was a modified form of implosive therapy (Stampfl and Levis, 1976) that is based upon the learning theory notion of extinction. This therapy involves a high degree of effortful participation on the part of the participant. The other therapy was based solely on the expenditure of physical effort and was not based upon any existing theory of therapy. Subjects were asked to jump rope, run in place, and perform other physical exercises. Subjects in each therapy condition first attempted to come as close as they could to a six-foot boa constrictor. Half of the subjects in each condition then made an informed choice about participating in the effortful therapy; the other half participated without such a choice. According to dissonance theory, the effort justification sequence should be invoked only under conditions of an informed choice (e.g., Linder, Cooper, and Wicklund, 1968). Therefore it was predicted that, on the basis of dissonance theory, effort justification would be invoked by either therapy since both implosive therapy and the newly created physical exercise therapy involved high degrees of effort—but this would occur only under conditions of high decision freedom. It was found that subjects in either form of therapy made significant improvement in the degree to which they could approach the boa constrictor—if and only if they participated under conditions of free choice. Neither the exercise therapy nor implosive therapy was effective in the absence of choice.

The basic approach was repeated in a conceptual replication. Participants were university students who were nonassertive. Cooper (forthcoming) assigned half of them to a physical exercise therapy and half to a standard therapy for lack of assertiveness (i.e., behavior rehearsal; see Salter, 1949). Once again, it was found that either form of therapy was effective in increasing participants' assertiveness, as long as it was engaged in voluntarily.

The conclusion from Cooper's experiments is that the voluntary expenditure of effort is at least one of the effective ingredients in psychotherapy, regardless of whether that effort forms part of a traditional therapy or whether it is improvised in a series of physical exercises. Exercise therapies or traditional therapies may be effective in promoting change, as long as they are engaged in voluntarily. However, several important questions remain unanswered. First, the notion that effort leads to positive changes in psychotherapy has yet to receive a direct test. This is because neither of Cooper's experiments used *variations* in effort as an independent variable. To the extent that effort justification is involved in psychotherapy, it should be shown that variations in the degree of effort will lead to variations in the degree of change that ensues.

Equally important is the question of duration of change that occurs as a result of effort justification. Both of Cooper's experiments involved single-session therapies with change measured immediately after the session. While this is a typical procedure in laboratory experiments involving attitude change, it is not parallel to the desired outcome of psychotherapy. Change as a result of therapy is anticipated to be more long lasting, and an assessment of the effort justification procedure should be based not so much on an immediate assessment as on its long term consequence. There are only a few reports in the dissonance literature of cognitive changes lasting well beyond the experimental session (Freedman, 1965). So the question of duration of consequences takes on enormous significance in assessing the appropriateness of the conceptual analysis that is based upon the psychology of effort justification.

The focus of the present experiment was on weight loss. This problem offers many advantages for an experimental study of effort justification procedures. First, there are objective and nonreactive measurements available for body weight that can be easily and repeatedly sampled. Second, the dependent measure thoroughly defines the criterion of a successful therapy, since weight loss represents the specific goal of treatment (Wollersheim, 1970). In addition, the number of different approaches to weight loss that exist in society attests to the fact that there is, at present, no agreed upon effective treatment for the loss of weight. The present experiment uses the concept of effort justification to effect such a treatment.

METHOD

Subjects

Subjects were recruited via newspaper ads for an "Experiment concerning possible methods of weight reduction." They were contacted by phone for scheduling and were assured that the procedure would be safe and would not include any medication. The following restrictions applied: First, only those 18 years of age and older were solicited. Second, men were excluded from the final subject pool (this decision was made after the initial response to the ads was overwhelmingly female). A third step, which was also taken to increase homogeneity, restricted allowable weight deviations to include only those women who were 10 to 20 percent above "desirable body weight" (according to the Metropolitan Life Insurance Company statistics [1959]). The lower limit was similar to that used by Schachter and his colleagues in their studies of obesity (Schachter, 1971); the upper limit decreased the probability of there being a physiological etiology and complication associated with overweightness (Olson, 1964). Fourth, no one receiving therapy or medication for their weight was included. Fifth, potential subjects who participated in athletic activities that would greatly increase weight by increasing muscle size rather than body fat were excluded (Schachter, Goldman, and Gordon, 1968).

Finally, only subjects who lived within twenty minutes' traveling time of the laboratory were included, since extraordinary distances could affect the manipulation of effort.

Sixty-eight subjects who fulfilled the above requirements began the experiment. Each was paid at the rate of one dollar per session. Fifteen subjects were lost through attrition. Chi-square analyses failed to reveal any relationship between experimental conditions and the decision to terminate.[1] In addition, data from one subject were omitted when she arrived at two consecutive sessions too late to complete the full procedure. Fifty-two subjects comprised the final sample.

Procedure

General overview. A 2 (level of effort) $\times$ 2 (level of choice) between-subjects design with an external control group was utilized. Subjects attended five sessions over a three week period. They attended two sessions during each of the first two weeks and one during the final week. To minimize extraneous influences on weight change, sessions for each subject were scheduled at the same time of day and the same days of the week throughout. Six months after the experimental sessions were completed, a final weighing session was conducted.

The choice variable. Subjects in the four experimental conditions were first met by a female experimenter who measured and recorded their weight.[2] She then administered the choice variable. Subjects in the *High Choice* conditions were told: "I have been instructed to advise you that although the procedures you will follow are perfectly safe and harmless, they may also be effortful and anxiety producing. If you like, you can stop the experiment now and you will be paid for this session. Would you like to continue?"

A Princeton University Informed Consent form was then given to be read and signed. This stated, in part, "I may withdraw my consent and discontinue participation in the project at any time." As the form was being read, the weigher emphasized, "As you notice on the form, you may withdraw your consent and stop at any time . . . you still have the prerogative to stop participation later. . . ."

Subjects in the *Low Choice* conditions were not asked whether they wanted to continue. Subjects were merely warned about the potential effort and anxiety and then told, "We'll go ahead and begin."

The second experimenter introduced himself and administered a "Life Pattern Questionnaire" concerning everyday eating and exercise patterns and other activ-

---

[1] Reported reasons for dropping included loss of interest, outside interference such as sudden job changes, and unexpected transportation problems, which prevented participation.

[2] Subjects were weighed in indoor clothing and without shoes.

ities that might be useful in interpreting weight change data. It was also administered to increase the perceived legitimacy of the procedure. All subjects were then
given a small booklet in which to monitor their eating over the three week period.
This too was partly to increase the perceived legitimacy of the procedure.

The second experimenter explained the rationale for the study by noting that
psychologists have frequently found strong correlations between heightened neurophysiological arousal and increased emotional sensitivity. The present researchers, he added, had been able to take advantage of this by presenting subjects with
various tasks designed specifically to increase this neurophysiological arousal and
thereby enhance emotional sensitivity in a way that helped lead to weight reduction. He then explained that although the procedure had been very successful in
the preliminary investigations, the precise reasons for the weight loss obtained were
still uncertain, and the present study aimed to make the process clearer.

The tasks were described as requiring much concentration and consequently being
neurophysiologically arousing and sometimes stressful. The subject was assured
that this arousal and stress would be brief and not last beyond any session. In
keeping with the cover story, all subjects were attached to a galvanic skin response (GSR) apparatus (ostensibly to measure their level of arousal) while performing the tasks described below.

The effort variable. Effort in the present experiment involved the degree of
difficulty of a variety of cognitive tasks. Effort was manipulated by varying task
difficulty and duration, the parameters chosen being established through pre-test-
ing. Subjects in the High Effort conditions worked for 20 minutes at a 3-channel
tachistoscope. Their task was to discriminate which of several near-vertical lines
presented sequentially was most vertical. Each line was visible for only 350 milliseconds. Those in the Low Effort condition worked for 3 minutes and were given
1 second to view each line.

Subjects then moved to a delayed-auditory feedback apparatus. High Effort subjects were given 30 minutes of recitation, in which they attempted to recite nursery rhymes, read a short story, and recite the U.S. Pledge of Allegiance with their
own voice reflected back to them via earphones at a delay of 316 milliseconds.
The delay was similar to that used by Zimbardo (1965) in his manipulation of high
effort. In addition, the voice of a woman attempting similar tasks during pretesting
was overlayed onto the recorder so that the subject not only had to contend with
the delay, but also with yet another voice. Low Effort subjects worked for only
10 minutes and the auditory delay was cut in half (158 milliseconds). This reduced
delay, accomplished by increasing tape speed, also rendered the voice distraction
incomprehensible and therefore less disruptive (Mackworth, 1970).

To avoid the potential confounding of session length with level of effort, subjects in the Low Effort condition, after completing the T-scope task, returned to
the waiting room to relax for 40 minutes before finishing the session. This pro-

cedure, which is similar to that used by Wicklund, Cooper, and Linder (1967), was explained to Low Effort subjects as being necessary to allow the arousal due to the T-scope task to dissipate before beginning the next task.

Upon finishing the final task, the subject completed a brief questionnaire concerning her impression of the session. Most importantly, she was asked, "In general, how effortful would you describe the hour as a whole?" This was followed by a 7-point response scale labeled "Very little" and "Very much" at the end points. The subject then returned to the weighing room for the final choice manipulation. For High Choice subjects, the first experimenter stated, "I'll remind you again that you can still stop the experiment and be paid for what you've done so far. Would you like to continue?" When the subject acknowledged that she wished to continue, she was given an appointment card containing the time and date of the next session. Low Choice subjects were merely given the appointment card without any mention of a new decision.

Sessions two through four.   The following three sessions were similar to the first. The Life Pattern Questionnaire was not administered during these sessions and the cover story was not repeated. The sessions contained T-scope and DAF tasks that were of the same durations as those of Session One, although the content of the visual discriminations and DAF reading tasks was altered to relieve possible boredom.

The final experimental session.   Session Five contained the final assessment of weight change during the experimental period and was conducted by the second experimenter. Unaware of the subject's choice condition, he weighed the subject and then informed her that her participation in the study was completed. He asked the subject to fill out a questionnaire that covered various weight-related topics and the subject's perceptions of the study. Crucial were two final manipulation checks: "How effortful would you describe the experiment as a whole?" (1 = very little; 9 = very much); and "How free did you feel *not* to continue with the experiment at any time?" (1 = not free to choose; 9 = very free to choose). Next the Life Patterns Questionnaire was readministered. Finally, the subject was carefully questioned about any suspicions she may have had about the purpose of the study, fully debriefed, and paid for her participation. The need for deception and the importance and possible implications of the study were discussed at length.

Control group.   To provide a baseline indication of normal weight fluctuation among subjects in our sample, ten subjects from those who responded to the advertisement were randomly assigned to a control condition. When contacted to begin, they were told that they would be unable to participate as originally planned due to a change in the procedure that meant using less participants. They were then asked to engage in a project "to determine normal female weight fluctuation over time." Control subjects also participated in five sessions, scheduled in a similar fashion as the experimental subjects. They were simply greeted and weighed during each session. The Life Patterns Questionnaire was administered at the first and fifth sessions. A shortened version of the final questionnaire was also administered at the fifth session. It was emphasized that this was not a weight reduction study and that subjects should therefore simply carry on their normal daily activities. Dieting was left to their discretion.

The final assessment.   Six months after the fifth session was completed, participants were contacted for a follow-up weighing. The subjects were unaware that they would be re-contacted. Forty-two of the 52 subjects were able to return; nine had since moved from the area or were unable to be reached; one had become pregnant. At this follow-up, subjects were given a copy of the results from the original experiment.

RESULTS

Checks on the Manipulations

The amount of effort involved in the subjects' participation was assessed in two ways. First, at the end of the four experimental sessions, subjects were asked to rate the effortfulness of the preceding hour (1 = very little; 7 = very much). A 2 (choice) $\times$ 2(effort) $\times$ 4 (sessions) repeated measures analysis of variance showed a significant main effect for Effort ($\bar{X}_{high} = 4.31$ vs. $\bar{X}_{low} = 3.14$), F (1, 37) = 10.21, $p < .005$. In addition, during the final session all subjects, including the control group, rated the effortfulness of the experiment as a whole (1 = very little; 9 = very much). Planned comparisons again confirmed that High Effort subjects rated the study as more effortful, F (1, 45) = 4.62, $p < .05$. Dunnett comparisons showed that only the High Effort–Low Choice groups differed significantly from the effortfulness ratings of the Control group ($p < .08$ and $p < .05$, respectively).

Results from the perceived choice manipulation checks revealed that the manipulation of this variable was not entirely effective. At the final session, the subjects were asked how free they felt not to continue with the experiment at any time. This primary measure of decision freedom was supplemented by three additional questions that related to perceived freedom (all scaled 1 = not free to choose; 9 = very free to choose). A multivariate analysis of variance on these questions revealed a marginal main effect for Choice, Wilkes Likelihood Ratio Exact F (4, 34) = 2.19, $p < .10$. Why this effect was only marginal can be seen by examining the primary choice measure. Of the 41 subjects in the experimental groups who responded to this question, 56 percent marked the highest level of choice. In fact, even 40 percent of the Low Choice subjects responded this way. This apparent ceiling effect was evident on all four questions. Although a univariate analysis of variance performed on the data from the primary choice measure showed that High Choice subjects felt significantly more freedom to discontinue the experiment than Low Choice subjects, F (1, 45) = 5.06, $p < .05$, the ceiling effect caused both the normality and homogeneity assumptions underlying the analysis of variance to be violated ($F_{max} = 13.04$, $p < .01$), which results in a positive bias in the test (Winer, 1971). The overwhelming proportion of extreme responses also rendered nonparametric analyses of little use, since the number of tied ranks was unacceptably high. We are thus left uncertain as to the statistical and psychological significance of the choice manipulation.

Weight loss: data from session five.   The amount of weight change at the immediate conclusion of the experimental sessions was measured in pounds.[3,4] As can be seen from Table 13.1, the effort justification hypothesis that there would be more weight loss in the High Effort–High Choice condition was not found. Planned comparisons revealed only a tendency toward a main effect for effort, F (1, 47) = 2.28, $p < .14$. Given the questionable effectiveness of the choice manipulation, these results are more understandable. If choice is considered high in all cells, then the effort justification prediction must be that subjects in the High Effort conditions would lose more weight than those in the Low Effort conditions. Although the evidence for this is marginal, additional data should be considered: Seven of the nine subjects who lost the most weight were from the High Effort conditions, while five of the six who gained the most weight were from the Low Effort conditions. In addition, t tests, adjusted for the number of possible comparisons, revealed that only in the combined High Effort conditions was weight deviation significantly different from zero, t (20) = 4.11, $p < .02$.

The equivocal success of the choice manipulation makes interpretations involving that variable difficult. To better understand the role of choice, an internal analysis was performed on the data. Specifically, the analysis used the subjects' report of how much freedom they experienced as the choice variable. Table 13.1 shows the weight lost by the subjects as a function of their *assigned* degree of effort and their *reported* experience of freedom, the latter variable obtained by a median split on the primary choice measure. A planned contrast shows a marginally significant difference between the High Effort–reported High Choice condition and the remaining three conditions, F (1,37) = 3.39, $p < .08$. In addition, only the High

---

[3] For clarity of presentation, the weight change analyses reported are based on difference scores. The crucial assumptions behind the use of difference scores, namely, that within-group regression coefficients relating initial to final weight are homogeneous and equal to 1.00, are both satisfied (b's for High Effort–High Choice; High Effort–Low Choice; Low Effort–High Choice; Low Effort–Low Choice; and the control were, respectively, 1.01, .95, 1.01, .99, and 1.02).

[4] Since the duration of this portion of the experiment was short and weight changes were expected to be small, data on each subject's menstrual cycle were also collected (during the fifth session). This was to ensure against any spurious weight change effects resulting from water retention (congestive dysmenorrhea) coinciding with menstruation. For example, if a subject began the experiment at or near menstruation, perhaps any weight loss obtained later would simply reflect a return to normal weight; likewise, legitimate losses could be concealed if the final session ended at menstruation. However, analyses of menstrual data showed that only a small number of subjects had either started or finished their sessions around menstruation and that these subjects were equally distributed across conditions.

TABLE 13.1
Mean Weight Change Over the Initial Five Sessions (in pounds)

Group	Unadjusted	Internal Analysis
High Effort–High Choice	− 1.75	− 2.25
High Effort–Low Choice	− 1.77	− 1.20
Low Effort–High Choice	− 1.16	− 1.33
Low Effort–Low Choice	− .45	0
Control	+ .17	—

Note: For the internal analysis, subjects were grouped according to High vs. Low *assigned* Effort and High vs. Low *reported* Choice (obtained by a median split of responses to the choice manipulation check).

Effort–reported High Choice condition showed weight deviations that were significantly different from zero, t (9) = 4.26, $p < .06$.

Weight loss: the final measure. Encouraged by the direction if not the magnitude of the results, we proceeded with the final dependent measure: the six month follow-up. As Table 13.2 indicates, dramatic differences had developed over the six month period. High Effort subjects had lost an average of 6.5 additional pounds. Low Effort subjects, on the other hand, showed a slight gain in weight. The difference was highly significant, whether measured as change from the fifth session, F (1, 30) = 20.30, $p < .001$, or the beginning of the experiment, F (1,30) = 20.45, $p < .001$. Of the 16 High Effort subjects who reported for the final measure, 15 had lost weight, whereas only 7 of the 18 Low Effort subjects and 4 of the 8 Control subjects lost weight. Once again, an internal analysis using a median split on the choice measure was performed. These results showed that subjects in the High Effort–reported High Choice group lost the most weight, al-

TABLE 13.2
Mean Weight Change With First Follow-up (in pounds)

Group	Over Follow-up Period	Overall	Internal Analysis: Over Follow-up Period	Internal Analysis: Overall
High Effort–High Choice	− 6.46a	− 7.93a	− 7.21a	− 9.75a
High Effort–Low Choice	− 6.61a	− 9.03a	− 5.03a	− 6.61a
Low Effort–High Choice	− .14b	− 1.47b	+1.10b	− .48b
Low Effort–Low Choice	+ 1.75b	+1.33b	+ .44b	+ .44b
Control	+ .94b	+ .94b	—	—

Note: "Over follow-up period" indicates weight change from the conclusion of the initial sessions to the follow-up; "overall" indicates weight change from the beginning of the experiment to the follow-up. Within each column, means with different subscripts differ significantly from one another ($p < .05$ by Newman-Keuls).

though this group did not differ significantly from those in the High Effort–reported Low Choice condition.

Weight loss: one more time.   One year from the date of the initial experimental session, subjects were contacted once again. By this time, many had either moved or otherwise changed situations so that they could no longer return in person for the weighing. These subjects weighed themselves at home and reported their weight by phone. As a result, data were obtained for 50 of the 52 original subjects. Analyses using only those subjects who returned to be weighed were similar to findings based on the full subject pool, so only the latter are reported. Once again, the main effect for effort was highly reliable, F (1, 45) = 7.79, $p < .01$. The average weight loss for high effort subjects one year from their participation in the experimental session was 6.7 pounds whereas low effort subjects lost only .34 pounds (Control $\overline{X} = -1.86$ pounds). Ninety percent of the High Effort participants were below their initial weight whereas only 48 percent of the Low Effort and 56 percent of the Control subjects were below their initial weight. Once again, an internal analysis revealed that the High Effort—reported High Choice group had lost the most weight. An *a priori* contrast showed this group to be significantly different from the other three, F (1, 36) = 5.26, $p < .05$.

DISCUSSION

The weight loss observed in the High Effort cells was both substantial and consistent. Since subjects were initially an average of 17 pounds overweight, High Effort subjects–who lost an average of 8.55 pounds by the first follow-up–were able to achieve a 50 percent reduction in excess weight. The chief issue to be addressed, then, is not whether positive therapeutic change occurred but rather whether effort justification processes can account for this change.

The results offer clear support for the role of effort in instigating the weight reduction. But the weakness of the choice manipulation makes a final interpretation less clear than would be desired. The notion of effort justification predicts weight loss for participants who engaged in a high degree of effort and who perceived their freedom to participate to be high. Since an overwehelming proportion of subjects perceived their freedom to be high, the main effects found for the effort variable are consistent with the effort justification approach.

In retrospect, the problem with the manipulation of choice could have been expected. Subjects came to the university from various locations in the community. They returned for several sessions. Attempting to convince Low Choice subjects that they were not at all free to discontinue the experiment would have strained ethical considerations as well as credibility. Rather, we remained silent about choice considerations to Low Choice subjects and instead emphasized the high degree of decision freedom to those subjects assigned to the High Choice conditions. Since freedom was apparently already assumed to be high by the participants, our manipulation did not have much differential impact. Because of this strong ceiling effect, even internal analyses could offer only slight assistance by showing tendencies for the High Effort–reported High Choice group to lose the greatest amount of weight.

What of those subjects who actually reported feeling low decision freedom? There were eight such subjects who checked points lower than the mid-point of the primary choice scale. For them, it would be predicted that, regardless of effort, weight loss would not occur. Indeed, six of the eight actually gained weight over the initial five sessions. The mean weight change for all eight was + .56 pounds. Seven returned for the first follow-up. Although their mean weight change was − 1.75 pounds, this was largely due to one highly aberrant subject who had lost 13 pounds. The remaining six were virtually unchanged ($\overline{X} = +.13$ pounds). By the second follow-up, overall weight change for the eight subjects reporting low choice was almost zero ($\overline{X} = -.06$ pounds).

The dramatic amount of weight change that occurred during the six month period following the experimental session stands in marked contrast to the marginal degree of weight change that occurred immediately after the experiment. Of course, weight change takes time and it is not surprising that only a few pounds could be changed during the three weeks of the sessions. But this weight loss was not only maintained over the half-year period, it was markedly increased by those subjects in the High Effort conditions. Why should this have happened? Long-term changes as a function of laboratory intervention are not commonplace in the literature.

Our answer, based upon the concept of effort justification, must still be in the form of speculation. It would appear that effort justification worked to increase the attractiveness of the *goal* of weight loss. There is no evidence that subjects in the High Effort conditions found that they could be successful at losing weight during the experimental sessions and continued their method over the long period of time prior to the follow-up measure. The data show little relationship between weight lost during the initial experimental sessions and weight lost by the first follow-up. Our initial statement of effort justification principles is consistent with this lack of relationship, for it is the goal state that effort renders more attractive. Just as Aronson and Mills's (1959) subjects who underwent effort viewed the goal of belonging to the discussion group with greater attractiveness, so too did our subjects come to view the goal of losing weight with more zeal and fervor. As a result, subjects in the High Effort conditions pursued that goal *regardless* of how successful they might have been during the five experimental sessions. Subjects in the Low Effort conditions, on the other hand, did not view the goal as being more attractive. Regardless of whether they had shown any weight loss during the five experimental sessions, they showed no consistent pattern of weight loss during the succeeding six months. Thus, the data are consistent with the notion that the goal of weight loss became more attractive to High Effort subjects who then pursued that goal more vigorously during the intervening six months.

Alternative explanations for the data may still arise due to the equivocal nature of the choice manipulation. One such explanation is that High Effort subjects may have formed differentially higher expectations about the therapy's potential outcome. These high expectations may have led to stricter adherence to some self-prescribed regimen that resulted in weight loss. Those who initially lost weight might have formed what Ross, Lepper, and Hubbard (1975) have referred to as antecedent—consequent explanations for why they were losing weight. However, it is difficult for an expectation-based hypothesis to explain why those initially showing little or no weight loss would later lose. As we have already mentioned, the data reveal that High Effort subjects lost weight during the six month period regardless of whether they had initially lost or gained. This factor is more consistent with the effort justification hypothesis for it appears to be the goal of weight loss that becomes more attractive.

But this speculation should only emphasize the need for more work in the area. In particular, the specific processes producing the changes need more attention. Perhaps an attributional alternative—a self-perception process, for example—could account for our findings.

When combined with the two experiments reported by Cooper (forthcoming), the present experiment adds to the confidence with which the concept of effort justification can be applied to psychotherapy. In the present experiment, the concept of effort was expanded to include cognitive mental tasks, and the data showed that variations in this type of effort led to systematic differences in the amount of weight that participants lost. Although the findings were of a small magnitude at the conclusion of the five experimental sessions, the amount of weight lost after a six month period by subjects who had undergone a highly effortful procedure was considerable. The extension over time adds important new dimensions to research on effort justification as well as to research on psychotherapeutic outcomes.

The application of effort justification principles to psychotherapy would seem to call for serious consideration on the basis of the present data. Some therapists may feel uneasy applying effort justification because the principles have become associated with the use of deception (Brehm, 1976). Such deception is not endemic to effort justification. In fact, effort justification could be relied upon by initially advising clients very truthfully of the effort and unpleasantness that lie ahead. As Freud (1929) stated, "when we take a neutrotic patient into psychoanalytic treatment . . . we point out the difficulties of the method to him, the long duration, the efforts and sacrifices it calls for . . ." (p. 15). Effort justification also stresses the client's personal responsibility for choosing to participate and remain in therapy. When viewed in this light, perhaps the effort justification principle emanating from the social psychological laboratory will become more palatable to therapists. We are not saying that effort justification is the only influence on therapy outcomes; nor are we necessarily advocating an "effort therapy." What we do believe is that effort justification influences outcomes. To the extent that therapists take advantage of this process they may enhance the efficacy of their therapies.

References

ARONSON, E., AND MILLS, J. The effects of severity of initiation on liking for a group. *Journal of Abnormal and Social Psychology,* 1959, **59,** 177–181.

BREHM, S. S. *The application of social psychology to clinical practice.* Washington, D. C.: Hemisphere, 1976.

COOPER, J. *Reducing fears and increasing assertiveness: The role of dissonance reduction.* Manuscript submitted for publication, 1979.

FESTINGER, L. *A theory of cognitive dissonance.* Stanford: Stanford University Press, 1957.

FRANK, J. D. *Persuasion and healing.* Baltimore: The Johns Hopkins Press, 1961.

FREEDMAN, J. L. Long-term behavioral effects of cognitive dissonance. *Journal of Experimental Social Psychology,* 1965, **1,** 145–155.

FREUD, S. *Introductory lectures on psychoanalysis.* New York: Norton, 1966.

GOLDSTEIN, A. P., HELLER, K., AND SECHREST, L. B. *Psychotherapy and the psychology of behavior change.* New York: Wiley, 1966.

LINDER, D. E., COOPER, J., AND WICKLUND, R. A. Pre-exposure persuasion as a result of committment to pre-exposure effort. *Journal of Experimental Social Psychology,* 1968, **4,** 470–482.

MACKWORTH, J. F. *Vigilance and attention: A signal detection approach.* Harmondsworth: Penguin, 1970.

METROPOLITAN LIFE INSURANCE COMPANY OF NEW YORK. New weight standards for men and women. *Statistical Bulletin,* **40,** 1959.

OLSON, R. E. Obesity. In H. F. Conn (ed.), *Current therapy.* Philadelphia: W. B. Saunders, 1964.

ROSS, L., LEPPER, M. R., AND HUBBARD, M. Perseverance in self-perception and social perception: Biased attributional processes in the debriefing paradigm. *Journal of Personality and Social Psychology,* 1975, **32,** 880–892.

SALTER, A. *Conditioned reflex therapy: The direct approach to the reconstruction of personality.* New York: Creative Age Press, 1949.

SCHACHTER, S. *Emotion, obesity, and crime.* New York: Academic Press, 1971.

SCHACHTER, S., GOLDMAN, R., AND GORDON, A. Effects of fear, food deprivation, and obesity on eating. *Journal of Personality and Social Psychology,* 1968, **10,** 91–97.

SLOANE, R. B., STAPLES, F. R., CRISTOL, A. H., YORKSTON, N. J., AND WHIPPLE, K. *Short-term analytically oriented psychotherapy vs. behavior therapy.* Cambridge, Mass.: Harvard University Press, 1975.

STAMPFL, T., AND LEVIS, D. Essentials of implosive therapy: A learning theory based psychodynamic behavioral therapy. *Journal of Abnormal Psychology,* 1967, **72,** 496.

STRONG, S. R. Social psychological approach to psychotherapy research. In S. Garfield and A. Bergin (eds.), *Handbook of psychotherapy and behavior change.* New York: Wiley, 1978.

WICKLUND, R. A., COOPER, J., AND LINDER, D. E. Effects of expected effort on attitude change prior to exposure. *Journal of Experimental Social Psychology,* 1967, **3,** 416–428.

WINER, B. J. *Statistical principles in experimental design* (2nd ed.). New York: McGraw-Hill, 1971.

WOLLERSHEIM, J. Effectiveness of group therapy based upon learning principles on the treatment of overweight women. *Journal of Abnormal Psychology,* 1970, **76,** 462–474.

ZIMBARDO, P. G. The effect of effort and improvisation on self-persuasion produced by role playing. *Journal of Experimental Social Psychology,* 1965, **1,** 103–120.

14

Dishonest Behavior as a Function of Differential Levels of Induced Self-Esteem

Elliot Aronson and David R. Mettee

After taking a personality test, Ss were given false feedback aimed at temporarily inducing either an increase in self-esteem, a decrease in self-esteem, or no change in their self-esteem. They were then allowed to participate in a game of cards, in the course of which they were provided with opportunities to cheat under circumstances which made it appear impossible to be detected. Significantly more people cheated in the low self-esteem condition than in the high self-esteem condition. A chi-square evaluating cheater frequency among the high self-esteem, the no information (no change in self-esteem), and the low self-esteem conditions was significant at the .05 level. The results are discussed in terms of cognitive consistency theory.

Recent theorizing and experimentation have suggested that a person's expectancies may be an important determinant of his behavior. Working within the framework of the theory of cognitive dissonance, Aronson and Carlsmith (1962) conducted an experiment in which subjects were led to develop an expectancy of poor performance on a "social sensitivity" test. The subjects then proceeded to perform beautifully. Aronson and Carlsmith found that these subjects subsequently changed their superior performance to an inferior one when retested over the same material. Similarly, Wilson (1965) found that subjects were significantly more attracted to a negative evaluator than to a positive evaluator if the negative evaluations were in accord with a strong performance expectancy which had led the subjects to

Reprinted with permission from the authors and *The Journal of Personality and Social Psychology,* Vol. 9, No. 2, 1968. Copyright 1968 by the American Psychological Association.

This experiment was supported by a National Science Foundation Graduate Fellowship (NSF-26-1140-3971) to David R. Mettee and by grants from the National Science Foundation (NSF GS 750) and the National Institute of Mental Health (MH 12357-01) to Elliot Aronson. Authors are listed in alphabetical order.

withdraw from a competitive event. Consistency theory thus has received some support in specific expectancies and performance directly related to these expectancies.

But what about more pervasive expectancies such as those about the self? Bramel (1962) showed some evidence for the impact of self-esteem on subsequent behavior. In his study he temporarily raised or lowered the subjects' self-esteem by providing them with positive or negative information about their personalities. He then allowed them to discover irrefutable negative information about themselves. The individuals who held low self-concepts were more willing to accept this information; that is, they were not as prone as people who had been induced toward high self-esteem to project this specific negative attribute onto others. These results are consistent with the work of Rogers (1951), who argued that negative or maladaptive responses occur as the result of being consistent with a negative self-concept, and that such responses can be altered only by first changing the self-concept in a direction consistent with adaptive responses.

The prediction being tested in the present experiment is in accord with the experiments cited above. In addition, it carries our interest in the self-concept one step further in the direction Rogers has taken—toward greater generalization. What Aronson and Carlsmith showed is that people try to behave in a highly specific manner which will coincide with a highly specific self-expectancy; that is, people who believe that they are poor in a "social sensitivity" test will take action aimed at performing poorly on the test. But does this generalize? If we feel low and worthless on one or two dimensions do we behave generally in low and worthless ways—even if the behavior is not directly and specifically related to the low aspects of the self-concept? For example, if a person is jilted by his girlfriend (and thus feels unloved), is he more apt to go out and rob a bank, kick a dog, or wear mismatched pajamas?

In the present experimental situation we are predicting just that. Concretely, individuals who are provided with self-relevant information which temporarily causes them to lower their self-esteem (but does *not* specifically make them feel immoral or dishonest) are more apt to cheat than those who are made to raise their self-esteem—or those who are given no self-relevant information at all (control condition). Similarly, people who are induced to raise their self-esteem will be less likely to cheat than the controls. This hypothesis is based upon the assumption that high self-esteem acts as a barrier against dishonest behavior because such behavior is inconsistent. In short, if a person is tempted to cheat, it will be easier for him to yield to this temptation if his self-esteem is low than if it is high. Cheating is not inconsistent with generally low self-esteem; it *is* inconsistent with generally high self-esteem.

METHOD

General Procedure

The subjects were led to believe that they were participating in a study concerned with the correlation between personality test scores and extra-sensory perception (ESP). They were told that their personalities would be evaluated with the self-esteem scales of the California Personality Inventory (CPI) and that their ESP ability would be ascertained with the aid of a modified game of blackjack. Before participating in the blackjack game, subjects took the personality test and received false feedback (either positive, negative, or neutral) about their personalities. During the blackjack game subjects were faced with the dilemma of either cheating and winning or not cheating and losing in a situation in which they were led to believe (erroneously) that cheating was impossible to detect. The opportunity to cheat occurred when the subjects were "accidentally" dealt two cards at once instead of one. The rightful card put the subject over 21 and ensured defeat, whereas the mistakenly dealt extra card, if kept, provided the subject with a point total that virtually assured victory. The card which the subject kept constituted the dependent variable.

Subjects

The subjects were 45 females taken from introductory psychology classes at the University of Texas, who were randomly assigned to one of three self-esteem conditions: high, low, and neutral. In actuality, 50 subjects were run; the result of five subjects were discarded because of suspicion. Three of these were in the low self-esteem condition; two were in the high self-esteem condition. The criteria for elimination were determined a priori and were followed rigidly throughout the experiment. It was made explicit that the experimenter had no preconceptions as to how personality traits might be related to ESP ability, but simply wanted to determine whether or not, for example, people who are easily angered have more ESP than calm people.

Personality Test

All subjects came to the first session together and were given the self-esteem scales of the CPI. The CPI was administered by a person who introduced himself as a member of the University Counseling Center staff. Subjects were told at this session that the experiment was concerned with ESP and personality characteristics. They were informed that this session was to determine the personality traits of the subjects, with ESP ability to be measured in the second session.

A shortened version of the CPI was used to evaluate the personalities of the subjects. This version contained only the six scales related to self-esteem and, for our purposes, constituted a measure of the subjects' chronic self-esteem. However, the *primary* experimental purpose of this text was merely to provide the opportunity and rationale for situationally manipulating the subjects' self-esteem via preprogrammed feedback regarding subjects' personality test results.

In order to separate the experimenter as much as possible from the personality evaluation aspects of the experiment, subjects were told by the experimenter that Miss Jacobs,[1] a member of the University of Texas Counseling Center staff, had kindly consented to administer the personality tests. It was emphasized that she would score the personality inventories and that the experimenter's access to their scores would not be on a name basis but via a complicated coding process. It was indicated that, as a matter of convenience, subjects would be given feedback regarding their personality tests when they came for the second session of the experiment. Following this, the CPI's were distributed, subjects completed them, and before leaving were assigned a time to return for the second session with three subjects assigned to each specific time slot.

Personality Score

In the second session subjects were tested in groups of three. Upon arrival for the second session, subjects were greeted by Miss Jacobs and told to be seated in an outer office which provided access to three adjoining offices. After all three subjects had arrived, Miss Jacobs handed each subject a manila envelope bearing appropriate Counseling Center insignia and assigned each subject to a different office where she was to go to read the results of her personality test. In addition, Miss Jacobs told the subjects that she had been given a sheet of instructions to deliver to them. The sheet of instructions was handed to the subjects along with the manila envelope.

The personality test results consisted of three standard feedbacks unrelated to subjects' performance on the CPI. Each of the three subjects present at any one specific time was randomly assigned feedback of either high self-esteem (HSE = positive), no self-esteem (NSE = neutral), or low self-esteem (LSE = negative) in content.

The HSE and LSE personality reports were parallel in content except, of course, for the nature of the evaluation. For example, a portion of the HSE report stated:

> The subject's profile indicates she has a stable personality and is not given to pronounced mood fluctuations of excitement or depression. Her stableness does not seem to reflect compulsive tendencies, but rather an ability to remain calm and level-headed

[1] The authors would like to express their appreciation to Sylvia Jacobs for her assistance in running the study.

in almost any circumstance. Her profile does suggest she might be rather impulsive concerning small details and unimportant decisions. This impulsive tendency is probably reflected in a lack of concern with material things. In addition, it appears that material things are important to the subject only insofar as they enable her to express her generosity, good nature, and zest for living. . . . [she] is intellectually very mature for her age.

The corresponding portion of the LSE report stated:

The subject's profile indicates that she has a rather unstable personality and is given to pronounced mood fluctuations of excitement or depression. Her instability seems to reflect compulsive tendencies and relative inability to remain calm and level-headed in circumstances which involve tension and pressure. Her profile does suggest she might be rather meticulous and careful concerning small details and when making unimportant decisions. In addition, it appears that material things are very important to the subject as an end in themselves. She appears to be a very selfish person who clings to material things as a source of personal gratification and as an emotional crutch.

In the NSE condition the subject was told that her report had not yet been evaluated due to a heavy backlog of work at the Counseling Center. Instead, she was presented with a sample profile which was described as a typical or average CPI profile such as one might find in a psychological textbook. The comments in this NSE report paralleled those of the HSE and LSE reports (e.g., ". . . fairly stable personality . . . occasionally experiences mood fluctuations . . ."). The contents of the rest of the reports was designed to be global in nature and contained comments concerning the person's ability to make friends, general impact of personality, and depth of thought. Since the dependent variable involved honesty, special care was taken to refrain from mentioning anything directly involving honesty. Similarly, nothing whatever was mentioned, either explicitly or implicitly about the person's moral behavior or "goodness" of conduct.

Warm-up Instructions

After reading their personality test results, subjects turned to their page of instructions. The instructions stated:

The purpose of this experiment is to correlate extrasensory perception ability with personality characteristics. In order to get a true measure of a person's ESP ability, it is necessary that one's mind be primed for thinking. In order to accomplish this, I am having you engage in a period of cerebral warm-up. It's not important what you think about—anything will do—but the crucial point is that you are to use your mind, warm it up by thinking. You need not concentrate intensely, just keep your mind active and filled with thought. A few minutes of this will suffice to prepare you for the ESP experiment.

The purpose of this "warm-up" period was to provide an opportunity for the impact of our manipulations to sink in. Since the subjects had just received an evaluation of their personalities we were quite confident that they would be thinking about this material.

Following the warm-up period, the subjects were sent, one at a time, at intervals of approximately 30 seconds, from the second floor of the psychology building to a room on the fourth floor. Here they were met by the experimenter who was unaware of which subject had received which type of personality feedback. The subjects were placed in one of three cubicles which isolated them from each other. Thus it was impossible for subjects to converse with one another between the time they received their personality results and the time the dependent measure was collected.

Apparatus

The apparatus consisted of four booths or cubicles; the front panels of three of the cubicles bounded an area 1 foot square. Access to this area was available via the fourth cubicle which had no front panel. The experimenter, when sitting in the fourth cubicle, was thus able to receive and dispense playing cards to the subjects in the other three cubicles via slots in their front panels. Subjects in their cubicles had no means of communicating with their fellow subjects nor could they see the experimenter due to plywood panels on both sides and at the top of their booths.

Inside each booth there were two slots and two toggle switches. One switch was designated to be turned on to indicate a "no" answer and the other to indicate a "yes" answer. One of the slots was horizontal with the base of the booth, and was situated in the middle of the front panel one-half inch above the cubicle base. The other slot was vertical and was the slot through which subjects returned cards to the experimenter.

The experimenter's cubicle contained three inclined sheet-metal slides leading down to the horizontal slots of the subjects' booths, three vertical slots from each subject's booth which were shielded so that subjects could not see into the experimenter's compartment, three scoreboards (one for each subject), six small light bulbs, each connected to one of the subjects' "yes" and "no" switches. The experimenter's booth also contained a small electric motor. The motor was functionless except for sound effects, and was activated periodically during the experimental session.

Experimental Materials

The experimenter had two decks of cards. One deck was used for the general game of cards to be played; the other deck was divided into three stacks of cards (one stack for each subject) with a prearranged sequence. These were the crucial hands which a subject would be dealt on those occasions when she would be given the opportunity to cheat. The prearranged stacks consisted of four hands of black-jack with four cards in each hand. The first two cards in all hands totaled approx-imately 11–13 points with the third card sending the point total over 21. The fourth card, if substituted for the third, always brought the point total to between 19 and 21 points.

Each subject's cubicle contained, in addition to the two toggle switches, a 15-watt light bulb and 10 fifty-cent pieces (or $5). The half-dollars were used as "chips" and potential reward in the experimental task.

Experimental Instructions

After the subjects were seated, the experimenter recited the instructions. He de-scribed the results of some previous experiments concerning ESP. Rhine's conclu-sion that some persons do indeed possess ESP was presented; however, *no* indi-cation was given that certain types of persons possess ESP while others do not. In order to make the cover story appear more credible, subjects were asked to read a recent newspaper article taped to a side panel of their cubicles, which told of how a young girl had been winning an astonishing number of raffle contests which, according to a research institute at Duke University, was perhaps due to her ESP powers. This article indicated that having ESP could be materially valuable, but no mention was made either in the article or in the instructions that having ESP was an intrinsically positive or valued trait to possess. Following this, subjects were told that their ESP ability was going to be evaluated in the context of a mod-ified game of blackjack. It was clearly explained that the game was to be played among the three subjects and that the experimenter was not participating in the game as a contestant. The subjects were then informed of the presence and neces-sity of a "card-dealing machine." The experimenter said that the cards were to be dealt by a machine "in order to insure against possible interference with your ESP due to another person handling the cards." In actuality, the machine was used in order to provide an opportunity for the subjects to cheat, and to make it easier for subjects to cheat since the machine apparently removed the experimenter from the situation; this will be described below. According to the experimenter's descrip-tion, the machine automatically dealt them a card on each round. In order to "stand pat," subjects had to switch on their "no" light so that the experimenter could divert the dealing machine from giving them a card on the next round. The use of light signals rather than verbal ones to inform the experimenter that the subject wanted to stand pat was justified by "the necessity of keeping talking at a mini-mum in order not to interfere with your ESP." The modifications and rules of the blackjack game were as follows:

Each subject was provided with $5 by the psychology department as a stake with which to play the game. Each subject was to "bet 50¢, no more or no less," on the outcome of each hand. Subjects were informed that at the conclusion of the experiment they would be allowed to keep all winnings over $5, but that subjects having less than $5 at the game's conclusion *would not* be required or obligated to make up the deficit out of their own pockets; this made it impossible for any subject to "lose" any of her own money.

We presented the following cover story regarding how ESP ability would be measured: The subjects were told to "concentrate for approximately 5 to 10 seconds before each hand," that is, to think about whether they would win or not win the upcoming hand. If they thought or felt they were going to win, they were to push the "yes" switch in front of them to the "on" position for 5 seconds, whereas if subjects felt they were not going to win the upcoming hand, they were to push the "no" switch to the "on" position for 5 seconds. The number of correct guesses *above chance* supposedly constituted a subject's ESP score. Prior to each guess, all subjects deposited a fifty-cent piece in the vertical slot, and it was explained to the subjects that on every hand they were trying to win a kitty with $1.50 in it.

A thorough explanation of the objectives of blackjack was given in order to equalize differential card game experience between the subjects. It was emphasized that point totals of 14–15 made the choice of whether or not to take another card especially difficult because of the high probability of going over 21 if another card was taken, and the low probability of winning by standing pat. The point totals of the various cards were also emphasized with the Ace counting only 11 points rather than 1 or 11 points, to simplify the experiment; also, all face cards were worth 10 points, and all numbered cards were worth their face values.

It was explained to subjects that:

> The card-dealing machine is a "home-made" affair. It has not as yet been perfected and at times makes mistakes. So far we've been able to iron out most of the mistakes, but occasionally the machine will deal two cards at a time instead of one. The machine is set up so that it deals from the top of the deck, as in a normal game of cards; when a person receives two cards, the top card is actually the card he should have gotten and the bottom card is the mistaken card. Now, the machine is fairly good, but at times will make this mistake. If it ever happens that you are dealt two cards, pick up the two cards and immediately return the bottom card to me by slipping it through the vertical slot. Remember, the top card is your card, the bottom card should be returned to me through the vertical slot.

Subjects were told that when all "no" switches were turned on, the experimenter would say "Game," which was a signal to subjects that they were to push all their cards through the vertical slot to the experimenter. Again, no talking was allowed under the pretext of not interfering with ESP. The experimenter then examined each player's hand, determined the winner, collected the three fifty-cent pieces and slid them down the slide of the winner. When the experimenter said "Begin," indicating that the next game was to start, the two subjects not having received any money were thus informed of their losing status.

Resumé of Procedural Instructions to the Subjects

The word "begin" was a signal to subjects that the previous hand was finished and that a new hand was to commence immediately. Following the word "begin," subjects deposited their fifty-cent pieces through the vertical slot, concentrated for 5 seconds, made a choice of the "yes" or "no" switches regarding their outcome expectancy on the upcoming hand, and the game began. The hand continued until all subjects had indicated they no longer wanted any more cards, at which time the experimenter said "game" and subjects then pushed all their cards through the vertical slot to the experimenter. This same procedure was repeated 35 times for all subjects. However, subjects were not aware as to exactly how many hands had been played at any given point in the experiment, nor did they know precisely when the game would end.

Dependent Measure

The dependent measure was the number of times during the experimental session that subjects kept the card they should have returned to the experimenter, thus enabling them to win the hand. The experiment was designed so as to present each subject with four opportunities to cheat during a session; thus, as far as each subject was concerned, the machine had mistakenly dealt two cards at once only four times in 35 hands. In reality, the experimenter dealt all cards but synchronized his dealing with the onset and termination of the machine-generated sound effects.

The method for providing subjects with a cheating opportunity was quite simple. However, complications arose in determining precisely *when* this opportunity should present itself. We decided to present a subject with an opportunity to cheat when the following conditions could be satisfied: *(a)* If she (the subject) had guessed "yes"; *(b)* when she possessed the same amount of money as the other subjects; *(c)* if she had not previously received more opportunities to cheat than the other subjects; *(d)* when her ESP hit rate was near chance level.

The separate stacks of "cheating cards" were prearranged so that the first two cards totaled 11–14 points, the third put the total over 21, and the fourth, if substituted for the third, brought the total to 19–21 points. Thus, when subjects received two cards at once they were faced with a dilemma: if they did not cheat, they would lose the hand; if they did cheat, they would almost certainly win. Cheating, therefore, was made relatively safe from exposure by having the cards dealt by a machine and also enabled subjects to net $1. The behavioral measure of cheating was whether or not subjects returned to the experimenter the card that was actually theirs and kept the bottom card which enabled them to win. The remaining two subjects in such a game were dealt cards from the general deck. If the subject had cheated, she was always declared the winner on that hand and was given the three fifty-cent pieces. This subject, of course, lost if she did not cheat.

The other two subjects, in this hand, were always dealt a hand less than 21 or at times one of them was dealt a hand exceeding 21. If a subject did *not* cheat, the other subject with the score closest to and under 21 was declared the winner. The experimenter used scoreboards to keep track of how often a subject had won a hand, how often each had been given a cheating opportunity, and whether or not a subject had cheated. The scoreboard also provided for an evaluation of trial effects.

Following the card game, all subjects were asked to be seated at a table in the same room. They then filled out a questionnaire consisting of a check on the experimental manipulations and several filler items. Subjects were then debriefed completely. The purpose of the experiment was explained and subjects were assured that the results of the personality test were preprogrammed and had no relationship to their actual test scores.

RESULTS AND DISCUSSION

Analyses of variance on continuous data were all nonsignificant, although the mean differences were of the order hypothesized (mean cheats: LSE = 1.87, NSE = 1.54, HSE = 1.07). Frequency analyses, however, produced significant chi-squares. Subjects were divided according to whether they never cheated or cheated on at least one occasion. Table 14.1 shows a 2 × 3 contingency table chi-square with $df = 2$. Note that 13 people who were given negative feedback cheated at least once, while there were only 6 cheaters among the positive-feedback subjects. The chi-square proved to be significant at beyond the .05 level ($\chi^2 = 7.00$, $p < .05$). Another chi-square was computed to evaluate the cheating differences between just the high and low self-esteem groups. This chi-square with $df = 1$ also proved to be significant after the correction for continuity had been made ($\chi^2 = 5.17$, $p < .03$).

Taken as a whole, the data indicate that whether or not an individual cheats is influenced by the nature of the self-relevant feedback he received. People who learned uncomplimentary information about themselves showed a far greater tendency to cheat (on at least one occasion) than individuals who received positive information

TABLE 14.1
Number of people cheating at least
once as a function of self-esteem

Condition	Cheat	Never cheat
LSE	13	2
NSE	9	6
HSE	6	9

Note: $\chi^2_{LHN} = 7.00$, $df = 2$, $p < .05$.
$\chi^2_{LH} = 5.17$, $df = 1$, $p < .03$.

about themselves. This suggests that individuals with low self-esteem are more prone to commit immoral acts than individuals with high self-esteem, at least when the immoral act is instrumental in producing immediate material gain. Moreover, the number of people cheating in each of the above groups fell on either side of the neutral condition. Although neither experimental condition was significantly different from the control, it is important to note that in the high self-esteem condition there was a greater trend toward honest behavior than in the control, whereas in the low self-esteem condition there was a greater trend toward cheating than in the control.

Our interpretation of the results hinges upon our contention that the manipulation employed in the experiment made subjects feel good about themselves or bad about themselves. In short, we contend that the feedback the subjects received had some impact (however temporary) upon their level of self-esteem. This interpretation is bolstered by our check on the manipulation which indicated that LSE subjects felt worse about themselves than either NSE or HSE subjects. However, since the manipulation check occurred subsequent to the card game, this difference might be due to the cheating behavior rather than the self-esteem manipulation.

There is some additional evidence which is also consistent with our interpretation. This involves the cheating behavior of subjects in terms of their *chronic levels* of self-esteem. Although it was not our intention to measure chronic self-esteem, one can extract a rough measure by looking at the self-concept scores on the CPI, which all subjects filled out as part of the cover story of the experiment. According to these measures, people of high, medium, and low self-esteem had been almost equally distributed among experimental conditions. The following data emerge: In the no feedback (NSE) condition, slightly more "low chronics" cheat than "high chronics." Also, as one might expect, the greatest percentage of cheaters falls among the low chronics who were given negative feedback; the smallest percentage of cheaters falls among the high chronics who were given positive feedback. The small number of subjects in each cell makes statistical analysis of these data unfeasible. The most one can say about these results is that they are consistent with our interpretation of the overall data on the basis of the experimental treatments themselves.

It should be emphasized that one of the unique aspects of this study is the nonspecific nature of the self-concept manipulation. The subjects were not told anything about themselves which would lead them to infer that they were moral-honest or immoral-dishonest people. Rather, they were told things designed to reduce their self-esteem in general. The social implications of these findings may be of some importance. Our results suggest that people who have a high opinion of themselves are less prone to perform any activities which are generally dissonant with their opinion. Similarly, it may be easier for a person with a low self-concept to commit acts of a criminal nature. Moreover, it may be that a common thread running through the complex variables involved in successful socialization (Sears, Maccoby, and Levin, 1957) is that of different development of self-esteem. Granted that most children become aware of what behavior is approved (moral) or disapproved (immoral), the development of high self-esteem in the individual may be crucial in his choosing a moral rather than an immoral mode of behaving.

This discussion is highly speculative to say the least. Further experimentation is necessary before the validity of our reasoning can be determined. One reason for the note of caution is the fact that it is difficult to perform an experiment involving complex human cognitions, emotions, and behavior which leaves us with a single, untarnished explanation for the results. This experiment is no exception. One conceivable alternative interpretation concerns aggressiveness: the subjects in the LSE condition, because they received a negative evaluation, may have been angry at the evaluator and, consequently, may have cheated as a way of punishing him. We attempted to eliminate this possibility in two ways: (a) We separated the experimenter who ran the ESP experiment from the evaluator (a member of the Counseling Center). Toward this end, the experimenter appeared ignorant of and uninterested in these evaluations; it was made to seem to be strictly between the subject and the Counseling Center. (b) The subjects were playing against other subjects rather than against "the house." Thus, when an individual cheated, she was clearly not hurting the experimenter; rather, she was unjustly taking money from a fellow college student.

Perhaps a more compelling explanation involves compensation. The subjects in the low self-esteem situation may, in effect, be saying "Well, I may not have done well on that personality test, but at least I'm going to see to it that I win some money." This explanation is quite different from the one that holds that it is easier to cheat because such behavior is consistent with feelings of low self-esteem. Note, however, that the "compensation" explanation applies only to the subjects in the low self-esteem condition. Thus, this alternative explanation would have been weakened if the subjects in the high self-esteem condition had cheated less than those in the control condition. But as the reader will recall, although these data were in a direction favoring consistency theory, they were not statistically significant. Thus, until further research is performed on this problem, compensation remains a possible explanation.

A final piece of unclarity should be mentioned. We predicted that people in the LSE condition would *cheat* and people in the HSE condition would *not cheat* because we felt that such actions would reflect a consistency between self-esteem and behavior. But *cheating* is merely one of many ways in which a person's behavior could show consistency with low or high self-esteem. For example, in the present situation a LSE subject could try to *lose;* being a loser might be considered as consistent with a low self-esteem. It should be noted that the experimenters took special pains to assure the subjects that they could not lose money. In this situation it seems reasonable to assume that being a loser simply means having bad luck—not being a bad person or even a poor person (financially). We selected cheating as our dependent variable because we felt that it is an unambiguously unethical piece of behavior. Although losing at cards is not pleasant, it does not seem as bad, especially since a loss of money is not involved. The data indicate that our reasoning was correct—that if a sizable number of LSE subjects had sought to lose, our data would have failed to reach significance.

References

ARONSON, E., AND CARLSMITH, J. M. Performance expectancy as a determinant of actual performance. *Journal of Abnormal and Social Psychology,* 1962, **62,** 178–182.

BRAMEL D. A. dissonance theory approach to defensive projection. *Journal of Abnormal and Social Psychology,* 1962, **64,** 121–129.

ROGERS, C. R. *Client-centered therapy.* Boston: Houghton-Mifflin, 1951.

SEARS, R. R., MACCOBY, E. E., AND LEVIN, H. *Patterns of child rearing.* Evanston, Ill.: Row-Peterson, 1957.

WILSON, D. T. Ability evaluation, postdecision dissonance, and co-worker attractiveness. *Journal of Personality and Social Psychology,* 1965, **1,** 486–489.

15

Dissonance and Alcohol: Drinking Your Troubles Away

Claude M. Steele, Lillian L. Southwick,
and Barbara Critchlow

Based on recent evidence supporting the assumption that cognitive dissonance is experienced as an unpleasant emotional state, and further evidence pertaining to the effects of drinking alcohol, it was predicted that among social drinkers, dissonance arousal would increase the amount of drinking and that drinking, in turn, would reduce dissonance and subsequent attitude change. This hypothesis was tested in the first two experiments by having subjects taste rate different brands of an alcoholic beverage—ostensibly to test taste discrimination but in fact to measure the amount of drinking—immediately after dissonance was aroused by having them write a counterattitudinal essay. The effect of drinking on dissonance reduction was assessed by measuring subjects' postattitudes immediately after the drinking task. Both experiments found that although dissonance arousal had little effect on the amount of drinking, whatever drinking occurred was sufficient to eliminate dissonance-reducing attitude change. The second experiment further established that these results occurred for light as well as heavy social drinkers. Evidence that the dissonance-reducing effect of drinking resulted from some effect of drinking alcohol was provided by the finding, in the second and third experiments, that neither water nor coffee drinking was sufficient to eliminate attitude change in this paradigm. Both the practical and theoretical implications are discussed. The practical implication is that some forms of alcohol abuse may evolve through the reinforcement of drinking as a means of reducing dissonance; the theoretical implication is that dissonance may be frequently reduced through behaviors that ameliorate the feelings of dissonance without involving cognitive change.

Cognitive dissonance theory reasons that behavior that contradicts an important belief about oneself will arouse dissonance, an uncomfortable state of psychological tension (Aronson, 1969; Zanna and Cooper, 1976). When this happens, attitude

Reprinted with permission of the authors and *Journal of Personality and Social Psychology*, Vol. 41, No. 5, 1981. Copyright 1981 by the American Psychological Association, Inc. This research was supported in part by Grant ROI AA02448-02 from the National Institute of Alcoholism and Alcohol Abuse to the first author and John P. Keating. The authors would like to express their special thanks to Robert Croyle, Bruce Morash, Thomas Liu, and Dorothy Steele for their generous assistance with several aspects of this research and to Andrew R. Davidson for his comments on an earlier draft of this manuscript.

change may occur in order to restore consistency between one's self-image and one's behavior, thereby reducing dissonance. Following this reasoning, dissonance reduction has typically been measured by changes in attitudes and behavior that are capable of reconciling the inconsistent cognitions. Despite the logic of this approach, it may have obscured the possibility that dissonance reduction may sometimes involve other responses that have no relationship to the inconsistent cognitions. Because it is an uncomfortable state of tension, dissonance may motivate behavior that is capable of alleviating the discomfort of dissonance, even when such behavior does not remedy the particular stimulus inconsistency.

Several recent experiments by Kidd and Berkowitz (1976), for example, showed that dissonance aroused by having subjects write a counterattitudinal essay increased subjects' helping behavior. Even though helping was unrelated to the dissonance-causing inconsistency, its ability to enhance subjects' self-esteems, and presumably make them feel better, may have made it a more likely response to dissonance. Similarly, Steele (1975) has suggested that dissonance arousal might also motivate behaviors such as drinking, smoking, or drug use that neither reduce inconsistency nor enhance self-esteem but simply elevate one's mood or reduce inhibitions. Thus, the first question addressed by the present research is whether dissonance can motivate a response that does not reconcile the provoking inconsistency yet is capable of alleviating its unpleasantness.

To address this question, the present research examines the possibility of a relationship between dissonance and drinking. Several known effects of drinking alcohol suggest that it might be a particularly effective way of eliminating the unpleasant feelings of dissonance. First, evidence on the "biphasic" physiological effects of alcohol (Himwich and Callison, 1972; McCollam, Burish, Maisto, and Sobell, 1980; Mello, 1968) indicates that during the initial phases of ingestion, small amounts of alcohol stimulate the central nervous system and cause positive affect. In describing their early reactions to drinking, subjects report feeling "alert and pepped up" and as having more energy (Naitoh and Docter, 1968). In larger amounts and during later phases of the drinking episode, alcohol acts as a depressant. Also, recent evidence indicates that this pattern of physiological effects is reflected in subjects' *expectations* about the effects of drinking. Two recent independent surveys (Brown, Goldman, Inn, and Anderson, 1980; Southwick, Steele, and Marlatt, 1981) revealed that during the initial phases of drinking, college drinkers expected alcohol to make them fell mildly euphoric, less inhibited, less self-conscious, and so forth. Negative effects such as poor coordination, hostility, and so forth, were expected to occur later in the drinking episode. Thus, regardless of how it is mediated, moderate drinking might relieve dissonance through its ability to induce positive affect.

Drinking may also relieve dissonance through its effect as a disinhibitory agent. Cognitive dissonance can be conceptualized as an internalized pressure toward psychological consistency and away from inconsistency. In this regard, alcohol has been found to reduce the pressure of internalized constraints against a variety of behaviors: aggression (e.g., Boyatzis, 1974; Shuntich and Taylor, 1972; Taylor and Gammon, 1975; Ziechner and Pihl, 1979), violent crime (Wolfgang, 1958), self-restrained eating behavior (Polivy and Herman, 1976a, 1976b), self-disclosure (Rohrberg and Sousa-Poza, 1976), gambling (Hurst, Radlow, Chubb, and Bagley, 1969), and risk taking (Teger, Katkin, and Pruitt, 1969). MacAndrew and Edgerton (1969) further argue that based on our general cultural expectations about the effects of alcohol, drinking is frequently used as an acceptable "time out" from the normative (internalized as well as externally imposed) regulation of behavior. Thus, whether mediated physiologically or cognitively, drinking may relieve the unpleasantness of dissonance through a disinhibitory effect of freeing the individual from the internalized pressure to maintain psychological consistency.

Finally, a possible dissonance-relieving effect of drinking is suggested by the conventional hypothesis that alcohol has tension-reducing properties (Conger, 1956). This hypothesis has been tested for many types of tension, but the findings have been inconsistent (cf. Cappell and Herman, 1972; Higgins, 1976). Nonetheless, one reason for the variability in support of the hypothesis may be that drinking reduces some forms of tension and not others. Thus, drinking may possibly reduce the discomfort of dissonance through a direct tension-reduction effect, which could be mediated either physiologically or cognitively.

If any one or all of these effects of drinking can reduce the unpleasant feelings of dissonance, then dissonance arousal might be expected to cause regular drinkers, especially, to increase their drinking as an expedient means of relieving these feelings.

Equally important is whether drinking, by relieving these feelings, will interfere with cognitive efforts to restore consistency. Several recent findings suggest that it might. The results of experiments by Cooper and Zanna and their associates (e.g., Cooper, Zanna, and Taves, 1978; Zanna and Cooper, 1974, 1976) show that in order for consistency-restoring attitude change to follow dissonance arousal, subjects must both experience negative arousal and also attribute this arousal to the dissonance-provoking act. Because drinking follows dissonance arousal by some time in the present research, it is unlikely that it would eliminate dissonance-reducing attitude change by causing subjects to misattribute their dissonant feelings to the act of drinking. Cooper et al. (1978), however, found that the normal tendency toward dissonance-reducing attitude change could also be eliminated by covertly giving subjects enough tranquilizer (phenobarbital) to directly alleviate the negative arousal associated with dissonance. Thus, if drinking alcohol, through any of the means described above, eliminates the unpleasantness of dissonance, it could also be expected to eliminate dissonance-reducing attitude change.

STUDY 1

The purpose of this experiment was to test the hypothesis that dissonance arousal would increase drinking and that drinking, in turn, would eliminate dissonance-reducing attitude change. To do this, subjects who were social drinkers were induced to write, under conditions of free choice, an essay either favoring or opposing an increase in tuition. Following a design developed by Cooper, Fazio, and Rhodewalt (1978), half of the subjects were then presented with a measure of their attitudes toward the increase. The administration of the attitude measure at this point enabled a test of the standard effect of dissonance on attitude change in this experiment; subjects in the dissonance condition should show consistency-restoring attitude change on this measure, whereas subjects in the consonant condition should not. Following the attitude measure, these subjects participated in a beer-tasting task ostensibly to assess their ability to discriminate among types of beer. If dissonance motivates drinking, these subjects should drink relatively little, since dissonance was either not aroused (as for the consonant subjects) or was reduced by prior attitude change (for the dissonance subjects). For the other half of the subjects, the drinking measure preceded the attitude measure, If dissonance motivates drinking, drinking should be relatively high for the dissonance subjects in this condition. If drinking, in turn, reduces the motivation to restore consistency, dissonance subjects should show relatively little attitude change in this condition.[1] These conditions lead to the prediction of an order of measurement by type of measure interaction in the dissonance condition. Thus the levels of drinking and attitude change depend on whether they were measured first or second.

Subjects' drinking habits might also affect their reaction to the variables in this experiment. Heavier social drinkers could have more experience and familiarity with the use of alcohol as a means of disinhibition and pleasure enhancement than moderate social drinkers. This, in turn, could make them more likely to drink in response to dissonance arousal. To examine this possibility, subjects' drinking habits were included as a factor in the present design.

---

[1]It will be noted that this drinking manipulation tests the effect of drinking, but not alcohol per se on dissonance reduction. A test of the effect of alcohol requires an administration procedure in which subjects are brought to the same intoxicating blood alcohol level at the time of the attitude measure without their being aware that they had consumed alcohol. Cooper, Zanna, and Taves (1978) used such a procedure to administer phenobarbital and amphetamine in their test of the role of arousal in mediating dissonance. In contrast, the drinking task in the present experiment allowed subjects to regulate their own dosage level, and the attitude measure was taken relatively soon (10 minutes and 15 minutes) after the task began. This task, it is reasoned, is sufficient to foster the specific effects of drinking that are hypothesized to reduce dissonance. Mediated by subjects' expectancies about the effects of drinking and physiological effects resulting from the amount of alcohol that subjects do consume, it is assumed that freely drinking alcohol in this task is mood-enhancing and disinhibiting in a way that the standard blood alcohol level procedure would not be and that it is thus generalizable to the effects of real life social drinking. Nonetheless, because this task allows subjects' blood alcohol levels to vary, this research cannot be viewed as a test of the effect of alcohol per se on dissonance.

METHOD

Subjects

An advertisement in the student newspaper was used to recruit subjects for a beer-tasting experiment who were (a) 21 years of age or older, (b) University of Washington students, and (c) experienced with alcohol. The advertisement offered $5 for participation and provided a number to call. Students who called were then screened for participation on the basis of their self-reported drinking habits. Using the definition that one drink equals one can of beer, one glass of wine, or one cocktail, only students who reported seven or more drinks per week were eligible for participation. These subjects were then divided into two groups, depending on whether they averaged between 7 and 10 or 10 and more drinks per week. Use of these selection and grouping criteria created groups whose drinking habits approximated what Cahalan, Cisin, and Crossley (1969) considered to be "moderate" and "heavy" levels of social drinking, respectively, based on their survey of national drinking practices. The selected subjects were then telephoned to schedule their appointments and were told not to eat for a period of 4 hours prior to their arrival time. In response to a question on the postexperimental questionnaire, subjects indicated that they had endured, on the average, 4.4 hours of food deprivation prior to the experiment. In this way, 64 subjects (47 males and 17 females) were recruited for this experiment, (32 moderate and 32 heavy social drinkers). The average number of drinks per week reproted by the moderate drinkers was 8.3 and 15.1 for the heavy drinkers.[2]

Upon arrival subjects were randomly assigned to conditions, with the restrictions that an equal number of moderate and heavy drinkers and an equal proportion of males and females be assigned to each of the four experimental conditions (dissonance-attitude measure first, dissonance-drinking measure first, consonance-attitude measure first, and consonance-drinking measure first) and a base-rate control condition in which subjects did not write an essay but did complete the attitude measure followed by the drinking measure. Over the course of the experiment, nine subjects (five in the dissonance conditions and four in the consonance conditions) had to be replaced because they refused to write the essay. Two subjects were signed up for each experimental session, with each one assigned to separate rooms. All experimental sessions were conducted between the hours of 3:00 P.M. and 6:00 P.M. because subjects are generally less willing to drink earlier in the day.

---

[2]It should be noted that potential subjects who called in were highly motivated to participate in this experiment. Because the newspaper ad specified that "experience with alcohol" was a requirement for participation, these self-reports of number of drinks per week may be somewhat inflated. Thus, our division between moderate and heavy social drinkers must be considered an approximate use of the Cahalan et al. criteria.

Procedure

The experimenter began the session by stating that the purpose of this study was to examine the relationship between biographical, attitude, and life-style factors and sensory discrimination, in particular, alcohol taste discrimination. It was explained that recent research had shown that people's personal characteristics, such as background and sociopolitical attitudes, affected their level of tactile and auditory discrimination. Examples were provided of how people made finer discriminations among the voice inflections of speakers with whom they agreed than of those with whom they disagreed, and that people from dense urban environments were better at making auditory discriminations against loud background noise, and so on. Subjects were then told that the experiment consisted of three parts: a Biographical, Attitude and Life-Style Survey; an Attitude-Activation Task (argument listing), which was ostensibly included to get subjects to recall and make salient their values and attitudes prior to the discrimination task; and a Beer Taste Discrimination Task. At this point subjects began work on the biographical-attitude questionnaire. This questionnaire was divided into three sections: the first asked for biographical and background information (e.g., number of siblings, region of country where born), the second was a 20-item survey of sociopolitical attitudes, and the third included 12 items that asked subjects about their current life-style (e.g., number of movies seen per month). Included in the sociopolitical section of the questionnaire was a pretest measure of subjects' attitudes, which asked them to indicate on a 20-point scale the degree of their agreement or disagreement with the statement, "Students at public universities should pay substantially more tuition than they currently pay." All experimenters were kept blind to the hypotheses and were required to conduct each condition an equal number of times.

Dissonance manipulation.   After completing the survey, subjects were introduced to the Attitude Activation Task. It was explained that the purpose of this task was to fully activate subjects' attitudes and values so that they would have maximum impact on their discrimination performance. The example was provided of how one's attitudes and beliefs become more prominent and clear during the course of an argument or debate. In this regard, it was explained that the activation task would involve participation in a real survey being conducted by the Faculty and Student Council on University Affairs. It was stated that the Council was preparing an orientation program for students entering the university during the subsequent fall quarter, which included the issue of a large increase in tuition at the University of Washington. In the interest of fully understanding the relevant arguments on both sides of the issue, the Council wanted students to list arguments favoring only one side. In the dissonance conditions, subjects were told that since the Council had already received so many arguments opposing an increase, it was now seeking arguments favoring the increase. Subjects were then told to take the next 10 minutes and list the strongest, most forceful arguments supporting a substantial increase in tuition at the University of Washington. In the consonance conditions, subjects were told that the Council was still seeking arguments that opposed a tuition increase, and in the 10 minutes allotted, the Council would like them to list the strongest, most forceful arguments that supported this position.

At this point, following a procedure developed by Cooper et al. (1978), subjects were handed a large manila envelope addressed to the Faculty and Student Council on University Affairs, which contained the materials for writing the arguments. To help keep this argument-listing task separate from the rest of the experiment, subjects were told to place their arguments in the envelope when they finished so that they could be sent directly to the committee. To create the perception that subjects had chosen to write the arguments—a necessary part of dissonance arousal—and to reinforce the public nature of their arguments, the first page in the envelope was an informed consent form that asked subjects to acknowledge (a) the nature of the task, (b) gave the Council the right to use the subject's arguments, and (c) informed the subjects that they would be paid $5 regardless of whether they wrote the arguments. The form also requested that the subjects check their choice of whether to write the arguments and whether they agreed to release them to the Council. Finally, the form required the subjects to print their names and the date and to provide a signature. Subjects in the experimental conditions then spent the next 10 minutes writing.

Order manipulation. For subjects in the attitude measure first condition, the last page of the materials in the manila envelope contained the postmeasure of subjects' attitudes toward a tuition increase. The questionnaire stated that, "In order to properly evaluate your essay, the Faculty and Student Council for University Affairs would like to know what your own opinion toward the issue is," and asked subjects to indicate, on a 31-point scale, the degree of their agreement or disagreement with the statement that "the University of Washington should raise student tuition by a substantial amount." For subjects in the drinking first condition, this measure was not included in their envelopes but was administered after they had participated in the taste-rating task (described below) by the experimenter, who claimed to have forgotten to include it in their envelopes. The posttest attitude measure differed from the pretest measure in the statement and scale that is used, and like all the materials included in the envelope, it was dittoed rather than mimeographed. These differences were included to make it more credible that the argument-listing task was separate from the other parts of the experiment.

Beer-tasting task. Either immediately after the attitude postmeasure in the attitude-measure first condition or immediately after the listing task in the drinking first condition, subjects were individually taken to separate rooms where the taste-rating task was set up. This task was adapted from a wine-tasting discrimination task used by Marlatt and his associates in several recent experiments (e.g., Higgins & Marlatt, 1975). Each subject was provided with three 16 ounce cans of cold beer of different brands (Schlitz, Budweiser, and Olympia) and three large, chilled mugs. While the subject turned away, the beers were poured into the mugs that were then labeled either *A, B,* or *C* and shifted around on the table so that the subject would not know which brand was in each mug. The task required subjects to discriminate among the beers by using a list of adjectives that included *sweet, bitter, malty, sour, tangy,* and so forth. The adjectives were typed on cards that were stacked upside down in a card box so that the subject could not tell how many cards were to be rated. Subjects had to rate which beer was the "most" and "least" for each adjective, and they were not told how long the task would last. As the experimenter left the room, he told each subject, "Feel free to drink as much beer as you need in order to make your best discriminations." After 15 minutes the task was stopped and the amount the subject drank was measured. Also, the number of sips subjects took throughout the task was recorded for each of the three 5 minute time periods by an observer who watched subjects through a one-way mirror. After the task, subjects' blood alcohol levels were measured with an intoxilizer to determine if the blood alcohol level of any subject had exceeded the .05 level. Although this level does not indicate intoxication, it does indicate a need for caution, and subjects who reached this level were to be driven home. No subjects in this experiment reached the .05 blood alcohol level.

RESULTS

The number of subjects in each analysis is the number of subjects who completed all necessary measures.

Pretest Attitudes

A 2 (dissonance versus consonance) × 2 (order: attitude first versus drinking first) × 2 (drinking habits: heavy versus moderate) analysis of variance on subjects' pretest attitudes revealed a main effect for drinking habits, $F(1, 39) = 8.81$, $p < .003$. As the means in Table 15.1 show, moderate drinkers were significantly more

TABLE 15.1
Pretest attitudes as a function of condition and drinking habits

Condition	Drinking habits	
	Heavy	Moderate
Dissonance-attitude 1st	2.5	9.0
Dissonance-drinking 1st	2.8	7.6
Consonance-attitude 1st	3.8	4.6
Consonance-drinking 1st	2.3	4.8
Base-rate control	7.3	6.3

Note: Twenty indicates greatest possible favorability toward a tuition hike.

favorable toward an increase in tuition than were heavier drinkers, especially in the dissonance conditions. These subjects did not experience the same level of dissonance arousal as heavier social drinkers and, in fact, probably experienced very little dissonance at all, since in the dissonance conditions their attitudes were toward the midpoint of the 20-point scale, indicating little opposition to the tuition increase. The absence of successful dissonance arousal for these subjects meant that the experimental hypotheses could be adequately tested only for heavier drinkers.

Drinking and Attitude Change as Reactions to Dissonance among Heavy Drinkers

Following an analysis used by Cooper et al. (1978), postattitude scores and the total milliters of beer consumed during the drinking task were converted to standardized scores in order to place both variables on a common scale of measurement. The overall Order × Measure interaction was significant for heavy drinkers, $F(1, 18) = 5.50, p < .03$. Tests of this interaction within the dissonance and consonance conditions separately revealed that as predicted, this interaction was significant in the dissonance conditions, $F(1, 18) = 8.16, p < .01$, but not in the consonance conditions. As can be seen in Figure 15.1, heavy drinkers in the dissonance conditions either favored the tuition hike relatively more and then drank relatively less in the attitude-measure first condition, or drank relatively more and then favored the tuition hike relatively less in the drinking first condition; they countered their dissonance with whichever opportunity came first, attitude change or drinking. In the consonance conditions there was no evidence of this tendency. The three-way interaction between dissonance arousal, order, and measure that reflects this pattern of results was marginally significant, $F(1, 18) = 3.5, p < .08$.

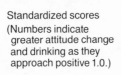

Standardized scores
(Numbers indicate greater attitude change and drinking as they approach positive 1.0.)

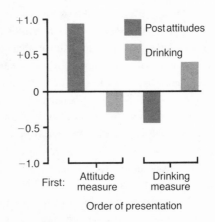

FIGURE 15.1
Postattitudes and drinking for heavy drinkers.

Effect of Dissonance on the Amount of Drinking

Although the pattern of data in Figure 15.1 suggests that dissonance may increase drinking, to fully establish this effect, the appropriate comparisons must be made separately. If dissonance increases drinking, then heavy drinkers in the dissonance-drinking first condition should drink more beer than those subjects in either the consonant-drinking first or control conditions. Although the means are in the predicted direction (in milliliters, the dissonance-drinking first, $M = 734$, the consonant-drinking first, $M = 542.5$, and the base-rate control, $M = 701.2$), neither the separate contrasts of each no-dissonance condition against the dissonance condition nor the contrast of both no-dissonance conditions combined against the dissonance condition reached significance (all $Fs < 1$). (The same pattern of results emerged when these contrasts were made with subjects' drinking habits removed as a covariate.) These data thus provide little support for the hypothesis that dissonance directly increases the amount of drinking even among heavy drinkers. It must be noted, however, that the within-cell variability in subjects' drinking during the rating task was extremely high.

Effect of Drinking on Cognitive Efforts to Reduce Dissonance

If the self-regulated drinking that subjects engaged in during the tasting task was sufficient to eliminate dissonance-reducing attitude change, then subjects in the dissonance-drinking first condition should show no more attitude change than subjects in the control condition or consonant-drinking first condition, and significantly less attitude change than was observed in the dissonance-attitude first condition.

Before presenting these results, however, it must first be established that dissonance produced significant attitude change in this experiment. To do this, postattitudes (with subjects' pretest scores removed as a covariate because of the high variability of initial attitudes in this experiment) in the dissonance-attitude first condition were contrasted against postattitudes in the consonant-attitude first condition and the control condition, both separately and together. (The adjusted means were dissonance-attitude first, $M = 15.4$, consonant-attitude first, $M = 24.8$, base-rate control, $M = 25.2$, where 31 indicates the strongest opposition to the tuition hike and, thus, the least attitude change.) These contrasts revealed that dissonance produced significantly more attitude change than was found in the control condition, $F(1, 9) = 5.20, p < .05$, marginally more change than was observed in consonant-attitude first condition, $F(1, 8) = 2.71, p < .14$, and significantly more change than was observed in both of these conditions combined, $F(1, 15) = 6.46, p < .03$. Dissonance-reducing attitude change was thus successfully replicated in this experiment. Was drinking sufficient to eliminate this change? The postattitudes for subjects in the dissonance condition who drank prior to the attitude measure ($M = 27.1$) did not differ significantly from those in the consonant-drinking first and control conditions ($Fs < 1$) and, in fact, were significantly less favorable than postattitudes in the dissonance-attitude first condition, $F(1, 8) = 5.20$, $p = .051$.[3]

DISCUSSION

This first experiment provides suggestive evidence of the hypothesized relationship between dissonance and drinking. For heavier social drinkers, drinking beer and attitude change served as interchangeable responses to dissonance arousal. The significant order of presentation by measure interaction for these drinkers within the dissonance conditions showed that they availed themselves of whichever response came first, beer drinking or attitude change, and that both responses were equally effective at reducing dissonance, as indicated by the reduced magnitude of these responses when they came second. Individual contrasts, however, revealed that these results provided significant support for only the second half of this hypothesis. Although drinking beer clearly eliminated dissonance-reducing attitude change, these contrasts did not reveal a significant effect of dissonance on the amount of beer consumed. Also, the forced exclusion of moderate drinkers from the analysis makes it impossible to generalize the results beyond the behavior of heavier social drinkers. With these limitations in mind, however, the present data suggest that drinking may be an effective means of reducing dissonance.

The fact that subjects in the dissonance-drinking first condition drank prior to the measurement of their attitudes, however, raises several possible alternative explanations of the dissonance-reducing effect of drinking in that condition. One of these can be addressed by data from the present experiment. Because the attitude measure followed drinking in this condition, the lack of attitude change may have resulted from subjects simply being too drunk to attend to the measure. Several factors, however, discredit this possibility. For one thing, although it is possible that subjects consumed enough beer during the 15 minute drinking task to induce modest affect enhancement and intoxication (cf. Ekman, Frankenhaeuser, Goldberg, Hagdahl, and Myrsten, 1964; McCollam et al., 1980), it is unlikely that they were drunk enough to ignore the attitude measure or its meaning. This point is corroborated by the fact that no subject in this experiment reached the .05 blood alcohol level (a .10 blood alcohol level is the legal criterion for intoxication in most states). Also, if drinking interfered with subjects' ability to attend to the postattitude measure, it might be expected that the correlation between pre- and posttest attitudes

---

[3]The results for moderate drinkers were also analyzed. As would be expected from the evidence showing that the essay favoring a tuition hike was not counterattitudinal for them, neither the order of presentation by type of measure interaction that was predicted in the dissonance conditions, nor the contrasts reflecting an effect of dissonance on drinking and an effect of drinking on dissonance, reached significance.

would be lower in the drinking-first conditions, where the postattitude measures were administered after subjects drank, than in the attitude-measure first condition, where attitudes were measured prior to drinking. These correlations, however, were .63 and .72, respectively. It is thus unlikely that subjects were too drunk to attend to the attitude measure in the drinking first condition. Nonetheless, several other alternative explanations of this effect remain that cannot be addressed by evidence available from the present experiment.

1. The greater time delay between dissonance arousal and attitude measurement experienced by subjects in the drinking first condition compared to subjects in the attitude first condition may have allowed dissonance to dissipate simply as a function of time, aside from any effect of drinking. Research that has assessed the effect of time on dissonance generally shows that it is the activity that fills the time, rather than time per se, that determines the persistence of dissonance (e.g., Crano and Messé, 1970; Freedman, 1965; Higgins, Rhodewalt, and Zanna, 1979; Walster and Berscheid, 1968). Nonetheless, to definitively eliminate this alternative explanation, it is necessary to determine whether the time delay that was confounded with drinking in the drinking first condition contributed to the lack of attitude change in that condition.

2. In addition to the time factor, the taste-rating task in the dissonance-drinking first condition may have been sufficiently involving to reduce dissonance by distracting subjects' attention away from their dissonant cognitions. Prior research (e.g., Zanna and Aziza, 1976) has shown that dissonance-reducing attitude change can be eliminated by participation in an involving task that presumably diverts attention away from the dissonant cognitions.

STUDY 2

A condition that replicated the general procedure of the dissonance-drinking first condition of Study 1 was included in this experiment both to test the replicability of the dissonance-reducing effect of drinking observed in Study 1 and to provide a second, more sensitive test of the effect of dissonance on the amount of drinking. With respect to this last aim, all subjects were given water prior to the drinking task, and vodka drinks were used in place of beer to reduce the extent to which variations in subjects' thirst contributed to variability in alcohol consumption. Changing the type of drinks used in the drinking task also tested whether the effects of drinking observed in Study 1 would generalize over different types of alcoholic beverages.

To test the time and distraction alternative explanations, the present experiment included another condition that exactly replicated the procedures of the dissonance-drinking first condition of Study 1 except that the taste-rating task required subjects to discriminate between types of water (tap and distilled) rather than brands of alcohol. If the dissonance-reducing effect of beer drinking in Study 1 resulted from the time and distraction involved in the rating task, then no significant attitude change should be observed in this condition.

This experiment also included a drinking-habits factor, and to prevent the exclusion of subjects because of inappropriate initial attitudes, only subjects with appropriate initial attitudes were treated as eligible for participation. The test of the drinking habits factor also involved change in strategy from Study 1: Light drinkers (people who drank no more than once a month) rather than moderate drinkers were used to constitute the lighter drinking level of the drinking habits factor. Because both moderate and heavy social drinkers selected by the Cahalan et al. (1969) norms drank relatively heavily (more than seven drinks per week), it was felt that the use of light drinkers would reveal more about the role of drinking habits in mediating the predicted effects. The inclusion of this factor led to a 2 (heavy vs. light drinking habits) × 3 (dissonance-alcohol, dissonance-water, and base-rate control, alcohol condition) design.

METHOD

Subjects

Potential subjects were identified through the administration of a Biographical, Attitude and Life Style questionnaire (used as the pretest in Study 1) to psychology classes and some fraternities at the University of Washington. Embedded in this 30-item questionnaire was a question assessing respondents' attitudes toward a tuition increase on an 11-point scale and two questions regarding drinking habits, one asking respondents for an estimate of the number of drinks they consumed per week, and the other asking them to check one of four verbal descriptions of their drinking that were based on the Cahalan et al. descriptions of drinking habit categories: never drink; drink at least once a year but less than once a month; drink at least once a month, no more than three drinks at a time; drink nearly every day or weekly, often more than five drinks at a time. Students over 21 were selected as eligible for participation if they opposed a tuition hike (a 4 or lower rating where 1 indicated strongest opposition) and either averaged more than 10 drinks per week (heavy drinkers) or drank at least once a year but less than once a month (light drinkers). Those meeting these criteria were telephoned and in most cases offered extra class credit, though some subjects were recruited from classes that did not give extra credit and were therefore asked to volunteer. Subjects were told during the phone contact that depending on the experimental condition they were assigned to, their participation could involve taste rating alcoholic drinks. The average number of drinks per week reported by the heavy drinkers was 15; by light drinkers, .7.

Upon arrival, subjects were randomly assigned to one of the three conditions — dissonance alcohol, dissonance water, base-rate alcohol control — with the restriction that an equal number of heavy and light drinkers be assigned to each group. Five subjects were dropped from the final analyses for not following directions, and an additional five females were randomly eliminated from the dissonance-alcohol condition in order to equalize the proportion of males to females in each of the six conditions, leaving a total of 10 females and 35 males in the entire experiment.

Procedure

As in Study 1, the experimenter began by explaining to all subjects that the purpose of this experiment was to examine the relationship between background attitude and life-style variables and sensory discrimination, in particular alcohol taste discrimination. Subjects assigned to the water condition were told that they would be participating in a control procedure to assess the effect of background attitudes, and so forth, on taste discrimination among nonalcoholic beverages. It was stated that their task would involve tasting different types of water. In the dissonance-alcohol and dissonance-water conditions, subjects were told that the experiment consisted of three parts: an attitude activation task, the ostensible purpose of which was to make their attitudes more salient before the task discrimination task but in fact provided the occasion for the counterattitudinal essay; a taste discrimination task (vodka drinks or water); and a final postexperimental questionnaire. For subjects in these conditions, the procedure of this experiment followed that of the "dissonance-drinking first" condition of Study 1, with several exceptions: (a) the Biographical Attitude Survey was not included as part of the procedure of this experiment, since subjects' pretest attitudes were measured earlier by the administration of this questionnaire in psychology classes; (b) immediately after the purpose of the experiment was explained, all subjects were instructed to drink from a glass of water, ostensibly to sensitize their taste buds but actually to reduce the effect of thirst on the amount consumed during the later rating task; (c) because subjects were not paid for their participation, the bogus form used to solicit subjects' consent to write the counterattitudinal essay assured subjects that they would not lose experimental credit if they did not write the essay; (d) all subjects completed a postexperimental questionnaire (described below) as the last part of the experiment. Subjects in the base-rate alcohol control condition quenched their thirst following the introduction, completed a 10-item attitude questionnaire in which the postattitude measure was embedded, and completed the alcohol taste-rating task and, finally, the postexperimental questionnaire.

Taste-rating task. The drinks in the alcohol condition were two mixed vodka drinks, each made with 2 ounces of vodka and 4 ounces of tonic water poured into glasses with no ice, which were marked *A* and *B*. The drinks differed in both the brand of vodka (Schenley or Gordon) and in the brand of tonic (Schwepps or Canada Dry). In the water condition, one glass contained 240 ml of tap water, the other the same amount of distilled water.

Procedures and instructions to the subject were identical to those of Study 1, with these exceptions: Subjects compared two rather than three drinks, the experimenter entered the room with the drinks already prepared, the number of sips taken by the subject was not observed (this measure was of little use in analyzing the data from Study 1), and the task was stopped after 10 minutes instead of 15. After the amount consumed was measured, each subject in the alcohol condition was asked his/her weight in order to determine if their blood alcohol had exceeded the .05 level; no subject reached this level.

Postexperimental questionnaire. In the dissonance-alcohol and disso- nance-water conditions, the postattitude measure (identical to that used in Study 1) was presented to the subject immediately after the drinking task. Subjects in all three conditions also completed a postexperimental questionnaire that was ostensi- bly being collected by the psychology department staff to help them evaluate stu- dents' reaction to their experimental participation. This questionnaire contained self-report items concerning the extent of perceived deception, and to measure subjects' affective reaction to the drinking task, five semantic differential items were included: pleasant – unpleasant, relaxing – tension producing, interesting – uninteresting, exciting – tedious, fun – no fun.

RESULTS

Analysis of subjects' initial attitudes toward the tuition hike revealed no significant differences as a function of either drinking habits or condition.

Effect of Drinking on Attitude Change

If drinking eliminated dissonance-reducing attitude change in Study 1 either because of the time or distraction involved in the taste-rating task, then attitude change in the dissonance-water condition of this experiment should not exceed that in the base-rate control condition. If, on the other hand, the dissonance-reducing effect of drinking in Study 1 was mediated by some aspect of alcohol consumption, then two effects should emerge in the present study: (a) The finding of no signifi- cant attitude change should be replicated in the dissonance-alcohol condition of this experiment, and (b) there should be significant attitude change in the disso- nance-water tasting conditions. Figure 15.2 shows that these latter predictions were generally supported, implicating some aspect of alcohol consumption as largely responsible for the dissonance-reducing effect of drinking in Study 1. Over all of the subjects, the main effect for treatments was highly significant, $F(2, 38) = 13.27$, $p < .001$. Condition contrasts further revealed, in support of the dissonance-reduc- ing effect of alcohol, that subjects in the water condition showed significant attitude change when compared with the base-rate condition, $t(42) = 4.14$, $p < .0001$, whereas subjects who drank alcohol during this task did not show change, $t(42) < 1$. Also, as implied by these results, dissonant-alcohol subjects changed their atti- tudes significantly less than dissonant-water subjects, $t(42) = 4.02$, $p < .0001$.

Of further interest, the dissonance-reducing effect of alcohol does not appear to depend on drinking habits. The drinking habits' main effect for postattitudes did

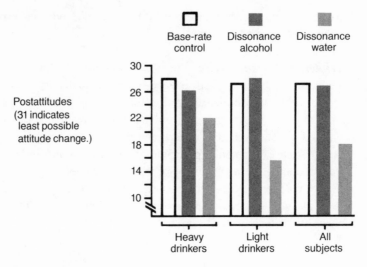

FIGURE 15.2
Postattitudes by condition and drinking habits.

not reach significance, $F(2, 38) = 1.65$, nor was there an interaction of this factor with experimental conditions. In fact, as can be seen from Figure 15.2, the pattern of condition means supporting the dissonance-reducing effect of alcohol is strongest for light drinkers. For these subjects, attitude change in the water condition is significantly greater than attitude change in the control condition, $t(20) = 4.79$, $p < .001$, whereas change in the alcohol condition was even slightly less than the control level. For heavy drinkers this pattern was not as strong. Attitude change in the alcohol condition was no greater than in the control condition, but change in the water condition, though in the predicted direction, was not significantly greater than the control level. (Because of the uncharacteristically large within-cell variance for heavy drinkers' postattitudes within the water condition, a logarithmic transformation was performed on the data, and the water versus control contrast was redone. Again it was only suggestive, $p < .14$.) Nonetheless, considering the strength of this contrast over all of the subjects, it is clear that task time and distraction are not solely responsible for the dissonance-reducing effect of alcohol observed in these two experiments, nor was it mediated by drinking habits.

Effects of Dissonance on Drinking

The possibility that a true effect of dissonance on drinking had been obscured in Study 1 by high within-cell variability in subjects' drinking was addressed in the present experiment by precautions to reduce the effect of thirst on drinking. If dissonance increases drinking, subjects in the dissonance-alcohol condition should drink more vodka drink than control subjects for whom dissonance was not aroused. As in Study 1, however, this contrast did not reach significance either when it was tested over all the subjects ($F < 1$) or for heavy and light drinkers separately ($ts = 1.07$ and 1.03 for heavy and light drinkers, respectively). The means were again in the expected direction for heavy drinkers (80.6 ml and 122.7 ml for the control and dissonance-alcohol conditions, respectively), but not for the light drinkers (49.4 ml and 51.6 ml for the control and dissonance-alcohol conditions, respectively). Efforts to reduce within-cell variance, which was again very high (the standard deviations ranged from 35.8 ml to 128.3 ml), through scale transformations and the use of subjects' habitual drinking levels as a covariate failed to improve these contrasts. Finally, data from both experiments were combined to test this hypothesis. The drinking scores for heavy drinkers in the dissonance-drinking first and control conditions of both Studies 1 and 2 were combined after they were standardized to place the two measures on the same scale. Although this procedure led to mean differences that were again in the predicted direction (.208 and .034 for the dissonance alcohol and control subjects, respectively), the condition contrast failed to reach significance ($t < 1$). Because of the high variability in drinking observed in this task, the inherent mildness of the dissonance that is aroused in experiments of this sort, and the small number of subjects in the critical conditions, the present test of the effect of dissonance on the amount of drinking must be considered conservative.

The Relationship between Task Enjoyment and Drinking: Was Affect Enhancement a Cause or an Effect of Dissonance Reduction?

Drinking was positively correlated with enjoyment of the drinking task (as measured by the postexperimental questionnaire items that assessed subjects' affective reactions to the task, e.g., pleasant–unpleasant, exciting–tedious, etc.) in the dissonance-alcohol condition, $r = .54$, $n = 15$, $p < .02$, but not in the no-dissonance control-alcohol condition, $r = .06$, $n = 14$. This pattern of correlations provides suggestive evidence (these correlations differed from each other at the $p < .09$ level of significance) against the interpretation, described in the introduction, that drinking would reduce dissonance by directly enhancing the level of subjects' affect (e.g., McCollam et al., 1980; Mello, 1968), thereby eliminating the unpleasantness of dissonance.[4] If drinking directly enhanced affect in this research, it should have

---

[4]It will be noted that to measure the effect of drinking on affect enhancement, the subjects' general moods should have been measured rather than their enjoyment of the specific drinking task. Nonetheless, because any measure of mood had to be taken after both the drinking task and the postattitude measure, it would have been impossible to know which of these factors—drinking or dissonance reduction through attitude change—was responsible for subjects' moods. Thus, to focus this measure on the effects of drinking, subjects were asked to complete the mood scales in specific reference to the drinking task.

been positively correlated with task enjoyment among no-dissonance, control subjects, as well as among dissonance subjects. The fact that it was not suggests that greater task enjoyment may have been an aftereffect of dissonance reduction that was brought about first by some other effect of drinking, such as a possible disinhibitory or tension-reducing effect.

It will also be noted that there were no condition differences on the task enjoyment measure ($F < 1$). Thus, although overall task enjoyment was equally strong in all three experimental conditions, it was linked to the amount of drinking only in the dissonance-alcohol condition.

DISCUSSION

The results of this study provided little support for the hypothesis that dissonance increases the amount of drinking, but they did replicate the finding of Study 1 that whatever amount of self-regulated drinking takes place during this task is enough to eliminate dissonance-reducing attitude change. Furthermore, beer and vodka drinks are apparently equally effective at accomplishing this effect. The finding of significant attitude change in the water-tasting condition also made it less credible that the time and distraction involved in the drinking task were themselves sufficient to cause the dissonance-eliminating effect. Also, the dissonance-reducing effect of drinking does not appear to depend on subjects' drinking habits; drinking eliminated attitude change for light as well as heavy social drinkers.

This finding for light drinkers also discredits another version of the distraction explanation. It might be argued that even through the water and alcohol conditions involved the same task, heavy drinkers were still more distracted from their dissonant cognitions in the alcohol condition because of their greater interest in and liking for alcohol. The results for the light drinkers, however, who were selected for not liking alcohol, indicate that this degree of distraction, if it occurred, is not necessary for drinking to eliminate dissonance. Furthermore, light drinkers in the dissonance-alcohol condition, where drinking eliminated dissonance, rated the drinking task as no more interesting and exciting on these postexperimental questionnaire items (the means were 17.0 and 17.1, respectively, on 31-point scales where larger numbers indicated less interest and excitement) than light drinkers in the dissonance-water condition (the same means were 10.25 and 16.0) , where attitude change was significant. Thus, a greater involvement in the alcohol-tasting task that might result from heavier drinking habits, or from the alcohol task being generally more interesting and exciting than the water task, is not necessary for drinking to eliminate dissonance. In fact, heavy drinkers in the dissonance-alcohol condition, where dissonance was eliminated, rated the tasting task as no more interesting ($M = 11.3$) and exciting ($M = 17.6$) than light drinkers did in the water condition, where attitude change was significant. (Heavy drinkers in the water condition, however, were sufficiently disappointed in the tasting substance as to rate the task more uninteresting, $M = 25.0$, and less exciting, $M = 21.3$.) Thus drinking alcohol is sufficient to eliminate dissonance in this paradigm regardless of whether subjects habitually like to drink alcohol or find the task particularly interesting or exciting.

Finally, correlational evidence indicated that subjects' affective reaction to the drinking task was positively related to alcohol consumption only for subjects who were already experiencing dissonance. This result suggests that drinking reduced dissonance not by directly inducing positive affect but by relieving the unpleasantness of dissonance through a direct disinhibitory or tension-reducing effect.

STUDY 3

Studies 1 and 2 support the notion that drinking alcohol can eliminate dissonance-reducing attitude change, regardless of one's drinking habits, and discredit trivializing alternative explanations that have to do with the task. At this point, however, another question arises: Does the dissonance-reducing effect of drinking in this research result from a type and magnitude of effect that is specific to drinking alcohol or from more general effects (e.g., mild affect enhancement, the above correlational evidence notwithstanding) that might result from drinking any moderately pleasurable substance in this drinking task, such as coffee, coca cola, and so forth? Study 3 examines this question by replicating the general procedure and design of Study 1 except that coffee tasting was used in place of beer tasting and subjects were required to be coffee drinkers. If coffee drinking, a presumably pleasurable activity for coffee drinkers, is sufficient to eliminate dissonance-reducing attitude change, then it could be concluded that the dissonance-reducing effects of drinking observed in Studies 1 and 2 were not mediated by effects that were specific to drinking alcohol but by effects that were at least common to coffee drinking as well. If, however, coffee drinking failed to reduce dissonance, then the dissonance-reducing effects of drinking cannot be attributed to the type or magnitude of effects that result from drinking coffee.

METHOD

The design of this experiment followed exactly that of Study 1 except that no base-rate control condition was included, nor was there a drinking habits factor. This left a 2×2 design in which the first factor was whether subjects wrote a dissonant or consonant essay (both under conditions of high choice), and the second factor was whether subjects drank before or after the measurement of their postattitudes.

As in Study 2, eligible subjects were identified through the administration of a

Biographical Attitudes and Life Style questionnaire in psychology classes, which included an item asking for the number of coffees respondents drank per day and an item assessing their attitudes toward a tuition increase (on an 11-point scale). Students who were opposed to a tuition hike and drank one or more cups of coffee per day were then telephoned and offered extra class credit for participation. Forty-one subjects (29 females and 12 males) were recruited in this fashion and randomly assigned to the experimental conditions, with the restriction that equal proportions of males and females be assigned to each condition. These subjects averaged 2.7 cups of coffee per day.

The procedure of this experiment followed that of Study 1 except that the taste-rating task involved the following changes: Subjects compared two blends of coffee rather than three beers, and subjects using cream and/or sugar were instructed to add equal amounts of each to each coffee. As in Study 2, the tasting task lasted for only 10 minutes.

RESULTS

Analysis of subjects' initial attitudes toward the tuition hike revealed no significant condition differences.

Figure 15.3 shows that more favorable attitudes toward the tuition hike resulted in the dissonance conditions than in the consonant conditions regardless of whether subjects drank coffee before or after the attitude measure. The dissonance main effect was statistically significant, $F(1, 40) = 4.15$, $p < .05$. Also, the dissonance condition in which subjects drank prior to the attitude measure produced significantly more attitude change than either the consonant-coffee first condition alone, $t(15) = 2.4$, $p < .03$, or both consonant conditions combined, $F(1, 29) = 6.02$, $p < .02$. Unlike drinking alcohol, drinking coffee had no dissonance-reducing effect on attitudes in this paradigm.

Also, mirroring the findings of the earlier studies, dissonance had no effect on the amount of coffee subjects consumed in this experiment. (The means in milliliters

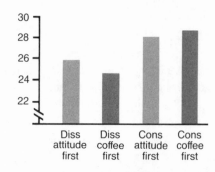

Postattitudes (31 indicates least possible attitude change.)

FIGURE 15.3
Postattitudes by condition.

were dissonance-coffee first = 198.18; dissonance-attitude first = 181.82; consonance-coffee first = 192.78; consonance-attitude first = 223.00.)

DISCUSSION

These results clearly show that the dissonance-reducing effect of self-regulated alcohol consumption does not generalize to self-regulated coffee consumption. This finding, of course, does not preclude the possibility that other substances would have a dissonance-reducing effect. It does make it less plausible, however, that the consumption of just any moderately pleasurable substance is sufficient to reduce dissonance.

Also, the inability of coffee drinking to eliminate dissonance-reducing attitude change among people who like coffee—taken together with the relevant findings from Study 2—supports the conclusion that the time and distraction inherent in the taste-rating task (including the distracting involvement that might result among people selected for liking the tasting substance) are not sufficient in themselves to account for the dissonance-reducing effect of drinking alcohol.

This result is also relevant to a misattribution explanation of the observed alcohol effect: Dissonant subjects who drank alcohol prior to the attitude measure in Studies 1 and 2 shomehow attributed their dissonant feelings to the drinking rather than to the dissonant essay and, thereby, felt less pressure to change their attitudes. As noted earlier, because the drinking task followed dissonance arousal by 10 minutes in these experiments, it is unlikely that subjects could have misattributed the arousal caused by the essay to the act of drinking. The finding that drinking coffee —another plausible source of arousal to which dissonance caused by the essay could have been attributed—also failed to eliminate dissonance-reducing attitude change provides empirical support for this reasoning.

GENERAL DISCUSSION

Perhaps the most important finding of the present research is that self-regulated drinking can very quickly reduce cognitive dissonance, as indicated by its elimination of dissonance-reducing attitude change within 10 minutes after drinking began. Studies 1 and 2 found that although dissonance arousal seemed to have little effect on the amount of drinking, whatever drinking subjects did was sufficient to eliminate dissonance, regardless of the subjects' drinking habits. Furthermore, the fact that neither water nor coffee consumption had a similar dissonance-reducing effect eliminates a variety of alternative explanations of the alcohol effect and supports the notion that drinking has a specific capacity to relieve this type of psychological distress. The connection between drinking and dissonance established in this research reinforces the view that normal social-psychological processes may play a role in the etiology of alcohol abuse. Psychological distresses of the sort aroused by the dissonance-producing procedures of this research are a common part of everyday life. Any situation in which one's behavior can be evaluated against an internalized value, attitude, or standard is a potential source of dissonance and related aversive feelings. Thus, the evidence that independent of drinking habits even light drinking can quickly alleviate this type of distress, and the fact that such distress is so common, suggests that some forms of alcohol abuse may evolve through the frequent reinforcement of drinking as a response to these tensions—tensions that in the present instance were the by-products of normal social–psychological processes of self-regulation.

Processes Underlying the Dissonance-Reducing Effect of Drinking

At the outset of this article, several processes that might be initiated by drinking alcohol (i.e., affect enhancement, psychological disinhibition, tension reduction) were discussed as possible mediators of any dissonance-reducing effect of drinking. Though the results of the completed experiments do not conclusively identify a single mediating process, they do enable us to exclude a number of possibilities— which by convergence helps to establish the nature of the observed effect—and to grade other possibilities as to their plausibility. We believe, for example, that in ruling out the trivializing explanations of this effect that have come to our attention, the present experiments establish that the dissonance-reducing effect of drinking is not an artifact of experimental procedures but results from some specific effect of drinking alcohol that is not dependent on drinking habits.

In evaluating the plausibility of the remaining explanations, the finding in Study 2 that task enjoyment was positively correlated with drinking in the dissonance condition but not in the control condition is perhaps most useful. This pattern of correlations suggests that drinking did not reduce dissonance by *directly* inducing enough positive affect to overwhelm the unpleasantness of dissonance but by somehow reducing dissonance first, which then resulted in affect enhancement as a result of dissonance reduction. Thus, through a process of elimination—never a completely satisfying process—the most plausible account of the observed drinking effect appears to be that drinking directly reduced dissonance through its capacity as a disinhibitory or tension-reducing agent. The more alcohol that dissonant subjects drank, the freer they may have felt from the internalized pressure to reestablish psychological consistency in this situation (i.e., a disinhibition effect), or the more they drank, the more they may have relaxed (e.g., Mayfield, 1968). In either case, drinking would be positively associated with task enjoyment. Future research can evaluate the extent of physiological and cognitive mediation of the observed drinking effect. (The "balanced placebo" design used by Ross, Krugman, Lyerly, and Clyde, 1962, and reviewed by Marlatt and Rohsenow, 1980, for example, was developed for this purpose.) This question aside, however, the dissonance-reducing effect of drinking seems most plausibly interpreted as resulting from the disinhibitory or tension-reducing effects of drinking alcohol.

Effects of Dissonance on Drinking

Throughout these experiments, dissonance failed to increase drinking. The apparent puzzle this presents is that if drinking disinhibits dissonance or reduces tension, why does dissonance not increase the amount of drinking? Several considerations are relevant. First, as noted earlier, the hypothesis that dissonance would increase drinking may have been insensitively tested in this research. Thus, before abandoning the hypothesis, future research should attempt to test it more sensitively. As importantly, although drinking may have reduced dissonance through the hypothesized mechanisms, some amount of experience with alcohol may be necessary before the drinking-dissonance reduction connection is learned well enough for dissonance to motivate greater drinking. This view is consistent with the finding that only heavier social drinkers in these experiments showed a consistent (though insignificant) tendency to drink more after dissonance arousal. Also, because even small amounts of drinking proved sufficient to reduce dissonance in these experiments, it is possible that dissonance was eliminated early in the drinking episode and simply did not persist long enough to motivate further drinking. Dissonance may have a greater effect on the choice to drink or not to drink than on the amount of drinking. This would be especially likely if it was the knowledge that one is drinking alcohol (as mediated by the expectations subjects have about the effects of drinking) rather than the pharmacological properties of alcohol, which mediated its dissonance-reducing effect. Future research will have to sort out these possibilities. For the present, however, these considerations suggest several reasons why drinking may have reduced dissonance without dissonance having had a strong effect on the amount of drinking in these experiments.

Theoretical Implications

Dissonance theory.   Together with the evidence that dissonance-reducing attitude change can be eliminated by a tranquilizing drug that relieves the unpleasantness of dissonance (i.e., Cooper et al., 1978), the present findings suggest that dissonance may be commonly reduced through activity that in some way alleviates the unpleasant feelings associated with the cognitive conflict. Thus, although dissonance theory itself has focused on cognitive change as the primary mode of dissonance reduction (i.e., a change in a behavioral or environmental cognitive element or the addition of new cognitive elements), the mounting evidence, of which the present research is a part, suggests that despite man's motivation for cognitive consistency, his perseverance for it may be easily usurped by immediate gratifications.

The tension-reduction hypothesis. The classic tension-reduction hypothesis of drinking (Conger, 1956) reasons that a state of tension from some source energizes the drinking response, which, in turn, is reinforced through alcohol's ability to reduce the tension. The present finding that drinking reduced dissonance is, therefore, consistent with this hypothesis, though the finding that dissonance had little effect on the amount of drinking is not. These results might thus be viewed as providing partial support for the hypothesis. It is important to note, however, that such support would be for only a specific version of the tension-reduction model, namely, that drinking, possibly through its disinhibitory effects, is capable of reducing cognitive dissonance. These results thus have little bearing on the global utility of the tension-reduction hypothesis.

A Practical Implication

As noted above, because of the ability of drinking to reduce dissonance, drinking may be easily learned as a response to this form of distress, especially when the individual has few alternative coping skills. In support of this reasoning, Marlatt and Gordon's (1979) analysis of the relapse episodes of a sample of alcoholics who were trying to abstain from drinking revealed that the most frequent precipitator of relapse (38 percent of the time) was "intrapersonal negative emotional states." An implication of this analysis, then, is that in order to reduce the appeal of drinking as an effective and fast-working response to these stresses, alcoholism prevention and treatment programs should emphasize training in alternative means of handling dissonance and related negative emotions as well as in drinking management and abstinence techniques.

References

ARONSON, E. The theory of cognitive dissonance: A current perspective. In L. Berkowitz (ed.), *Advances in experimental social psychology.* Vol. 4. New York: Academic Press, 1969.

BOYATZIS, R. E. The effect of alcohol consumption on the aggressive behavior of men. *Quarterly Journal of Studies on Alcohol,* 1974, **35,** 959–972.

BROWN, S. A., GOLDMAN, M. S., INN, A., AND ANDERSON, L. R. Expectations of reinforcement from alcohol: Their domain and relation to drinking patterns. *Journal of Consulting and Clinical Psychology,* 1980, **48,** 419–426.

CAHALAN, D., CISIN, I. H., AND CROSSLEY, H. M. *American drinking practices.* Monograph No. 6. New Brunswick, N.J.: Rutgers Center of Alcohol Studies, 1969.

CAPPELL, H., AND HERMAN, C. P. Alcohol and tension reduction: A review. *Quarterly Journal of Studies on Alcohol,* 1972, **33,** 33–64.

CONGER, J. J. Alcoholism: Theory, problem and challenge. II. Reinforcement theory and the dynamics of alcoholism. *Quarterly Journal of Studies on Alcohol*, 1956, **17**, 296–305.

COOPER, J., FAZIO, R. H., AND RHODEWALT, F. Dissonance and humor: Evidence for the undifferentiated nature of dissonance arousal. *Journal of Personality and Social Psychology*, 1978 **36**, 280–285.

COOPER, J., ZANNA, M. P., AND TAVES, P. A. Arousal as a necessary condition for attitude change following induced compliance. *Journal of Personality and Social Psychology*, 1978, **36**, 1101–1106.

CRANO, W. D., AND MESSÉ, L. A. When does dissonance fail? The time dimension in attitude measurement. *Journal of Personality*, 1970, **38**, 493–508.

EKMAN, G., FRANKENHAEUSER, M., GOLDBERG, L., HAGDAHL, R., AND MYRSTEN, A. Subjective and objective effects of alcohol as functions of dosage and time. *Psychopharmacologia*, 1964, **6**, 399–409.

FREEDMAN, J. L. Long-term behavioral effects of cognitive dissonance. *Journal of Experimental Social Psychology*, 1965, **1**, 145–155.

HIGGINS, E. T., RHODEWALT, F., AND ZANNA, M. P. Dissonance motivation: Its nature, persistence, and reinstatement. *Journal of Experimental Social Psychology*, 1979, **15**, 16–34.

HIGGINS, R. L. Experimental investigations of tension reduction models of alcoholism. In G. Goldstein and C. Neuringer (eds.), *Empirical studies of alcoholism.* Cambridge, Mass.: Ballinger, 1976.

HIGGINS, R. L., AND MARLATT, G. A. Fear of interpersonal evaluation as a determinant of alcohol consumption of male social drinkers. *Journal of Abnormal Psychology*, 1975, **84**, 644–651.

HIMWICH, H. E., AND CALLISON, D. A. The effects of alcohol on evoked potentials of various parts of the central nervous system of the cat. In B. Kissin and H. Begletter (eds.), *The biology of alcoholism.* Vol. 2. New York: Plenum, 1972.

HURST, P. M., RADLOW, R., CHUBB, N. C., AND BAGLEY, S. K. Effects of alcohol and D-amphetamine upon mood and volition. *Psychological Reports*, 1969, **24**, 975–987.

KIDD, R. K., AND BERKOWITZ, L. Effects of dissonance arousal on helplessness. *Journal of Personality and Social Psychology*, 1976, **33**, 5, 613–622.

MACANDREW, C., AND EDGERTON, R. B. *Drunken comportment: A social explanation.* Chicago: Aldine, 1969.

MARLATT, G. A., AND GORDON, J. R. Determinants of relapse: Implications for the maintenance of behavior change. In D. Davidson (ed.), *Behavioral medicine: Changing health lifestyles.* New York: Brunner/Mazel, 1979.

MARLATT, G. A., AND ROHSENOW, D. J. Cognitive processes in alcohol use: Expectancy and the balanced placebo design. In N. K. Mello (ed.), *Advances in substance abuse: Behavioral and biological research.* Greenwich, Conn.: JAI Press, 1980.

MAYFIELD, D. G. Psychopharmacology of alcohol: I. Affective change with intoxication, drinking behavior, and affective state. *Journal of Nervous and Mental Disease*, 1968, **146**, 314–321.

McCOLLAM, J. B., BURISH, T. G., MAISTO, S. A., AND SOBELL, M. B. Alcohol's effects on physiological arousal and self-reported affect and sensations. *Journal of Abnormal Psychology*, 1980, **89**, 224–233.

MELLO, N. K. Some aspects of behavioral pharmacology of alcohol. *Proceedings of the American Cellular Neuropsychopharmacology*, 1968, **6**, 787–809.

NAITOH, P., AND DOCTER, R. F. Electroencephalographic and behavioral correlates of experimentally induced intoxication with alcoholic subjects. *International Congress Alc. Alesm. Abst. 28th*, 1968, 76–77.

POLIVY, J., AND HERMAN, C. P. The effects of alcohol on eating behavior: Disinhibition or sedation? *Addictive Behavior*, 1976, **1**, 121–125.

POLIVY, J., AND HERMAN, C. P. Effects of alcohol on eating behavior: Influence of mood and perceived intoxication. *Journal of Abnormal Psychology*, 1976, **85**, 601–606.

ROHRBERG, R. G., AND SOUSA-POZA, J. F. Alcohol, field dependence and dyadic self-disclosure. *Psychological Reports*, 1976 (Dec.) **39**, (3, Pt. 2), 1151–1161.

Ross, S., Krugman, A. D., Lyerly, S. B., and Clyde, D. J. Drugs and placebos: A model design. *Psychological Reports,* 1962, **10,** 383–392.

Shuntich, R., and Taylor, S. The effects of alcohol on human physical aggression. *Journal of Experimental Research in Personality,* 1972, **6,** 34–38.

Southwick, L. L., Steele, C. M., and Marlatt, G. A. Alcohol-related expectancies: Defined by phase of intoxication and drinking experience. *Journal of Consulting and Clinical Psychology,* 1981, **49,** 713–721.

Steele, C. M. Name-calling and compliance. *Journal of Personality and Social Psychology,* 1975, **31,** 361–369.

Taylor, S., and Gammon, C. The effects of type and dose of alcohol on human aggression. *Journal of Personality and Social Psychology,* 1975, **32,** 169–175.

Teger, A. I., Katkin, E. S., and Pruitt, D. G. Effects of alcoholic beverages and their congener content on level and style of risk taking. *Journal of Personality and Social Psychology,* 1969, **11,** 170–176.

Walster, E., and Berscheid, E. The effects of time on cognitive consistency. In R. P. Abelson et al. (eds.), *Theories of cognitive consistency: A sourcebook.* Chicago: Rand McNally, 1968.

Wolfgang, M. E. *Patterns in criminal homicide.* New York: Wiley, 1958.

Zanna, M. P., and Aziza, C. On the interaction of repression-sensitization and attention in resolving cognitive dissonance. *Journal of Personality,* 1976, **44,** 577–593.

Zanna, M. P., and Cooper, J. Dissonance and the pill: An attribution approach to studying the arousal properties of dissonance. *Journal of Personality and Social Psychology,* 1974, **29,** 703–709.

Zanna, M. P., and Cooper, J. Dissonance and the attribution process. In J. H. Harvey, W. J. Ichkes, and R. F. Kidd (eds.), *New directions in attribution research.* Vol. 1. Hillsdale, N.J.: Erlbaum, 1976.

Zeichner, A., and Pihl, R. O. Effects of alcohol and behavior contingencies on human aggression. *Journal of Abnormal Psychology,* 1979, **88,** 153–160.

V

HUMAN AGGRESSION

16

The Effects of Observing Violence

Leonard Berkowitz

Experiments suggest that aggression depicted in television and motion picture dramas, or observed in
actuality, can arouse certain members of the audience to violent action.

An ancient view of drama is that the action on the stage provides the spectators
with an opportunity to release their own strong emotions harmlessly through identification with the people and events depicted in the play. This idea dates back at
least as far as Aristotle, who wrote in *The Art of Poetry* that drama is "a representation . . . in the form of actions directly presented, not narrated; with incidents arousing pity and fear in such a way as to accomplish a purgation of such
emotions."

Aristotle's concept of catharsis, a term derived from the Greek word for purgation, has survived in modern times. It can be heard on one side of the running
debate over whether or not scenes of violence in motion pictures and television
programs can instigate violent deeds, sooner or later, by people who observe such
scenes. Eminent authorities contend that filmed violence, far from leading to real
violence, can actually have beneficial results in that the viewer may purge himself
of hostile impulses by watching other people behave aggressively, even if these
people are merely actors appearing on a screen. On the other hand, authorities of
equal stature contend that, as one psychiatrist told a Senate subcomittee, filmed
violence is a "preparatory school for delinquency." In this view emotionally im-
mature individuals can be seriously affected by fighting or brutality in films, and
disturbed young people in particular can be led into the habit of expressing their
aggressive energies by socially destructive actions.

Until recently neither of these arguments had the support of data obtained by controlled experimentation; they had to be regarded, therefore, as hypotheses, supported at best by unsystematic observation. Lately, however, several psychologists have undertaken laboratory tests of the effects of filmed aggression. The greater control obtained in these tests, some of which were done in my laboratory at the University of Wisconsin with the support of the National Science Foundation, provides a basis for some statements that have a fair probability of standing up under continued testing.

First, it is possible to suggest that the observation of aggression is more likely to induce hostile behavior than to drain off aggressive inclinations; that, in fact, motion picture or television violence can stimulate aggressive actions by normal people as well as by those who are emotionally disturbed. I would add an important qualification: such actions by normal people will occur only under appropriate conditions. The experiments point to some of the conditions that might result in aggressive actions by people in an audience who had observed filmed violence.

Second, these findings have obvious social significance. Third, the laboratory tests provide some important information about aggressive behavior in general. I shall discuss these three statements in turn.

Catharsis appeared to have occurred in one of the first experiments, conducted by Seymour Feshbach of the University of Colorado. Feshbach deliberately angered a group of college men; then he showed part of the group a filmed prizefight and the other students a more neutral film. He found that the students who saw the prizefight exhibited less hostility than the other students on two tests of aggressiveness administered after the film showings. The findings may indicate that the students who had watched the prizefight had vented their anger vicariously.

That, of course, is not the only possible explanation of the results. The men who saw the filmed violence could have become uneasy about their own aggressive tendencies. Watching someone being hurt may have made them think that aggressive behavior was wrong; as a result they may have inhibited their hostile responses. Clearly there was scope for further experimentation, particularly studies varying the attitude of the subjects toward the filmed aggression.

Suppose the audience were put in a frame of mind to regard the film violence as justified—for instance because a villain got a beating he deserved. The concept of symbolic catharsis would predict in such a case that an angered person might enter vicariously into the scene and work off his anger by thinking of himself as the winning fighter, who was inflicting injury on the man who had provoked him.

FIGURE 16.1
Typical experiment tests reaction of angered man to filmed violence. Experiment begins with introduction of subject *(white shirt)* to a man he believes is a co-worker but who actually is a confederate of the author's. In keeping with pretense that experiment is to test physiological reactions, student conducting the experiment takes blood-pressure readings. He assigns the men a task and leaves; during the task, the confederate insults the subject. Experimenter returns and shows filmed prizefight. Confederate leaves; experimenter tells subject to judge a floor plan drawn by confederate and to record opinion by giving confederate electric shocks. Shocks actually go to recording aparatus. The fight film appeared to stimulate the aggressiveness of angered men. (Photographs by Gordon Coster.)

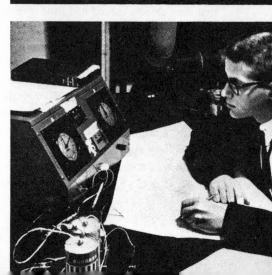

Instead of accepting this thesis, my associates and I predicted that justified film aggression would lead to stronger rather than weaker manifestations of hostility. We believed that the rather low volume of open hostility in the Feshbach experiment was attributable to film-induced inhibitions. If this were so, an angered person who saw what appeared to be warranted aggression might well think he was justified in expressing his own hostile desires.

To test this hypothesis we conducted three experiments. Since they resulted in essentially similar findings and employed comparable procedures, I shall describe only the latest. In this experiment we brought together two male college students at a time. One of them was the subject; the other was a confederate of the experimenter and had been coached on how to act, although of course none of this was known to the subject. Sometimes we introduced the confederate to the subject as a college boxer and at other times we identified him as a speech major. After the introduction the experimenter announced that the purpose of the experiment was to study physiological reactions to various tasks. In keeping with that motif he took blood-pressure readings from each man. Then he set the pair to work on the first task: a simple intelligence test.

During this task the confederate either deliberately insulted the subject—for example, by remarks to the effect that "You're certainly taking a long time with that" and references to "cow-college students" at Wisconsin—or, in the conditions where we were not trying to anger the subject, behaved in a neutral way toward him. On the completion of the task the experimenter took more blood-pressure readings (again only to keep up the pretense that the experiment had a physiological purpose) and then informed the men that their next assignment was to watch a brief motion picture scene. He added that he would give them a synopsis of the plot so that they would have a better understanding of the scene. Actually he was equipped with two different synopses.

To half of the subjects he portrayed the protagonist of the film, who was to receive a serious beating, as an unprincipled scoundrel. Our idea was that the subjects told this story would regard the beating as retribution for the protagonist's misdeeds; some tests we administered in connection with the experiment showed that the subjects indeed had little sympathy for the protagonist. We called the situation we had created with this synopsis of the seven-minute fight scene the "justified fantasy aggression."

The other subjects were given a more favorable description of the protagonist. He had behaved badly, they were told, but this was because he had been victimized when he was young; at any rate, he was now about to turn over a new leaf. Our idea was that the men in this group would feel sympathetic toward the protagonist; again tests indicated that they did. We called this situation the "less justified fantasy aggression."

Then we presented the fiom, which was from the movie *Champion;* the seven-minute section we used showed Kirk Douglas, as the champion, apparently losing

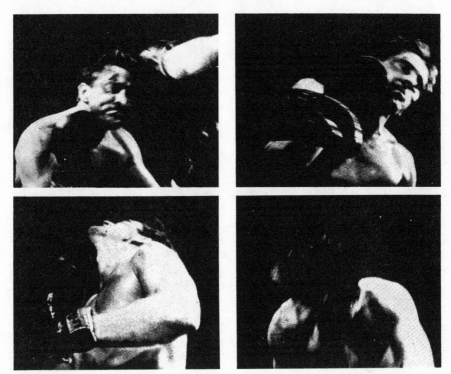

FIGURE 16.2
Filmed aggression shown in author's experiments was from the motion picture *Champion* and included these scenes in which Kirk Douglas receives a bad beating. Watchers had been variously prepared; after showing, they were tested for aggressive tendencies. (Astor Pictures, Inc.)

his title. Thereafter, in order to measure the effects of the film, we provided the subjects with an opportunity to show aggression in circumstances where that would be a socially acceptable response. We separated each subject and accomplice and told the subject that his co-worker (the confederate) was to devise a "creative" floor plan for a dwelling, which the subject would judge. If the subject thought the floor plan was highly creative, he was to give the co-worker one electric shock by depressing a telegraph key. If he thought the floor plan was poor, he was to administer more than one shock; the worse the floor plan, the greater the number of shocks. Actually each subject received the same floor plan.

The results consistently showed a greater volume of aggression directed against the anger-arousing confederate by the men who had seen the "bad guy" take a beating than by the men who had been led to feel sympathy for the protagonist in the film (Fig. 16.3). It was clear that the people who saw the justified movie violence had not discharged their anger through vicarious participation in the aggression but instead had felt freer to attack their tormenter in the next room. The motion picture scene had apparently influenced their judgment of the propriety of

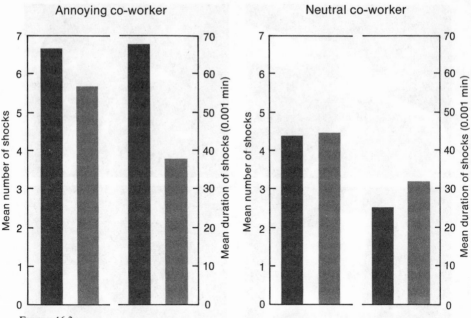

FIGURE 16.3
Responses of subjects invited to commit aggression after seeing prizefighting film varied according to synopsis they heard beforehand. One *(dark gray)* called Douglas' beating deserved; the other *(light gray)* said it was undeserved. After the film the subjects were told they could give electric shocks to an annoying or neutral co-worker based on his "creativeness" in doing a task. Seeing a man receive what had been described as a well-deserved beating apparently lowered restraints against aggressive behavior.

aggression. If it was all right for the movie villain to be injured aggressively, they seemed to think, then perhaps it was all right for them to attack the villain in their own lives—the person who had insulted them.

Another of our experiments similarly demonstrated that observed aggression has little if any effectiveness in reducing aggressive tendencies on the part of an observer. In this experiment some angered men were told by another student how many shocks they should give the person, supposedly in the next room, who had provoked them. Another group of angered men, instead of delivering the shocks themselves, watched the other student do it. Later the members of both groups had an opportunity to deliver the shocks personally. Consistently the men who had watched in the first part of the experiment now displayed stronger aggression than did the people who had been able to administer shocks earlier. Witnessed aggression appeared to have been less satisfying than self-performed aggression.

Our experiments thus cast considerable doubt on the possibility of a cathartic purge of anger through the observation of filmed violence. At the very least, the findings indicated that such a catharsis does not occur as readily as many authorities have thought.

Yet what about the undoubted fact that aggressive motion pictures and violent athletic contests provide relaxation and enjoyment for some people? A person who was tense with anger sometimes comes away from them feeling calmer. It seems to me that what happens here is quite simple: He calms down not because he has discharged his anger vicariously but because he was carried away by the events he witnessed. Not thinking of his troubles, he ceased to stir himself up and his anger dissipated. In addition, the enjoyable motion picture or game could have cast a pleasant glow over his whole outlook, at least temporarily.

The social implications of our experiments have to do primarily with the moral usually taught by films. Supervising agencies in the motion picture and television industries generally insist that films convey the idea that "crime does not pay." If there is any consistent principle used by these agencies to regulate how punishment should be administered to the screen villain, it would seem to be the talion law: an eye for an eye, a tooth for a tooth.

Presumably the audience finds this concept of retaliation emotionally satisfying. Indeed, we based our "justified fantasy aggression" situation on the concept that people seem to approve of hurting a scoundrel who has hurt others. But however satisfying the talion principle may be, screenplays based on it can lead to socially harmful consequences. If the criminal or "bad guy" is punished aggressively, so that others do to him what he has done to them, the violence appears justified. Inherent in the likelihood that the audience will regard it as justified is the danger that some angered person in the audience will attack someone who has frustrated *him,* or perhaps even some innocent person he happens to associate with the source of his anger.

Several experiments have lent support to this hypothesis. O. Ivar Lövaas of the University of Washington found in an experiment with nursery school children that the youngsters who had been exposed to an aggressive cartoon film displayed more aggressive responses with a toy immediately afterward than a control group shown a less aggressive film did. In another study Albert Bandura and his colleagues at Stanford University noted that pre-school children who witnessed the actions of an aggressive adult in a motion picture tended later, after they had been subjected to mild frustrations, to imitate the kind of hostile behavior they had seen.

This tendency of filmed violence to stimulate aggression is not limited to children. Richard H. Walters of the University of Waterloo in Ontario found experimentally that male hospital attendants who had been shown a movie of a knife fight generally administered more severe punishment to another person soon afterward than did other attendants who had seen a more innocuous movie. The men in this experiment were shown one of the two movie scenes and then served for what was supposedly a study of the effects of punishment. They were to give an electric shock to someone else in the room with them each time the person made a mistake on a learning task. The intensity of the electric shocks could be varied. This other person, who was actually the experimenter's confederate, made a constant number of mistakes, but the people who had seen the knife fight gave him more intense punishment than the men who had witnessed the nonaggressive film. The filmed violence had apparently aroused aggressive tendencies in the men and, since the situation allowed the expression of aggression, their tendencies were readily translated into severe aggressive actions.

These experiments, taken together with our findings, suggest a change in approach to the manner in which screenplays make their moral point. Although it may be socially desirable for a villain to recieve his just deserts at the end of a motion picture, it would seem equally desirable that this retribution should not take the form of physical aggression.

The key point to be made about aggressiveness on the basis of experimentation in this area is that a person's hostile tendencies will persist, in spite of any satisfaction he may derive from filmed violence, to the extent that his frustrations and aggressive habits persist. There is no free-floating aggressive energy that can be released through attempts to master other drives, as Freud proposed, or by observing others as they act aggressively.

In fact, there have been studies suggesting that even if the angered person performs the aggression himself, his hostile inclinations are not satisfied unless he believes he has attacked his tormentor and not someone else. J. E. Hokanson of Florida State University has shown that angered subjects permitted to commit aggression against the person who had annoyed them often display a drop in systolic blood pressure. They seem to have experienced a physiological relaxation, as if they had satisfied their aggressive urges. Systolic pressure declines less, however, when the angered people carry out the identical motor activity involved in the aggression but without believing they have attacked the source of their frustration.

I must now qualify some of the observations I have made. Many aggressive motion pictures and television programs have been presented to the public, but the number of aggressive incidents demonstrably attributable to such shows is quite low. One explanation for this is that most social situations, unlike the conditions in the experiments I have described, impose constraints on aggression. People are usually aware of the social norms prohibiting attacks on others, consequently they inhibit whatever hostile inclinations might have been aroused by the violent films they have just seen.

Another important factor is the attributes of the people encountered by a person after he has viewed filmed violence. A man who is emotionally aroused does not necessarily attack just anyone. Rather, his aggression is directed toward specific objectives. In other words, only certain people are capable of drawing aggressive responses from him. In my theoretical analyses of the sources of aggressive behavior I have suggested that the arousal of anger only creates a readiness for aggression. The theory holds that whether or not this predisposition is translated into actual aggression depends on the presence of appropriate cues: stimuli associated with the present or previous instigators of anger. Thus if someone has been insulted, the sight or the thought of others who have provoked him, whether then or earlier, may evoke hostile responses from him.

An experiment I conducted in conjunction with a graduate student provides some

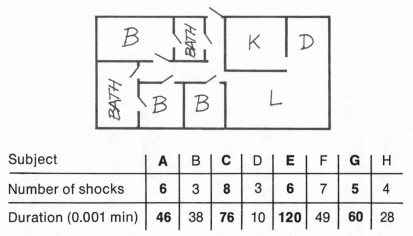

Subject	A	B	C	D	E	F	G	H
Number of shocks	**6**	3	**8**	3	**6**	7	**5**	4
Duration (0.001 min)	**46**	38	**76**	10	**120**	49	**60**	28

FIGURE 16.4
Task by annoying co-worker supposedly was to draw a floor plan. Actually, each subject saw the floor plan shown here. The subject was asked to judge the creativeness of the plan and to record his opinion by pressing a telegraph key that he thought would give electric shocks to the co-worker; one shock for a good job and more for poor work. Responses of eight subjects who saw prizefight film are shown; those in boldface type represent men told that Douglas deserved his beating; those in lightface type, men informed it was undeserved.

support for this train of thought. People who had been deliberately provoked by the experimenter were put to work with two other people, one a person who had angered them earlier and the other a neutral person. The subjects showed the greatest hostility, following their frustration by the experimenter, to the co-worker they disliked. He, by having thwarted them previously, had acquired the stimulus quality that caused him to draw aggression from them after they had been aroused by the experimenter.

My general line of reasoning leads me to some predictions about aggressive behavior. In the absence of any strong inhibitions against aggression, people who have recently been angered and have then seen filmed aggression will be more likely to act aggressively than people who have not had those experiences. Moreover, their strongest attacks will be directed at those who are most directly connected with the provocation or at others who either have close associations with the aggressive motion picture or are disliked for any reason.

One of our experiments showed results consistent with this analysis. In this study male college students, taken separately, were first either angered or not angered by *A*, one of the two graduate students acting as experimenters. *A* had been introduced earlier either as a college boxer or as a speech major. After *A* had had his session with the subject, *B*, the second experimenter, showed the subject a motion picture: either the prizefight scene mentioned earlier or a neutral film. (One that we used was about canal boats in England; the other, about the travels of Marco Polo.)

We hypothesized that the label "college boxer" applied to *A* in some of the

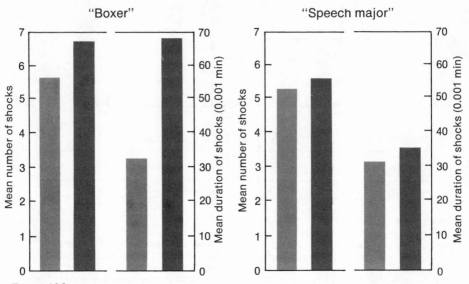

FIGURE 16.5
Co-workers introduction also produced variations in aggressiveness of subjects. Co-worker was introduced as boxer or as speech major; reactions shown here are of men who were angered by co-worker and then saw either a fight film *(dark gray)* or neutral film *(light gray)*. Co-worker received strongest attacks when subjects presumably associated with fight film.

cases would produce a strong association in the subject's mind between *A* and the boxing film. In other words, any aggressive tendencies aroused in the subject would be more likely to be directed at *A* the college boxer than at *A* the speech major. The experiment bore out this hypothesis. Using questionnaires at the end of the session as the measures of hostility, we found that the deliberately angered subjects directed more hostility at *A*, the source of their anger, when they had seen the fight film and he had been identified as a boxer. Angered men who had seen the neutral film showed no particular hostility to *A* the boxer. In short, the insulting experimenter received the strongest verbal attacks when he was also associated with the aggressive film. It is also noteworthy that in this study the boxing film did not influence the amount of hostility shown toward *A* when he had not provoked the subjects.

A somewhat inconsistent note was introduced by our experiments, described previously, in "physiological reactions." Here the nonangered groups, regardless of which film they saw, gave the confederate more and longer shocks when they thought he was a boxer than when they understood him to be a speech major (see Fig. 16.6). To explain this finding I assume that our subjects had a negative attitude toward boxers in general. This attitude may have given the confederate playing the role of boxer the stimulus quality that caused him to draw aggression from the angered subjects. But it could only have been partially responsible, since the

Unangered subjects

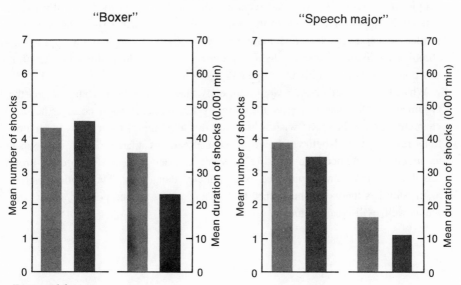

FIGURE 16.6
Similar test, varied by the fact that the co-worker behaved neutrally toward the subjects and therefore presumably did not anger them, produced these reactions. The greater number of shocks given to the co-worker introduced as a boxer than to the one introduced as a speech major apparently reflected a tendency to take a generally negative attitude toward persons identified as boxers.

insulted subjects who saw the neutral film gave fewer shocks to the boxer than did the insulted subjects who saw the prizefight film.

Associations between the screen and the real world are important. People seem to be emotionally affected by a screenplay to the extent that they associate the events of the drama with their own life experiences. Probably adults are less strongly influenced than children because they are aware that the film is make-believe and so can dissociate it from their own lives. Still, it seems clear from the experiments I have described that an aggressive film can induce aggressive actions by anyone in the audience. In most instances I would expect that effect to be short-lived. The emotional reaction produced by filmed violence probably dies away rather rapidly as the viewer enters new situations and encounters new stimuli. Subjected to different influences, he becomes less and less ready to attack other people.

Television and motion pictures, however, may also have some persistent effects. If a young child sees repeatedly that screen heroes gain their ends through aggressive actions, he may conclude that aggression is desirable behavior. Fortunately screenplays do not consistently convey that message, and in any event the child is exposed to many other cultural norms that discourage aggression.

As I see it, the major social danger inherent in filmed violence has to do with the temporary effects produced in a fairly short period immediately following the film. For that period, at least, a person—whether an adult or a child—who had just seen filmed violence might conclude that he was warranted in attacking those people in his own life who had recently frustrated him. Further, the film might activate his aggressive habits so that for the period of which I speak he would be primed to act aggressively. Should he then encounter people with appropriate stimulus qualities, people he dislikes or connects psychologically with the film, this predisposition could lead to open aggression.

What, then, of catharsis? I would not deny that it exists. Nor would I reject the argument that a frustrated person can enjoy fantasy aggression because he sees characters doing things he wishes he could do, although in most cases his inhibitions restrain him. I believe, however, that effective catharsis occurs only when an angered person perceives that his frustrater has been aggressively injured. From this I argue that filmed violence is potentially dangerous. The motion picture aggression has increased the chance that an angry person, and possibly other people as well, will attack someone else.

References

BANDURA, ALBERT, ROSS, DOROTHEA, AND ROSS, SHEILA A. Imitation of film-mediated aggressive models. *Journal of Abnormal and Social Psychology,* Vol. 66, No. 1, pp. 3–11; January, 1963.

BERKOWITZ, LEONARD. *Aggression: a social psychological analysis.* McGraw-Hill Book Company, 1962.

BERKOWITZ, LEONARD, AND RAWLINGS, EDNA. Effects of film violence on inhibitions against subsequent aggression. *Journal of Abnormal and Social Psychology,* Vol. 66, No. 5, pp. 405–412; May, 1963.

SCHRAMM, WILBUR, LYLE, JACK, AND PARKER, EDWIN B. *Television in the lives of our children.* Stanford University Press, 1961.

WALTERS, RICHARD H., THOMAS, EDWARD LLEWELLYN, AND ACKER, C. WILLIAM. Enhancement of punitive behavior by audio-visual displays. *Science,* Vol. 1236, No. 3519, pp. 872–873; June 8, 1962.

17

The Facilitation of Aggression by Aggression: Evidence Against the Catharsis Hypothesis

Russell G. Geen, David Stonner, and Gary L. Shope

Ninety male subjects were either attacked or treated in a more neutral manner by a male confederate. On a subsequent maze-learning task, one third of the subjects shocked the confederate, one third observed as the experimenter shocked the confederate, and one third waited for a period of time during which the confederate was not shocked. Finally, all subjects shocked the confederate as part of a code-learning task. Subjects who had been attacked and had shocked the confederate during the maze task delivered shocks of greater intensity on the code task than did subjects in the other two conditions, and the former subjects also experienced a greater reduction in diastolic blood pressure than did the latter. The results contradict the hypothesis of aggression catharsis and are discussed in terms of feelings of restraint against aggressing that a subject experiences after committing an aggressive act.

The idea of aggression catharsis is popular both in conventional wisdom and in psychological writing. Certainly it seems to make sense that attacking an enemy allows the attacker to purge himself of hostility and anger toward his victim. Furthermore, in both psychoanalytic theory (e.g., Schafer, 1970) and the frustration–aggression hypothesis (Dollard et al., 1939), the notion is expressed that aggression leads to a release of aggressive drive or instigation, so that further acts of violence are rendered less probable. This point of view stipulates a fairly simple cause–effect model: Instigation produces aggressive arousal, which motivates an aggressive response, which in turn reduces arousal. The end result is a declining level of aggression.

Largely because of this cause–effect assumption, however, the catharsis hy-

Reprinted with permission from the authors and *The Journal of Personality and Social Psychology,* Vol. 31, No. 4, 1975. Copyright 1975 by the American Psychological Association. This study was supported by National Science Foundation Grant GS 2748 to the first author.

pothesis has encountered troubles on empirical, methodological, and theoretical levels. Although some investigators have reported results that support the notion of catharsis (e.g., Doob, 1970; Feshbach, 1955; Thibaut and Coules, 1952), many others have made findings that contradict it (e.g., deCharms and Wilkins, 1963; Freeman, 1962; Kahn, 1966; Mallick and McCandless, 1966). Still other studies not directly concerned with catharsis have reported a tendency for subjects to *increase* progressively the intensity of electric shocks administered to an antagonist over a series of trials (Buss, 1966b; Geen, 1968). One reason for the lack of conclusiveness that characterizes data on aggression may be caused by more than just lowered aggressive instigation. If the experimental situation is structured in such a way, for example, that aggressive behavior produces feelings of guilt or anxiety in the subject, intensity of aggression may decline not because of reduced instigation but because of increased restraints (e.g., Baron, 1971; Geen, 1970). Perhaps subtle differences in experimental conditions in some of the studies cited above led to varying degrees of inhibitions against aggressing, thus producing the conflicting results. Finally, the catharsis hypothesis can be questioned on theoretical grounds: Because reduction in aggressive arousal may actually represent the termination of an aversive state, it is not at all obvious why such a reduction should reduce the probability of occurrence of the aggressive response that precedes it. On the contrary, it can be argued that lowered aggressive arousal may actually reinforce aggression (Berkowitz, 1970).

Some of the matters discussed here may have been involved in a study by Doob and Wood (1972), which was reported while the present experiment was in progress. In their experiment, Doob and Wood found apparent evidence for aggression catharsis: Angry subjects who aggressed against an antagonist twice were less punitive in their second attack than similarly angered subjects who aggressed only once. A slight tendency toward reduced aggression was also observed among subjects who, prior to their own aggressing, had witnessed the experimenter attacking their adversary. Although Doob and Wood did not present their findings as clearcut evidence for catharsis and suggested that the results could have been due to a process of restoration of equity, they did not mention as a possible explanation increased restraints against aggressing in the groups showing reduced aggression. Their procedures contained two features that could have led to strong inhibitions against aggressing following aggressive responses, however. First, the confederate who sustained the subject's attacks was always a female, and the subjects were both male and female. Buss (1966a) has shown that both male and female subjects, and especially females, tend to reduce the intensity of attacks on a female victim when they suspect that they have hurt her. Doob and Wood's subjects may have believed that they were causing some pain to the confederate and thus held back in later attacks on her. Second, the confederate aroused the subject to attack through verbal badgering and insult, whereas the subject's mode of attacking the confederate was at all times electric shock. It is not difficult to imagine the subject concluding that the confederate had more than paid for her insults by being shocked one time and thus holding back on hurting her further. The subject may have still harbored considerable resentment toward the antagonist but may have been restrained from aggressing by a fear of guilt over further unwarranted aggression. The experimental situation was therefore structured in such a way that aggression could have produced stimuli that inhibited subsequent aggression by the subject. Finally, the Doob and Wood study did not include any measure of arousal, which could have shed some light on the state of aggressive drive characteristic of the subject before and after aggression.

The present experiment, which was designed and begun before the appearance of the Doob and Wood (1972) article, adventitiously avoided some of the potential sources of inhibition implicit in that study. Subjects were all males, as was the confederate. Buss (1966a) has shown that males do not show marked decreases in aggression against other males over a series of shocks, even when they have evidence that the shocks are causing suffering. In addition, subjects were aroused to aggression by electric shocks and later retaliated with shocks. Finally, systolic and diastolic blood pressure were measured at several points during the session. Otherwise the study was similar to the Doob–Wood experiment. The subject was either attacked or not attacked by a confederate; he then either attacked the confederate himself, observed the confederate being attacked by the experimenter, or waited for a period of time during which the confederate was not attacked. Finally subjects in every condition attacked the confederate.

METHOD

Subjects

Subjects were 90 males recruited from an introductory psychology class, with each student receiving a point toward his course grade for participating. Subjects were randomly assigned to the six treatment conditions.

Procedure

The experimenter described the experiment to the subject, and to a confederate who posed as a subject, as one on the effects of punishment on learning. It was explained that one of the men, to be selected by chance, would perform on two types of learning problem, one involving maze learning and the other deciphering of a code, with threat of punishment as a motive in each case. By means of a rigged lottery, the confederate was designated as the one to work on the learning problem under threat of punishment. The two men were thereupon isolated from each other in separate booths which allowed auditory, but not visual, communication. The subject's booth contained the panel of an aggression machine, with 10 shock buttons. The experimenter then said that before the learning problems could begin, it was necessary to discover the degree of similarity that existed between the "learner" and the other person (the subject), who would at some points in the experiment play the role of a teacher; this was justified by the assertion that similarity between teacher and learner is an important variable in the learning process. The first measurement of blood pressure was made just after these instructions had been given.

The subject was then given a set of 12 cards, each bearing an attitude statement on some controversial issue. The subject was instructed to read the statement loud and state whether he agreed or disagreed with it. The confederate was instructed to express agreement with the subject by saying "I agree" and to register disagreement by giving the subject an electric shock. The rationale for this treatment was that the effects of disagreement between people are most acute when the situation is a highly stressful one, such as threat of shock would produce. The confederate then proceeded to give the subject either (a) two mild shocks following 2 of the statements and no shocks after the other 10 statements (no attack condition) or (b) fairly strong shocks following 8 statements, two mild ones after 2 other statements, and no shocks after the remaining 2 statements, and no shocks after the remaining 2 statements (attack condition). Previous studies (e.g., Geen and Stonner, 1973) have shown that these treatments arouse reliably different levels of anger in subjects. Blood pressure was again measured just after the shocks had been given.

The experimenter then announced that the first task, maze learning, would begin and called the attention of the subject and the confederate to a console in the room visible from both cubicles. Among a panel of light on this console were a red one, a green one, and a white one. The experimenter explained that the onset of the white light signaled the beginning of a trial. Each trial was said to consist of a run through an electronically simulated maze by the learner designate (the confederate), who would express his decision at each of the choice points by pressing one of two switches. If the learner made a correct choice, the green light on the console would go on, but if he made an error the red one would go on. Two thirds of the subjects were told that the red light signaled that the learner was to receive a shock, after which he would go on to the next choice point. The remaining third were told nothing about shocks (and thus comprised the no shocks condition). Of those subjects who were informed that a shock was to be given for each choice-point error, half were told that the experimenter would give the shock (experimenter shocks condition), whereas the other half were instructed to deliver a constant-intensity shock themselves (subject shocks condition). The subject was told to give shocks by pressing the number 5 button on his panel. Ten trials were run, in which the confederate made 20 errors according to a prearranged schedule. A third measure of blood pressure followed.

The experimenter then explained that the second and final learning task would begin. Again, the confederate was given a learning task, this one consisting of translating an alphabetical code into numerals according to an abstract principle, which the learner was required to discover. The subject was instructed to present trigrams from a list by pressing combinations of switches from among five on his panel labeled A through E. The subject was informed that he would learn only whether the learner had correctly encoded the information or whether he had made an error. A correct response was said to govern the onset of a green light on the subject's panel and an error to control a red light on the same panel. The subject was instructed to give the confederate a shock of any intensity from 1 to 10 by pressing the appropriate button on his panel; he was told that the intensity of the shock increased across the series of buttons. The confederate then proceeded to make 10 errors over a series of 16 trials, following a standard schedule. The intensity and duration of each shock supposedly given to the confederate were measured electronically. The fourth and final measurement of blood pressure was made just after the shock series had ended.

After completing the fourth measurement of blood pressure, the experimenter asked the subject and the confederate to fill out a short questionnaire on which they could express their overall impressions of the experiment. The questionnaire consisted of several ratings scales on which were described (a) the subject's mood— how happy, angry, excited, and anxious he felt; (b) how much he liked the confederate; (c) how much he had held back and restrained himself in shocking the confederate on the code-learning task. Each scale consisted of a continuous 100-mm line, which the subject checked. Finally the experimenter interviewed each subject in order to discover whether the deceptions had been successful. A total of 11 subjects were able to at least approximately verbalize the true nature of the study. These subjects were eliminated from analysis and replaced by other subjects randomly drawn from the same population. The true nature of the experiment was fully discussed by the experimenter and the subject at the end of each session.

RESULTS

Effectiveness of Attack

Ratings of anger made by the subject just after being shocked by the confederate were used as the basis for inferring the effectiveness of the attack manipulation. Subjects who had received 10 shocks reported a mean anger rating of 55.1 (on a 100-mm scale on which 0 represented maximum anger and 100 represented minimum anger), whereas subjects who had received 2 shocks reported a mean rating of 78.3. This difference was significant, F (1, 88) = 13.88, $p < .01$. The treatment used to induce different levels of instigation to aggression was therefore judged to be successful.

Shock Intensity

The mean intensities of shocks given by subjects to the confederate during the code-learning task are reported in Table 17.1. An analysis of variance showed main effects for the attack variable, $F(1, 84) = 26.10$, $p < .001$, and the variable regarding treatment of the confederate during the maze-learning task, $F(2, 84) = 5.93$, $p < 01$. The latter of the two main effects was due primarily to an interaction between the two variables, $F(2, 84) = 3.56$, $p < .05$. Comparisons among means further reveal that subjects in the attack–subject shocks condition gave *stronger* shocks than did those in both the attack—no shocks and the attack—experimenter shocks conditions.

TABLE 17.1
Mean intensities of shocks given by subjects on the code-learning task

Treatment of confederate on code-learning task	Treatment of subject by confederate	
	Attack	No attack
Subject shocks	6.65_a	3.92_{bc}
Experimenter shocks	4.13_{bc}	3.62_c
No shocks	5.20_b	3.20_c

Note: Cells having common subscripts are not significantly different at the .05 level by a Duncan multiple-range test.

Blood Pressure

Data regarding systolic and diastolic blood pressure were analyzed separately. Although systolic pressure failed to manifest any differences across conditions, treatment effects on diastolic pressure were found. The data were analyzed as a set of three change scores per subject: (a) from baseline to just after the attack–no attack treatment, (b) from that treatment to just after the first attack on the confederate by the subject or experimenter (or the absence of such an attack), and (c) from that treatment to just after the subject himself attacked. These are labeled, respectively d_{12}, d_{23}, and d_{34} in Table 17.2, which shows the relevant change scores.

No differences were found across the five conditions in baseline diastolic pressure, $F(5, 84) < 1.00$. Analysis of the d_{12} scores indicated that subjects in the attack condition experienced a greater elevation in diastolic pressure (in millimeters of mercury, Hg) than those in the no attack condition (5.92 mm Hg vs. -1.67 mm Hg: $F(1, 84) = 29.12$, $p < .01$. The analysis of the d_{23} scores revealed that subjects in the attack–subject shocks condition experienced a decrease in diastolic pressure to slightly below baseline level, while those in the attack–experimenter shocks condition manifested a slight increase, and those in the attack–no shocks condition revealed a slight decrease. Subjects in the no attack condition showed a moderate decrease in blood pressure. These data produced a significant interaction between the two variables, $F(2, 84) = 6.82$, $p < .01$. Analysis of the d_{34} scores showed main effects for both the attack, $F(1, 84) = 10.33$, $p < .01$, and treatment of confederate, $F(2, 84) = 3.15$, $p < .05$, variables. Both effects are due primarily to the fact that subjects in the attack–experimenter shocks condition experienced a greater drop in diastolic pressure than subjects in any other condition.

Questionnaire Data

Two of the items on the final questionnaire yielded significant differences across conditions.[1] One asked the subject the following question: "On the basis of what you know about the other subject in the experiment, what are your general feelings about him?" The subject replied to this item by marking a 100-mm line anchored at the low end by the statement "I liked him very much" and at the high end by the statement "I didn't like him at all." The results showed that subjects in the attack–subject shocks condition expressed significantly greater dislike for the confederate than did those in the attack–experimenter shocks condition, 45.8

TABLE 17.2

Mean diastolic blood pressure changes for the six treatment conditions

Treatment condition	Mean diastolic pressure change		
	$d_{12}{}^{a}$	$d_{23}{}^{b}$	$d_{34}{}^{c}$
Attack–subject shocks	$+6.66_a$	-8.80_a	-1.20_b
Attack–experimenter shocks	$+5.53_a$	$+1.83_c$	-10.67_a
Attack–no shocks	$+5.56_a$	-1.30_{bc}	-6.57_{ab}
No attack–subject shocks	-3.27_b	-1.87_{bc}	-1.07_b
No attack–experimenter shocks	-1.83_b	-1.80_{bc}	-3.53_b
No attack–no shocks	$+.10_b$	-6.60_{ab}	$-.07_b$

Note: Cells having common subscripts within time periods are not significantly different at the .05 level by a Duncan multiple-range test.
[a] Refers to the change in diastolic pressure during the time period from baseline to just after the attack–no attack treatment.
[b] Refers to the change in diastolic pressure during the time period from the attack–no attack treatment to just after the first attack on the confederate by the subject or experimenter (or the absence of such an attack).
[c] Refers to the change in diastolic pressure during the time period from just after the first attack on the confederate by the subject or experimenter (or the absence of such an attack) to just after the subject himself attacked.

_____

[1] None of the other scales from the final questionnaire yielded significant differences across conditions. These scales are not discussed further.

vs. 32.3; $t(18) = 2.73$, $p < .05$, and slightly more dislike than those in the attack–no shocks condition, 45.8 vs. 36.1; $t(18) = 1.98$, $p < .10$. Thus, in verbally expressed hostility, as in physical aggression, subjects were more intense after aggressing than after not aggressing. In response to the questionnaire item "How much did you hold back in shocking the other subject during the code-learning task," subjects again marked a 100-mm line extending from the statement "I held back a lot" (at the low end) to "I didn't hold back at all" (at the high end). The results indicated that subjects in the attack–subject shocks condition reported feeling less restrained than those in both the attack–experimenter shocks, 70.13 vs. 29.20; $t(18) = 4.93$, $p < .01$, and attack–no shocks, 70.13 vs. 48.21; $t(18) = 3.22$, $p < .01$, conditions. This indicates that the attack–subject shocks condition, relative to the other two attack conditions, lowered inhibitions against aggression in the subjects.

DISCUSSION

The findings of the present experiment provide no support for the hypothesis of aggression catharsis and suggest instead that when an experimental situation is arranged to minimize restraints against aggression, the opposite of catharsis occurs. The study also shows that aggressive behavior was followed by a reduction in blood pressure for all three groups of previously attacked subjects: Men in the subject shocks condition experienced decreased pressure after the maze task in which they aggressed, whereas men in the experimenter shocks and no shocks conditions experienced decreased pressure following the learning task, in which they aggressed for the first time. The finding that reduced arousal in the subjects of the attack–subject shocks condition was in turn followed by a high level of aggression rather than a low level casts doubt on the simple cause–effect assumption mentioned earlier in this article. Activity that we may call "physiological catharsis" does follow aggression, as previous reports have shown (e.g., Hokanson and Burgess, 1962), but this cardiovascular change is associated with relatively strong aggressiveness. The results reported here are consistent with those of Kahn (1966), who showed that verbal aggression was followed by both reduced arterial pressure and a high degree of verbal hostility. Our findings, like Kahn's, suggest that future studies of aggression catharsis should include measures of both behavioral aggression and physiological arousal, since we cannot simply assume that decreased arousal brings about a lowered instigation to aggress.

Two explanations of the findings of the present study seem plausible. One is that aggression, by lowering arousal, causes a person to feel relatively at ease. The person may interpret his relaxed condition as being due to the aggression he has just committed and may therefore experience a weakening of socialized restraints against further violence. Our data support this interpretation by showing that aggression is followed by a lowering of motivation to hold back in shocking.

The other plausible explanation of our data is that subjects in the attack–subject shocks condition, having aggressed once against the confederate on the maze-learning task, felt compelled to behave consistently by continuing to attack him on the subsequent code-learning phase of the experiment. Doob and Wood (1972) have proposed that such consistency-seeking behavior is enhanced by a high degree of similarity between the first occasion for aggression and the second, such as similarity in the means of aggression used by the subject. Thus, the fact that the subject used electric shock to attack the confederate on both occasions in the present study could have facilitated the subject's need to be equally violent in both. The data from this experiment also showed that men in the attack–subject shocks condition tended to express greater verbal hostility on the final questionnaire than did those in the other two attack groups. This, too, may indicate that subjects acted out of a need to maintain consistency. Perhaps committing violence against another person creates a feeling of need to justify the violent act, and one means of justification is derogation of the victim (see Glass, 1964). The data from this experiment do not allow us to test directly whether the subject's need for consistency underlies the failure to find evidence of catharsis, however, and this possibility remains an intriguing subject for future research.

References

BARON, R. A. Aggression as a function of magnitude of victim's pain cues, level of prior anger arousal, and aggressor–victim similarity. *Journal of Personality and Social Psychology,* 1971, **18,** 48–54.

BERKOWITZ, L. Experimental investigations of hostility catharsis. *Journal of Consulting and Clinical Psychology,* 1970, **35,** 1–7.

BUSS, A. H. The effect of harm on subsequent aggression. *Journal of Experimental Research in Personality,* 1966, **1,** 249–255. (a)

BUSS, A. H. Instrumentality of aggression, feedback, and frustration as determinants of physical aggression. *Journal of Personality and Social Psychology,* 1966, **3,** 153–162. (b)

deCHARMS, R., AND WILKINS, E. J. Some effects of verbal expression of hostility. *Journal of Abnormal and Social Psychology,* 1963, **66,** 462–470.

DOLLARD, J., DOOB, L., MILLER, N. E., MOWRER, O. H., AND SEARS, R. R. *Frustration and aggression.* New Haven, Conn.: Yale University Press, 1939.

DOOB, A. N. Catharsis and aggression: The effect of hurting one's enemy. *Journal of Experimental Research in Personality,* 1970, **4,** 291–296.

DOOB, A. N., AND WOOD, L. Catharsis and aggression: The effects of annoyance and retaliation on aggressive behavior. *Journal of Personality and Social Psychology,* 1972, **22,** 156–162.

FESHBACH, S. The drive-reducing function of fantasy behavior. *Journal of Abnormal and Social Psychology,* 1955, **50,** 3–12.

FREEMAN, E. Effects of aggressive expression after frustrations on performance: A test of the catharsis hypothesis. Unpublished doctoral thesis, Stanford University, 1962.

GEEN, R. G. Effects of frustration, attack, and prior training in aggressiveness upon aggressive behavior. *Journal of Personality and Social Psychology,* 1968, **9,** 316–321.

GEEN, R. G. Perceived suffering of the victim as an inhibitor of attack–induced aggression. *Journal of Social Psychology,* 1970, **81,** 209–215.

GEEN, R. G., AND STONNER, D. Context effects in observed violence. *Journal of Personality and Social Psychology,* 1973, **25,** 145–150.

GLASS, D. C. Changes in liking as a means of reducing cognitive discrepancies between self-esteem and aggression. *Journal of Personality,* 1964, **32,** 531–539.

HOKANSON, J. E., AND BURGESS, M. The effects of three types of aggression on vascular process. *Journal of Abnormal and Social Psychology,* 1962, **64,** 446–449.

KAHN, M. The physiology of catharsis. *Journal of Personality and Social Psychology,* 1966, **3,** 278–286.

MALLICK, S. K., AND MCCANDLESS, B. R. A study of catharsis of aggression. *Journal of Personality and Social Psychology,* 1966, **4,** 591–596.

SCHAFER, R. Requirements for a critique of the theory of catharsis. *Journal of Consulting and Clinical Psychology,* 1970, **35,** 13–17.

THIBAUT, J. W., AND COULES, J. The role of communication in the reduction of interpersonal hostility. *Journal of Abnormal and Social Psychology,* 1952, **47,** 770–777.

18

Peacetime Casualties: The Effects of War on the Violent Behavior of Noncombatants

Dane Archer and Rosemary Gartner

Violence by the State is strangely absent from most discussions of the problem of violence. Books about aggression, for example, often treat topics ranging from hormones to homicidal criminals without mentioning capital punishment, the shooting of looters, the beating of protesters, or even that most impressive form of "official" violence: war. This last omission is particularly astonishing, since the unrivaled mortality of wars is both well documented (Singer and Small, 1972) and reasonably contemporary—including the killing of at least 46,000 young American men between 1963 and 1973, in addition to the much great death toll of other nations involved in the Vietnam War.

Why, then is official violence nearly invisible in discussions of murder and aggression? The most obvious explanation is that wars and other forms of official violence are unique in that they wear the mantle of governmental legitimacy. When aircraft bomb a village, when the United States CIA hires assassins to kill foreign leaders, when a policeman shoots a looter, when a prison firing squad kills a convicted murderer, and when National Guardsmen use lethal weapons to break up a protest, the killings that occur are the result of orders (see Marx, 1970). These orders originate in a hierarchical organization; they are issued by appointed or elected officials and are carried out collectively by uniformed deputies who perform the actual killing. In each case the resulting homicides are described as necessary to

Reprinted (with minor changes) with permission from the authors and Jossey-Bass, Inc., Publishers, from I. L. Kutash, S. B. Kutash, and L. B. Schlesinger (eds.), *Violence: Perspectives of Murder and Aggression*, San Francisco: Jossey-Bass, 1978, 219–232.

This project and the development of the 110-Nation Comparative Crime Data File were supported by NIMH Grant MH 27427 from the Center for Studies of Crime and Delinquency and by a Guggenheim fellowship to the first author. Responsibility for the findings and interpretations in this paper belongs, of course, to the authors alone.

accomplish some official objective—to stop the spread of an alien ideology, to stem the destruction of private property, to deter potential murderers of the future, or to control opponents of governmental policy. Official violence may also be justified as retaliatory—a response to "illegal" violence or the alleged threat of illegal violence. Official killings, therefore, differ from illegal violence in that they result from governmental orders, are usually performed by several agents acting collectively, and are justified as instrumental to some higher purpose. These differentiating features are the basis of the legitimation of official violence.

This process of legitimation is clearly successful in the sense that large numbers of citizens appear to regard government violence as acceptable and unproblematic. Evidence of a public mandate for official violence is easily found. In a 1968 survey, for example, 57 percent of a national sample agreed with the statement "Any man who insults a policeman has no complaint if he gets roughed up in return" (Gamson and McEvoy, 1972, p. 336). The public mandate for official violence includes even the extreme act of homicide. In another 1968 survey, by the Gallup organizaton, respondents were asked to react to the fact that the mayor of a large city had ordered police to shoot looters on sight during race riots; 61 percent of the men interviewed said that this was the "best way" to deal with the problem of looting. In 1969 roughly half of the American people (48 percent) also thought that shooting was the best way to handle student protests on campus (Kahn, 1972, p. 48). This tolerance of official violence may explain the persistent public perception of riots and rioters as violent—despite the fact that the number of people killed by authorities during civil disorders has consistently exceeded the number of people killed by rioters by approximately ten to one (Couch, 1968).

Public support for official violence is so pervasive that the definition of violence is itself affected. In a 1969 survey, for example, 30 percent of a national sample said that "police beating students" was *not* an act of violence, and an astonishing 57 percent said that "police shooting looters" was *not* an act of violence (Blumenthal et al., 1972, p. 73). The semantics of the label *violence,* therefore, clearly reflect the perceived legitimacy of the actor and not merely the nature of his or her act. The same 1969 survey asked respondents what violent events were of greatest concern to them. Even though the survey occurred during the Vietnam War, only 4 percent of those interviewed mentioned war.

One consequence of the legitimation of official violence, therefore, is the creation of a reservoir of public support. For at least large numbers of Americans, it is clear that even the extreme act of homicide is not regarded as violence—if the homicide carries the stamp of government authority. Widespread deference to government violence may explain the intensity of public reaction to the concept of war crimes, including the trial of Lieutenant William Calley during the Vietnam War (Kelman and Lawrence, 1972). In view of citizen support for official killing, it is perhaps not surprising that public discussions of violence are generally limited to "criminal" violence.

The invisibility of official violence in scientific discussions of aggressions, however, is somewhat more disturbing. Since social scientists are also citizens, it is possible that they too have been captured by the process of legitimation and that they—like many other citizens—no longer think of governmental homicides as violence. For example, wars are generally discussed in political terms in scholarly circles, and political opponents of a war are the only ones who label wartime killings as "murders." Since social scientists are trained to try to regard behavior in a value-free manner, however, it is curious that there has not been more critical discussion of those acts of homicide that happen to have government auspices. Perhaps we are all, scientists included, socialized to accept the State's monopoly on legitimate violence—and perhaps this socialization influences our curiously selective use of the label *violence.*

This tendency to see official violence as legitimate seems limited, however, to the acts of our own government. As an extreme example, wartime practices in Nazi Germany had official government auspices and seem to have enjoyed legitimacy among most contemporary German citizens—including many scientists and intellectuals—but not among citizens of other nations. Even though socialization leads us to defer to governmental violence, therefore, a kind of moral ethnocentrism may restrict this deference to only the acts of our own government. Official acts of violence may enjoy perceived legitimacy inside a government's borders but not outside them.

There may also be structural explanations for the scientific neglect of official violence, however, and this possibility is particularly worrisome. Since social scientists operate in an institutional context, governments may be exerting a disproportionate influence on research agendas for the study of violence. The structure of scientific research itself may therefore create a lamentable "blind spot" concerning governmental violence. Unhappily, case studies of this scientific blind spot are easy to find. For example, when Governor Ronald Reagan of California announced his controversial proposal to establish a Center for the Study of Violent Behavior at UCLA, Earl Brian, the state's secretary of health and welfare, held a press briefing to describe the mission of the proposed center. When asked by a reporter whether the forms of violence studied by the violence center would include war, Dr. Brian replied, "I hadn't thought about war."

Another fascinating illustration of the way official violence can be defined to preclude scientific consideration is described in Short's (1975) excellent account of the history of the National Commission on the Causes and Prevention of Violence. Since Short served as codirector of research (with Marvin E. Wolfgang) for this important presidential commission, his narrative is as well informed as it is interesting. The national commission initially adopted a neutral definition of violence, one that concentrated on the nature of violent acts themselves: "the threat or use of force that results, or is intended to result, in the injury or forcible restraint or intimidation of persons, or the destruction or forcible seizure of property" (Short, 1975, p. 68). Early in the life of the commission, this neutral definition of violence prompted commission researchers to cast a wide net. The commission's *Progress Report,* for example, outlined the scope of the investigation: "There is no implicit value judgment in this definition. The maintenance of law and order falls within it, for a policeman may find it necessary in the course of duty to threaten or use force, even to injure or kill an individual. Wars are included within this definition, as is some punishment of children. It also includes police brutality, the violence of the Nazis, and the physical abuse of a child" (Commission *Progress Report,* p. 3; cited in Short, 1975, p. 68). Even though official violence, such as wars and police brutality, was clearly included in the commission's initial agenda, this emphasis all but disappeared in the commission's subsequent research. For example, by the time its *Final Report* was issued, the commission was concentrating on "all illegal violence" (Short, 1975, p. 69). The insertion of the adjective *illegal* is of pivotal importance, since the actions of governments are—by definition—seldom perceived as illegal. This change of emphasis in the work of a prestigious national investigation had the effect of shifting attention to the acts of "deviant" individual criminals, members of rioting mobs, and assassins. At the same time, this rather traditional focus on illegal violence excluded the acts of governments from consideration—even though the commission conducted its investigations at the very height of the Vietnam War. In fact, Short indicates that some commission researchers had suggested creating a task force on war to go along with the seven task forces on other topics. This idea was abandoned, according to Short, because of the "potentially explosive nature of such a direct focus on war in general and on the conflict in Indochina in particular" (p. 71). Even though much of the commission's research was excellent and even of permanent value, it is still troublesome that the commission was apparently influenced to concentrate on those acts of violence of greatest concern to government—a concern that did not, of course, include the violent acts of government itself.

For all these reasons, discussions of the problem of violence—in both popular and scientific publications—have centered almost exclusively on illegal or "deviant" violence. This prejudice is clearly reflected in language. The term *murder* is almost always reserved for the acts of individual criminals, and even the more neutral word *homicide* is almost never used to describe killings committed by government officials. During wartime, for example, governments use terms like *casualties, body counts,* or even just *losses* to refer to actions that are really—despite these pastel euphemisms—violent homicides. When officials take lives in peacetime, words like *execution* again camouflage the fact that a killing takes place. Only the opponents of a government's violence refuse to participate in this banal vocabulary. Opponents of a war, for example, accuse government officials of "murderous" policies and refer to soldiers as "killers."

We believe that the scientific neglect of official violence needs to be remedied. It is clear that "deviant" violence has in the past monopolized the attention of researchers interested in violence. We believe that this monopoly is scientifically shortsighted, for two reasons. First of all, acts of official violence may turn out to be particularly pernicious, since they are the only forms of violence that carry the prestige and authority of the State. Second, many important and exciting questions about official violence await answers: What kinds of citizens support government violence most strongly? What kinds of government justifications of violence (such as wars or executions) are most effective in legitimating such violence? Do young children regard government violence, such as wartime killing, as "wrong"? At what developmental age are children socialized to accept government violence? Are people at various stages of "moral development" differentially supportive or critical of official violence? Do violent "criminals" support government violence more strongly than a matched group of noncriminals? How do the deputies who carry out official violence justify their behavior to themselves? What kinds of arguments are most effective with juries in death penalty cases? Are Americans more supportive or tolerant of official violence than citizens of relatively nonviolent societies like England? This research agenda is clearly both rich and relatively unexplored.

CONSEQUENCES OF VIOLENCE BY THE STATE

For the past several years, we have been interested in the consequences of violence by the State. Specifically, we wondered whether the most impressive form of official violence, war, tends to increase the level of violence in a society after the war is over. There are rather compelling theoretical reasons to suspect that wars might produce a legacy of postwar violence. For example, there is now incontrovertible evidence that social learning or "modeling" mediates many forms of aggression and violence (Bandura, 1973). Although most research in this area has used either experiments or causal regression techniques to assess the effects of watching media violence, "modeling" theory also appears to provide the best explanation of the apparently contagious patterning of specific murder methods, airplane hijackings, and terrorism (Bandura, 1973, pp. 101–107). The basic tenet of social learning theory is that acts of media violence and real violence can provide a model or script that increases the likelihood of imitative violence. In addition, research indicates that aggressive models appear to be most influential when they are seen as rewarded for their aggression.

If the violent acts of real or fictional individuals can compel imitation, it seems to us very possible that official violence like war could also provide a script for the postwar acts of individuals. Wars, after all, carry the full authority and prestige of the State, and wars also reward killing in the sense that war "heroes" are decorated and lionized in direct proportion to the number of their wartime homicides. Wars also carry, of course, objectives and rationalizations unique to each war—securing the Crimea, humbling the Boers, stopping fascism, deterring communism, and so on. But what all wars have in common—when they are stripped of their idiosyncratic circumstances—is the unmistakable moral lesson that homicide is an acceptable, or even praiseworthy, means to certain ends. It seems likely that this moral lesson will not be lost on at least some of the citizens in a warring nation. Wars, therefore, contain in particularly potent form all the ingredients necessary to produce imitative or modeled violence: great numbers of violent homicides, official auspices and legitimation, and conspicuous praise and rewards for killing and killers.

Even though there are theoretical reasons to expect wars to provide an imitated model for postwar killing, a rigorous test of this legitimation hypothesis is fraught with complications. At first glance, there are a number of intriguing observations in support of the hypothesis. For example, during the Vietnam War the United States murder and nonnegligent manslaughter rate more than doubled—from 4.5 (per 100,000 persons) in 1963 to 9.3 in 1973 (Archer and Gartner, 1976b). This single case of an astonishing increase cannot be regarded as persuasive evidence of a violent legacy of war, of course, because there were many other social and demographic changes in the United States during this decade. A definitive test of the legitimation hypothesis requires a large number of cases, so that the changes in homicide rates in many warring societies can be assessed. In addition, it is important to disentangle the two very different questions included in the legitimation hypothesis: (1) the empirical question of whether, in general, homicide rates do increase after wars and (2) the more interpretive question of what wartime variables could actually cause such increases.

Although writers and social scientists have long suspected that war leaves a violent legacy, a general test of this suspicion has not been possible in the past. The central obstacle to such a test has been the unavailability of the necessary records on national rates of homicide over time. Without a large archive of annual homicide rates for many nations, it has not been possible to test the legitimation hypothesis. Over a period of three years, we have collected a Comparative Crime Data File (CCDF) containing annual rates of homicide and other offenses for 110 nations and 44 major international cities for the period 1900–1970 (see Archer and Gartner, 1976b; see also Archer et al., 1978; Archer and Gartner, 1977).

The creation of the CCDF makes it possible for the first time to attempt a general answer to the question of whether violence by the State increases subsequent violence by individuals. Our basic research design involves a comparison of homicide rate changes in combatant societies with the changes in a control group of noncombatant nations. This controlled comparison is designed to guard against the possibility that homicide increases might be universal over a given period. Only if wars produce consistent increases in postwar homicide rates, therefore, will the changes in combatant nations differ from those in noncombatant nations during the same years. Since wartime mobilization renders the data uninterpretable for war years themselves, we chose as the two periods of comparison the five years prior to a war and the five years after (Archer and Gartner, 1976b). Records of national participation in wars were obtained from an encyclopedic inventory published by Singer and Small (1972). The rich historical data in the CCDF made it possible to examine homicide rate changes after fourteen wars and, since several nations were involved in some wars, after a total of fifty "nation-wars" (one nation in one war). These changes, and the changes in thirty control cases uninvolved in these wars, are shown in Table 18.1.

TABLE 18.1
Homicide Rate Changes in Combatant and Control Nations After World War I, World War II, the Vietnam War, and Eleven Other Wars

			Homicide Rate Change (%)				
	Decrease	%	Unchanged (<|10%|)	%	Increase	%	
Combatant Nations	Australia (I)	−23	England (I)	−5	Belgium (I)	24	
	Canada (I)	−25	France (I)	4	Bulgaria (I)	22	
	Hungary (I)	−57	S. Africa (I)	−1	Germany (I)	98	
	Finland (II)	−15	Australia (VN)	7	Italy (I)	52	
	N. Ireland (II)	−83	Korea (VN)	6	Japan (I)	12	
	U.S. (II)	−12	Philippines (VN)	9	Portugal (I)	47	
	India (1962 Sino-Ind)	−14	Egypt (1956 Sinai)	−2	Scotland (I)	50	
	Israel (1956 Sinai)	−58	Egypt (1967 6-Day)	−4	U.S. (I)	13	
	Italy (1896 It-Eth)	−15	France (1884 Sino-Fr)	0	Australia (II)	32	
	Italy (1935 It-Eth)	−44	India (1965 2nd Kash)	6	Denmark (II)	169	
			Japan (1904 Russo-Jap)	−9	England (II)	13	
			Japan (1932 Manch)	−8	France (II)	51	
					Italy (II)	133	
					Japan (II)	20	
					Netherlands (II)	13	
					New Zealand (II)	313	
					Norway (II)	65	
					Scotland (II)	11	
					S. Africa (II)	104	
					New Zealand (VN)	50	
					Thailand (VN)	14	
					U.S. (VN)	42	
					Hungary (1956 Russo-H)	13	
					Israel (1967 6-Day)	14	
					Japan (1894 Sino-Jap)	15	
					Jordan (1967 6-Day)	35	
					Pakistan (1965 2nd Kash)	13	
Control Nations	Norway (I)	−37	Ceylon (I)	8	Finland (I)	124[a]	
	Ceylon (II)	−19	Chile (I)	−3	Thailand (I)	112[a]	
	Chile (II)	−67	Netherlands (I)	−2	Colombia (II)	34	
	El Salvador (II)	−20	Thailand (1932 Manch)	7	Sweden (II)	14	
	Ireland (II)	−22	Ceylon (1962 Sino-Ind)	−4	Turkey (II)	12	
	Switzerland (II)	−42			Canada (VN)	11	
	Thailand (II)	−17			England (VN)	23	
	Burma (VN)	−17			Taiwan (VN)	37	
	Indonesia (VN)	−23			Ceylon (1965 2nd Kash)	11	
	Japan (VN)	−23					
	Austria (1956 Russo-H)	−13					
	Burma (1965 2nd Kash)	−13					
	France (1896 Italo-Eth)	−13					
	Switzerland (1935 It-Eth)	−22					
	Turkey (1956 Sinai)	−33					
	Turkey (1967 6-Day)	−19					

[a]Finland and Thailand are both included as control cases for World War I because these two nations were not identified as combatants by Singer and Small (1972). However, the wisdom of classifying these nations as noncombatants during this period is far from clear. Finland underwent something of an "internal" or civil war in 1918, and Thailand actually sent troops to the Allied cause during World War I.

Source: Archer and Gartner (1976b).

Table 18.1 *(continued)*

		Homicide rate change (%)			
Decrease	%	Unchanged (<|10%|)	%	Increase	%
Burma (VN)	− 17			Taiwan (VN)	37
Indonesia (VN)	− 23			Ceylon (1965 2nd Kash)	11
Japan (VN)	− 23				
Austria (1956 Russo-H)	− 13				
Burma (1965 2nd Kash)	− 13				
France (1896 Italo-Eth)	− 13				
Switzerland (1935 It-Eth)	− 22				
Turkey (1956 Sinai)	− 33				
Turkey (1967 6-Day)	− 19				

[a]Finland and Thailand are both included as control cases for World War I because these two nations were not identified as combatants by Singer and Small (1972). However, the wisdom of classifying these nations as noncombatants during this period is far from clear. Finland underwent something of an "internal" or civil war in 1918, and Thailand actually sent troops to the Allied cause during World War I.
Source: Archer and Gartner (1976b).

Our analysis demonstrates that warring nations were more likely to experience homicide rate increases than nations not involved in war. A majority of the combatant nations experienced homicide rate increases of at least 10 percent, while a majority of the uninvolved nations experienced homicide rate *decreases* of greater than 10 percent. Many of the homicide rate increases in warring nations were very large; in several cases the nation's prewar homicide rate more than doubled. The legitimation hypothesis is, therefore, consistent with the results in Table 18.1. Wars do produce a postwar legacy of increased homicide rates.

Even though the difference between combatant and noncombatant nations is striking, this comparison is actually a very conservative test of the legitimation hypothesis. Along with many other changes, wars produce dramatic changes in the age and sex structure of a nation's population. The numbers of young men killed in twentieth-century wars have often been staggering. At the end of World War I, for example, out of every thousand men who were between 20 and 45 at the war's outbreak, 182 died in France, 166 in Austria, 155 in Germany, 101 in Italy, and 88 in Britain (von Hentig, 1947, p. 349). Since young men are universally overrepresented in homicide offense rates (Wolfgang and Ferracuti, 1967), these wartime losses remove from a nation's population precisely those who are statistically most likely to commit homicide in the postwar years. In addition, the postwar baby booms, particularly after large wars, also tend to reduce postwar homicide rates by inflating the population denominator on which they are calculated.

The implications of these conservative artifacts can be summarized easily: If one corrects for the wartime depletion of young men and the surge of postwar babies, warring nations experienced homicide increases even greater than those

shown in Table 18.1. For example, France is shown as having a negligible homicide rate increase of 4 percent after World War I and is classified as an unchanged
case in Table 18.1. If one controls for France's appalling wartime losses, however, the observed postwar homicide rate can only be considered as an impressive
increase. The fact that combatant nations still showed increases much more frequently than the noncombatant nations in Table 18.1, despite the conservative effect of these demographic changes, is therefore particularly impressive.

The nation-wars in Table 18-1 varied considerably in length of wartime involvement and number of men killed. The warring nations in our study, therefore, received different "doses" of war. In terms of the legitimation hypothesis, nations
with the greatest wartime mortality are exposed to a larger scale of legitimate killing than other nations. Using the information provided by Singer and Small (1972),
we were able to classify nation-wars according to the amount of war experienced
by each. The classifications we used was whether a nation had fewer or more than
five hundred battle deaths per million prewar population. This control procedure
produces even more dramatic results and again runs counter to what one would
expect from population changes alone. The nations that were heavily impacted by
wartime killing suffered postwar homicide increases with particular consistency;
79 percent of these nations experienced homicide rate increases, and only 21 percent had decreases. We interpret this result as an internal validation of the overall
finding in Table 18.1, since the nations most likely to show homicide increases
are precisely those nations that experienced the largest "amounts" of war.

The answer to the first of the two questions prompted by the legitimation hypothesis is therefore unambiguous. Warring nations are more likely than nonwar-
ring nations to experience postwar surges in individual acts of homicide. The increases occur despite massive wartime losses of young men and are particularly
common among nations whose wartime involvement is unusually deadly. It is still
a separate matter, of course, to test the second half of the legitimation hypothesis:
What is it about wars that actually produces these increases in postwar violence?
For example, there has been a long history of speculation about whether veterans
of wars are more likely than other citizens to commit acts of violence (Archer and
Gartner, 1976a). This fear appears to surface after each war. For example, attorney Clarence Darrow (1922, p. 218) attributed post-World War I crime increases
to the returned veterans, whom he described as "inoculated with the universal
madness," and Lifton (1973, p. 32) made this more recent prediction about Vietnam veterans: "Some are likely to seek continuing outlets to a pattern of violence
to which they have become habituated, whether by indulging in antisocial or criminal behavior or by offering their services to the highest bidder." The basic idea
behind this speculation is that the military socializes men to be both more accepting of and more proficient at killing and that this experience may increase the likelihood that veterans will use violence even after the war is over.

A number of spectacular case studies appear to lend substance to the image of
the violent veteran. For example, some soldiers who rode with Quantrill's guer-

rillas during the American Civil War became well known for their postwar law-lessness; they were Jesse and Frank James and the Younger brothers. There are also many recent examples. In 1949 a combat veteran named Howard Unruh went on a rampage in Camden, New Jersey, and killed twelve people with a souvenir pistol—Unruh had won marksman and sharpshooter ratings during World War II. During the Vietnam War, a soldier named Dwight Johnson killed about twenty enemy soldiers and was awarded the Congressional Medal of Honor; several months after his return home to Detroit, he was shot and killed while trying to rob a grocery store. Case studies prove very little by themselves, of course, and the more general question is whether the acts of violent veterans can explain the homocide increases we have observed in postwar societies. Since most nations, including the United States, do not maintain records on the military experience of persons arrested for homicide, it is difficult to determine whether veterans are overrepresented in homicide statistics. Even if the number of homicides committed by veterans were known, it is not clear what rate we would compare this with, since veterans and nonveterans differ in many ways in addition to military experience. An indirect approach to this question involves comparisons between two types of veterans, combat and noncombat; and recent research has not found unusually high rates of offenses or violence among combat veterans (Borus, 1975). In the absence of firm evidence, the image of the violent veteran may be more myth than reality. The persistence of the myth is probably due, in part, to the nature of the training that soldiers receive. Civilians may have deep-seated misgivings about teaching soldiers to kill, and they may fear that it is easier to unleash violence than to rein it in again.

We have found evidence that the hypothesis of the violent veteran, despite its popularity, cannot explain the postwar homicide increases observed in warring nations. The key answer to this question is our finding that postwar increases in violence have occurred among groups who could not have been combat veterans. During the Vietnam War, for example, violence increased precipitously for United States men and women; between 1963 and 1973 homicide arrests increased 101 percent for men and 59 percent for women. We have also obtained data from other nations on postwar offense rates for men and women, and it is clear that postwar rates of violence increase for both sexes. As a final bit of evidence, postwar homicide rates also increase among all age groups, and not just among the young veteran age cohort.

In summary, postwar increases in homicide rates are both common and pervasive. They occur after large and small wars, in victorious as well as defeated nations, in nations with improved postwar economies and nations with worsened economies, among veterans and nonveterans, among men and women, and among offenders in several age groups. We believe that these homicide rate increases cannot be explained by artifacts or by other social changes that happen to coincide with wars. Instead, we think that this finding reveals a potential linkage between the violence of governments and the violence of individuals. This linkage is mediated, we believe, by a process of legitimation in which wartime homicide becomes a high-status, rewarded model for subsequent homicides by individuals. Wars provide concrete evidence that homicide, under some conditions, is acceptable in the eyes of a nation's leaders. This wartime reversal of the customary peacetime prohibition against killing may somehow influence the threshold for using homicidal force as a means of settling conflict in everyday life. There is even some independent evidence of the mechanisms by which governmental violence can permeate the private lives of citizens in warring societies; for example, Huggins and Straus (1975) have shown that fictional violence in children's literature reaches a dramatic maximum in war years.

Even though social scientists have in the past amassed impressive experimental evidence that violence can be produced through imitation or modeling, they have in general neglected the possibility that government—with its vast authority and resources—might turn out to be the most potent model of all. This powerful influence of governments on private behavior seems to be what Justice Louis Brandeis had in mind when he wrote in 1928: "Our government is the potent, the omnipresent teacher. For good or ill, it teaches the whole people by its example. Crime is contagious. If the government becomes a lawbreaker, it breeds contempt for the law."

The astonishing neglect of official violence in the social sciences has almost certainly resulted from the curious tendency of both citizens and scientists to avoid labeling the acts of governments—including even the extreme act of homicide—as violence. This deference to the legitimacy of governments has resulted in the near omission of wars and other forms of official homicide from discussions of violence. The new finding that wars cause surges in postwar homicide rates suggests that this omission is lamentable and that the violent acts of individuals may in part be catalyzed by the violent times in which governments cause them to live.

References

ARCHER, D., AND GARTNER, R. The myth of the violent veteran. *Psychology Today*, Dec. 1976, 94–96, 110–111. (a)

ARCHER, D., AND GARTNER, R. Violent acts and violent times: A comparative approach to postwar homicide rates. *American Sociological Review*, 1976, **41**, 937–963. (b)

ARCHER, D., AND GARTNER, R. Homicide in 110 Nations: The development of the comparative crime data file. *International Annals of Criminology*, 1977, **16**, 109–139.

ARCHER, D., GARTNER, R., AKERT, R., AND LOCKWOOD, T. Cities and homicide: A new look at an old paradox. In R. F. Tomasson (ed.), *Comparative studies in sociology*. Vol. 1. Greenwich, Conn.: JAI Press, 1978 (now *Comparative Social Research*).

BANDURA, A. *Aggression: A social learning analysis*. Englewood Cliffs, N.J.: Prentice-Hall, 1973.

BLUMENTHAL, M. D., KAHN, R. L., ANDREWS, F. M., AND HEAD, K. B. *Justifying violence.* Ann Arbor: University of Michigan Press, 1972.

BORUS, J. F. The reentry transition of the Vietnam veteran. *Armed Forces and Society,* 1975, **2**, 97–114.

COUCH, C. Collective behavior: An examination of stereotypes. *Social Problems,* 1968, **15**, 310–321.

DARROW, C. *Crime and its causes.* New York: Crowell, 1922.

GAMSON, W. A., AND McEVOY, J. Police violence and its public support. In J. F. Short and M. E. Wolfgang (eds.), *Collective violence,* Chicago: Aldine, 1972.

HUGGINS, M. D., AND STRAUS, M. A. Violence and the social structure as reflected in children's books from 1850 to 1970. Paper presented at 45th annual meeting of the Eastern Sociological Society, 1975.

KAHN, R. L. Violent man: Who buys bloodshed and why. *Psychology Today,* June 1972. **6** (1), 46–48, 83.

KELMAN, H. C., AND LAWRENCE, L. H. American response to the trial of Lt. William L. Calley. *Psychology Today,* June 1972, **6**(1), 41–45, 78–81.

KOHLBERG, L. The cognitive-developmental approach to socialization. In D. A. Goslin (ed.), *Handbook of socialization theory and research.* Chicago: Rand McNally, 1969.

MARX, G. T. Civil disorder and the agents of social control. *Journal of Social Issues,* 1970, **26**, 19–57.

SHORT, J. F. The National commission on the causes and prevention of violence: Reflections on the contributions of sociology and sociologists. In M. Komarovsky (ed.), *Sociology and public policy: The case of presidential commissions.* New York: Elsevier, 1975.

SINGER, J. D. AND SMALL, M. *The wages of war, 1816–1965: A statistical handbook.* New York: Wiley, 1972.

TURIEL, E., AND ROTHMAN, G. R. The influence of reasoning on behavioral choices at different stages of moral development. *Child Development,* 1972, **43**, 741–756.

VON HENTIG, H. *Crime: Causes and conditions.* New York: McGraw-Hill, 1947.

WOLFGANG, M. E., AND FERRACUTI, F. *The subculture of violence.* London: Tavistock, 1967.

19

Deindividuation and Anger-Mediated Interracial Aggression: Unmasking Regressive Racism

Ronald W. Rogers and Steven Prentice-Dunn

A factorial experiment investigated the effects of deindividuation, anger, and race-of-victim on aggression displayed by groups of whites. Deindividuating situational cues produced an internal state of deindividuation that mediated aggressive behavior. Deindividuation theories were extended by the finding that the internal state of deindividuation was composed not only of the factors Self-Awareness and Altered Experience, but also Group Cohesiveness, Responsibility, and Time Distortion. As predicted, nonangered whites were *less* aggressive toward black than white victims, but angered whites were *more* aggressive toward blacks than whites. Interracial behavior was consistent with new, egalitarian norms if anger was not aroused, but regressed to the old, historical pattern of racial discrimination if anger was aroused. This pattern of interracial behavior was interpreted in terms of a new form of racism: regressive racism.

Mob violence that has occurred since the time of the Roman republic has been attributed typically to short-term economic motives and political issues (cf. Rude, 1964). Economic and political motives, however, were inadequate to explain the torture, mutilation, and burning that frequently occurred in outbursts of interracial violence. Lynch mobs convinced social scientists that "the fundamental need was for a better understanding of the causes underlying the resort to mob violence" (Southern Commission on the Study of Lynching, 1931, p. 5). The major purpose of the present experiment was to examine interracial aggression within a group context, especially a context conducive to deindividuation.

Reprinted with permission from the authors and *Journal of Personality and Social Psychology*, 1981, Vol. 41, No. 1, 63–67. Copyright 1981 by the American Psychological Association, Inc.

The authors gratefully acknowledge Kevin O'Brien, Henry Mixon, George Smith, and Rod Walls for their assistance in collecting the data.

Deindividuation is a process in which antecedent social conditions lessen self-awareness and reduce concern with evaluation by others, thereby weakening restraints against the expression of undesirable behaviors (e.g., Diener, 1977; Zimbardo, 1970). Prentice-Dunn and Rogers (1980) provided the first confirmation of deindividuation theory's major assumption that deindividuating situational cues produce an internal state of deindividuation that mediates the display of aggressive behavior. The deindividuating cues lowered self-awareness and altered cognitive and affective experiences. This deindividuated state weakened restraints against behaving aggressively that are normally maintained by internal and external norms of social propriety. In the present study, therefore, we hypothesized that deindividuating situational cues would produce more aggression than individuating cues, and that an internal state of deindividuation would mediate the effects of deindividuating cues on antisocial behavior.

Many problematic forms of interracial conflict occur in group contexts. The major contribution to our understanding of interracial aggression has come from the Donnersteins' research program (cf. Donnerstein and Donnerstein, 1976), which has focused on situations involving one aggressor and one victim; no published studies have examined interracial agression displayed by a *group* of whites toward a black individual. The present experiment examined interracial aggression in a group setting in which angry aggressors were deindividuated. This social situation approximates many naturalistic situations.

Studies of interracial aggression have consistently shown that the strength of aggression directed toward a different-race victim varies as a function of, for example, potential censure (Donnerstein and Donnerstein, 1973), threatened retaliation (Donnerstein, Donnerstein, Simon, and Ditrichs, 1972), and the victim's expression of suffering (Baron, 1979; Griffin and Rogers, 1977). Donnerstein and Donnerstein (1976) have reported that, in a variety of conditions, white subjects manifest less direct aggression toward black than white victims. Griffin and Rogers (1977) interpreted their white subjects' more lenient treatment of blacks than whites in terms of "reverse discrimination" (cf. Dutton, 1976): To avoid appearing prejudiced, whites treated blacks more favorably (i.e., less aggressively) than they treated whites.

Reverse discrimination is the overt manifestation of white people's viewing themselves as egalitarian and feeling threatened by the prospect of appearing prejudiced. Blacks would not be expected to display reverse discrimination, and studies of blacks' aggression have confirmed they do not (Wilson and Rogers, 1975). Both blacks' and whites' behavior, however, can be traced to the same underlying source: Both races seem to be "reacting against the older, traditional patterns for their races" (Griffin and Rogers, 1977, p. 157).

For whites, the historical pattern of appropriate behavior toward blacks was racial discrimination and inferior treatment. Although whites may have negative attitudes on several specific issues such as blacks' economic gains (Ross, Vanneman, and Pettigrew, 1976) and race riots (Davis and Fine, 1975), survey data indicate that the new norm is an egalitarian view of the races (Brigham and Wrightsman, in press; Campbell, 1971; Taylor, Sheatsley, and Greeley, 1978). This new norm is especially prevalent among college students. Surveys at the university where the present study was conducted confirmed that the current norm among white students is an unprejudiced, egalitarian view of the races (Rosenberg).[1] Theoretically, reverse discrimination is a product of this relatively new egalitarian view of blacks (Dutton, 1976).

For blacks, the historical pattern of appropriate interracial behavior was to inhibit aggression toward whites and to displace it to fellow blacks. The new norms favor more militancy, antiwhite attitudes, and overt hostility toward whites (Caplan, 1970; Wilson and Rogers, 1975). The new norms for blacks and whites represent dramatic departures from the deep-rooted values of the past. Both races have been found to act on these new norms if they are not emotionally aroused by a verbal insult. Thus, blacks are more aggressive towards white than black targets (Wilson and Rogers, 1975), and whites are more aggressive toward white than black targets (Griffin and Rogers, 1977).

But what happens to behavior based on these new norms if the aggressors are insulted? Baron (1979) reported a three-way interaction effect among race-of-victim, insult, and pain cues. An examination of the conditions comparable to those to be studied in the present experiment (i.e., Baron's no-pain-cues condition) indicated that when white subjects were not insulted, black victims received less aggression than white victims (i.e., reverse discrimination); if insulted, the level of aggression expressed toward blacks increased, but did not significantly differ from the level directed toward the white victims.

Since we wish to understand interracial aggression in general and not merely whites' behavior toward blacks, let us also examine blacks' aggression toward whites. To interpret the interracial aggressive behavior of blacks, Wilson and Rogers (1975) suggested that emotional arousal produced a regression to a chronologically earlier mode of responding. The data from that experiment were interpreted as evidence that blacks' behavior could be understood as a product of "the conflict between new militant norms and the residue of oppression" (p. 857). Anger-mediated aggression should not be as firmly under the cognitive control of new norms of in-group solidarity and pride. The black students had been exposed to the traditional values of in-group rejection and out-group preference for many years before the appearance of the Black Power movement. They undoubtedly retained some residual symptoms. Thus, when they became emotionally aroused, the new values, which had not been fully internalized, gave way to the older, more traditional pattern. Similarly, the young white adults in the present study had been exposed during their socialization to the older tradition of belief in black inferiority.

The foregoing considerations converge to suggest an interaction between race-of-victim and insult variables. If whites are not angered, we predicted that they

---

[1]Rosenberg, J. *Racial attitudes of undergraduate students.* Unpublished manuscript, University of Alabama, 1980.

would display reverse discrimination, directing weaker attacks against blacks than against whites. If angered, we hypothesized that whites would regress to the traditional pattern, displaying more aggression toward blacks than toward whites.

One class of deindividuation theories suggests that victim characteristics (e.g., different race) become less salient under conditions of deindividuation. For Festinger, Pepitone, and Newcomb (1952), the defining characteristic of deindividuation is that individuals are not paid attention to as individuals. As elaborated by Zimbardo (1970), deindividuated behavior is not under the controlling influence of usual discriminative stimuli; it is "unresponsive to features of the situation, the target, the victim" (p. 259). Based upon this theoretical position, any differential treatment of different-race victims should vanish when group members become deindividuated. On the other hand, Diener's (1980) theory of less extreme forms of deindividuation postulates that crowd members are *more* responsive to external stimuli as a result of the focus of attention shifting away from the self. It is plausible to infer that any differential treatment of different-race victims should be enhanced by deindividuation. Therefore, the present study was designed to test these two alternative predictions.

Verbal attack, or insult, is a potent and well-established antecedent of aggression in dyadic situations involving one aggressor and one victim (see review by Baron, 1977); however, the effects of anger in a group context have not been investigated. Virtually all studies of deindividuation and aggression have involved unprovoked aggression. Yet, anger adds an important theoretical and applied dimension to our understanding of mob violence. It was hypothesized that insult, or anger arousal, would facilitate the expression of aggression among members of small groups.

One class of deindividuation theories postulates that deindividuated behavior is not influenced by usual discriminative stimuli. It may be derived from these theories that prior insult would have less impact on deindividuated than individuated group members. On the other hand, it may be derived from Diener's (1980) theory that "because self-regulation is minimized or eliminated the deindividuated person is *more* susceptible to the influences of immediate stimuli, emotions [e.g., anger], and motivations" (p. 211, italics added).

There are several limitations to deriving predictions of interaction effects from these two classes of deindividuation theories. First, neither theory explicitly states how the variables of insult or different-race victim would interact with deindividuation. Thus, other interpretations are possible. Second, the form of the interaction effect may vary as a function of the strength of the deindividuation state. The present study is certainly not an experimentum crucis, but perhaps it can shed light on the interaction of deindividuation and anger-mediated interracial aggression.

METHOD

Design and Subjects

A $2 \times 2 \times 2$ factorial design was employed with three between-subjects manipulations: (a) deindividuating cues versus individuating cues, (b) white versus black victim, and (c) no insult versus insult. Ninety-six male introductory psychology students participated in the experiment to earn extra credit. Twelve subjects were randomly assigned to each cell.

Apparatus

The shock apparatuses were modified Buss aggression machines connected to a polygraph. Each of the four aggression machines had 10 pushbutton switches that could be depressed to deliver "shocks" of progressively increasing intensity. Of course, shocks were not actually delivered. A Grason-Stadler noise generator (Model 901A) was used to produce white noise in the deindividuating cues condition.

Procedure

The procedure was highly similar to one we had used previously (Prentice-Dunn and Rogers, 1980). Subjects arrived in groups of five; four were naive participants and one was our assistant. The study was explained as a combination of two experiments. The subjects had signed up for an experiment entitled "Behavior Modification" and were to be tested together. Our assistant, ostensibly another introductory psychology student, had volunteered for a study labeled "Biofeedback." After the experimenter determined who had volunteered for each topic, the biofeedback subject was sent to another room to receive detailed instructions for the biofeedback study.

After hearing explanations of the concepts of behavior modification and biofeedback, the four white behavior modifiers were told that the response of interest in both studies was heart rate. It was indicated that the biofeedback subject would be attempting to maintain his heart rate at a predesignated, high level. Whenever his heart rate fell below the predetermined level, the behavior modifiers would administer an electric shock. The purpose of having groups of four behavior modifiers was explained as an attempt to establish a laboratory analog of a ward at the local state hospital where behavior modifiers actually worked in small groups.

We explained to all subjects that they received their extra credit points for simply showing up and that they could discontinue at any time. Each subject was asked if he had any questions about or any objections to the use of electric shock. All questions were answered and no one declined to participate. In addition, writ-

ten informed consent was obtained. Two mild sample shocks were administered
to the behavior modifiers (i.e., the subjects) via finger electrode. The shocks were
from Switches 4 (.3 mA) and 6 (.45 mA) on the aggression machine and each
lasted for 1 sec. These samples were administered to convince the subjects that
the apparatus really worked and to give them some idea of the shocks they would
be delivering.

The behavior modification subjects were then taken to an adjoining room, seated
at aggression machines with partitions that blocked observation of others' responses (thus, responses were experimentally independent), and given instructions
for operating the shock apparatus. Each time the biofeedback subject's heart rate
fell below the predetermined level, a signal light would be illuminated on their
panels. It was explained that the higher the level chosen and the longer the switch
was depressed, the stronger the shock administered would be. The "shock" received by the biofeedback subject was alleged to be the average of the intensities
and durations selected by the four behavior modifiers.

The final instruction given to the subjects was that any of the 10 shock switches
would be sufficient for the purposes of the experiment. It was explained that the
equipment had been designed with different shock intensities because we had not
known how strong the shocks would have to be to increase heart rate. We explained we had discovered that the different shocks all had equal effects on the
biofeedback subject's heart rate, so the naive subjects could choose any intensity
they wished on each trial. These instructions were designed to eliminate any potential altruistic motivation, and they made clear that use of the lowest possible
intensity on every trial would fulfill the requirements of the experiment. Use of
any intensity greater than "1" would only result in additional pain to the biofeedback subject.

Each group was presented with 20 signal lights over the course of the experiment. The interval between the appearance of any two signals was initially chosen
randomly, ranging from 20–50 sec. The intervals were then held constant across
subsequent trials.

The experimenter then left to bring the biofeedback subject from a waiting room
to the experimental room. The doors were left open, so that subjects heard the
final instructions given by the experimenter to the biofeedback subject about his
role. Thus, the naive subjects would easily hear, but not see, their future victim.

Experimental Manipulations

The first manipulation attempted to differentiate maximally between deindividuating situational cues and individuating ones. In the *deindividuating* cues condition, the experimenter did not address subjects by name. They were informed
that the shocks they used were of no interest to the experimenter and that he would

not know which intensities and durations they selected (anonymity to the experimenter). Subjects were further informed that they would not meet or see the biofeedback subject (anonymity to the victim). The experimenter indicated that he assured full responsibility for the biofeedback subject's well-being (no responsibility for harm-doing). Finally, white noise was played at 65 dB (SPL) in the dimly lit room under the guise of eliminating any extraneous noise from the hall or other experimental rooms (arousal). Prentice-Dunn and Rogers (1980) have shown that such manipulations decrease the subjects' feelings of identifiability and self-awareness.

In the *individuating* cues condition, the subjects wore name tags and were addressed on a first-name basis. As in Zimbardo's (1970) study, the "unique reactions" of each subject were emphasized, and the experimenter expressed his interest in the shock intensities and durations used by the subjects. Subjects were informed that they would meet the biofeedback subject on completing the study. It was emphasized that the biofeedback subject's well-being was the responsibility of each individual behavior modifier. The room was well-lit and no white noise was broadcast.

A second independent variable, race of victim, was manipulated through the use of four experimental assistants, two whites and two blacks. Assistants were assigned to the treatment cells randomly, with the exception that they appeared an equal number of times in each treatment combination. Analyzing this "assistants" factor as an additional variable in the factorial design yielded no main or interaction effects. Thus, the data from the assistants of each race were pooled in the analyses reported below.

The third independent variable was introduced when the behavior modifiers overheard a conversation in an adjoining room between the experimenter and the biofeedback subject. This conversation took place immediately after the naive subjects received their instructions. This insult manipulation was operationalized as a series of questions and answers between the experimenter and the biofeedback subject (i.e., our assistant). Care was taken that the insulting remarks applied to all of the subjects and were devoid of any racial content or connotation. In the *insult* condition, the biofeedback subject, when asked if he objected to the behavior modifiers shocking him, responded that the equipment looked complicated and he wondered if people who appeared as dumb as the behavior modifiers did could follow instructions properly. When the experimenter reiterated the biofeedback subject's option to withdraw from the experiment, the biofeedback subject answered that he hoped the behavior modifiers were not as stupid as they appeared. Finally, when asked by the experimenter if he knew the behavior modifiers, the biofeedback subject said that he didn't know them personally, but that he knew their type; he could tell that they thought they were "hot stuff." In the *no insult* condition, the biofeedback subject simply stated that he had no objections to these particular behavior modifiers shocking him.

Postexperiment Session

Following the last shock trial, subjects completed a questionnaire containing manipulations-check items (10-point Likert rating scales) and 17 items tapping an internal state of deindividuation. A second questionnaire assessed suspicions about the experiment. Five subjects suspected that shocks were not actually delivered, two subjects correctly guessed the race-of-victim hypothesis, and four subjects believed that the insult manipulation was staged. These subjects were deleted from the data analyses. The data in this unequal *n* design were analyzed with the complete least squares model recommended by Overall, Spiegel, and Cohen (1975) because this model meets the criterion of estimating the same parameters as those estimated in an orthogonal design. After each experimental session, each subject was thanked and given a full debriefing that was based on Mills' (1976) recommendations. Finally, a questionnaire was given to each student in a stamped envelope addressed to the Psychology Department's Committee on Ethics. These anonymous responses were returned by 58% of the subjects. 98% of the respondents understood why the deception had been necessary and did not resent it. One subject (2% of the sample) indicated that he did not think the deception was necessary, and another subject stated that his participation had not been voluntary. Neither of these two subjects explained their responses in the space provided on the questionnaire. Fortunately, both of these students, along with every other respondent (100% of the sample), indicated that (a) the experiment should be allowed to continue and (b) that they would be willing to participate in another similar experiment.

RESULTS

Aggression

Deindividuating cues. A multivariate analysis of variance was performed on the sums of the shock intensity scores and the shock duration scores. This analysis yielded a main effect associated with the situational cues manipulation, Wilks's lambda (Λ) = .843, $F(2, 152) = 6.79$, $p < .001$. Univariate analyses of variance disclosed main effects for the situational cues variable on both shock intensity, $F(1, 77) = 12.01$, $p < .001$, and on shock duration, $F(1, 77) = 5.37$, $p < .03$. Compared to subjects in the individuating cues condition, subjects exposed to the deindividuating cues used higher shock intensities ($Ms = 5.3$ and 6.4, respectively) for longer durations ($Ms = 1.6$ & 2.8 sec, respectively). Neither of the two predicted 2-way interaction effects with deindividuating cues were significant ($Fs < 1$).

Insult and race. To determine if the insult manipulation had been successful, an analysis of variance was performed on the sum of the two items assessing

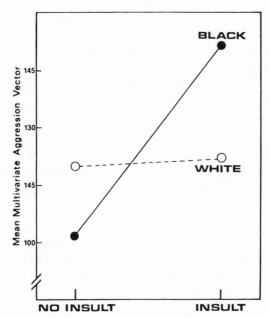

FIGURE 19.1
Aggression as a function of insult and race of victim.

anger toward the victim. The only significant effect was that the insulted groups ($M = 5.1$) expressed much more anger than the noninsulted groups ($M = 1.9$), $F(1, 77) = 29.71$, $p < .0001$. A multivariate analysis of variance indicated that the insulted groups expressed more aggression toward the victim than the noninsulted groups, $\Lambda = .904$, $F(2, 152) = 3.97$, $p < .02$. More importantly, the multivariate analysis of variance revealed a Race × Insult interaction effect, $\Lambda = .912$, $F(2, 152)$ 3.59, $p < 03$. The aggression centroids are shown in Figure 19.1. Univariate analyses of variance yielded identical interaction effects for both the intensity data, $F(1, 77) = 4.63$, $p < .04$, and the duration data, $F(1, 77) = 4.57$, $p < .04$. A Duncan's multiple-range test ($p < .05$) yielded an identical pattern of results for both measures. As may be seen in Figure 19.1, when white subjects were not insulted, they expressed less aggression toward black than white victims; however, if the white subjects were insulted, they expressed more aggression against black than white victims. There were no other significant main or interaction effects on the aggression data.

Internal State of Deindividuation

Prior research (Diener, 1979; Prentice-Dunn and Rogers, 1980) had established that a subjective state of deindividuation is composed of at least two factors, Self-Awareness and Altered Experience. The 17 items in the Prentice-Dunn and Rogers

(1980) study that loaded greater than .4 on only one of these factors were included in the postexperiment questionnaire. Although the present study differed from the former one in that half of the subjects had been emotionally aroused by an insult and in that half aggressed against a different-race victim, an initial principal-axes factor analysis with varimax rotation yielded two factors highly similar to those previously reported. However, five eigenvalues were greater than unity, hence five factors were rotated to a varimax solution. This solution accounted for an additional 26% of the variance. The factor loadings that are greater than or equal to .4 on only one of the factors are shown in Table 19.1. This factor structure, although not predicted, is readily interpretable. The first factor, which is composed of four items from the original Self-Awareness factor, is highly similar to Fenigstein, Scheier, and Buss' (1975) and Buss' (1980) concept of Public Self-Consciousness (coefficient alpha = .72). The items loading on this factor are highly similar to Fenigstein et al.'s items (e.g., "I'm concerned about what other people think of me.") The original Altered Experience factor decomposed into the second, third, and fourth factors. The second factor might still be labeled Altered Experience

TABLE 19.1
Factor loadings on the rotated factors

Factor	Loading
Factor 1 (Public Self-Awareness)	
Concerned with what experimenter thought of me	.78
Concerned with what victim thought of me	.78
Concerned with what other group members thought of me	.71
Felt self-conscious	.50
Factor 2 (Altered Experience)	
Emotions were different from normal	.69
Thinking was somewhat altered	.65
Felt aroused	.59
Factor 3 (Group Cohesiveness)	
Liked other group members	.70
Session was enjoyable	.65
Willing to volunteer for another study with same group	.77
Factor 4 (Time Distortion)	
Thoughts were concentrated on the moment	.69
Time seemed to go quickly	.67
Factor 5 (Responsibility)	
I had responsibility for harm-doing	.83
Responsibility was shared	.84

Note. These five factors accounted for 13%, 14%, 13%, 9%, and 10% of the variance, respectively.

(alpha = .62). The third factor represents another facet of altered experiencing, but refers specifically to Time Distortion (alpha = .35). The fourth factor is composed of items traditionally used to measure cohesiveness, thus, it is labeled Group Cohesiveness (alpha = .61). The final factor might be labeled Responsibility (alpha = .69); it reflected the degree to which all group members, including the subject himself, had responsibility for harmdoing. The remaining two items loaded on more than one factor. The question assessing a feeling of togetherness among group members loaded on both the Group Cohesion and Altered Experience factors. Finally, the item measuring inhibition loaded on both the Public Self-Awareness and Altered Experience factors.

To replicate our previous study, we asked whether the original two factors could discriminate between the individuating and deindividuating cues conditions. The sums of each subject's raw scores on the Self-Awareness items and the Altered Experience items were submitted to a multivariate analysis of variance. A significant difference was obtained between the two situational cues conditions, $\Lambda = .882$, $F(2, 164) = 5.34$, $p < .006$. Subjects in the deindividuating cues condition had higher scores on the Altered Experience factor $M = 65.8$) than subjects in the individuating cues condition ($M = 62.8$), but lower scores on the Self-Awareness factor ($Ms = 33.1$ & 38.5, respectively). A multivariate analysis of variance comparing the two situational cues conditions on the five factors also yielded a significant main effect, $\Lambda = .801$, $F(5, 79) = 3.91$, $p < .004$.

Deindividuation and Aggression

The correlation matrix among the original two factors of the internal state of deindividuation and shock intensity was decomposed into components corresponding to paths. This analysis attempted an exact replication of previous findings that there was a causal path from Altered Experience and Self-Awareness to shock intensity. The coefficients were .29 and $-.20$, respectively, confirming the previous structural model.

There was no strong a priori theoretical model that postulated relationships among our three independent variables, the five components of an internal state of deindividuation, and the dependent measures of shock intensity and duration. Therefore, no path analysis was performed. However, a multiple regression analysis was performed using the five factors to predict aggression (sum of shock intensity and duration). All of the relationships were in the expected direction, but the only statistically significant predictor of aggression was Group Cohesiveness (standardized partial regression coefficient = .25, $p < .05$).

DISCUSSION

Deindividuation

The results of the present experiment replicated and extended previous findings (Prentice-Dunn and Rogers, 1980) that deindividuating situational cues produce an internal state of deindividuation that mediates aggression. We had suggested previously that different configurations of stimulus conditions might produce different factor structures of the deindividuation state. Although the present data are only suggestive, the Self-Awareness and Altered Experience components may be differentiated into more precise aspects of deindividuation. First, the original Self-Awareness factor may reduce to Public Self-Awareness, which focuses on reactions of others to oneself (Buss, 1980; Fenigstein et al., 1975). The self-awareness literature has demonstrated that reductions in self-awareness are associated with increased feelings of deindividuation (Ickes, Layden, and Barnes, 1978) and with increased aggression (Scheier, Feningstein, and Buss, 1974). Diener, Lusk, De-Four, and Flax (1980) pointed out that self-awareness is not a unitary phenomenon, but may have several facets.

Our original Altered Experience factor may be differentiated into three independent factors. One factor contained only items assessing altered thinking and altered feelings, accompanied by increased arousal. Although we shall retain the label Altered Experience for this factor, it is interesting to note its similarity to Fenigstein et al.'s (1975) and Buss' (1980) concept of private self-awareness. Private self-awareness refers to the *process* of being conscious of one's thoughts and feelings, whereas the Altered Experience factor refers to a *product* of that awareness. Another interesting facet of the original Altered Experience factor may be found in the Time Distortion factor, which reflected a concentration on the "here and now" that seemed to pass more rapidly than it actually did. Perhaps the most readily interpretable new factor to emerge was Group Cohesiveness, which was the best predictor of aggressive behavior. Finally, the data suggested that feelings of responsibility may be another component of the subjective state of deindividuation.

It is important to emphasize that these data only suggest the existence of these additional factors for several reasons: (a) the ratio of subjects to items was only 5:1; (b) there were a small number of items per factor (2–4); and (c) partially due to the small number of items, the mean reliability coefficient was a modest .60. Future research that overcomes these problems is obviously required before we can confidently accept these additional components of a deindividuated state. Taken together, the available data indicate that the internal state of deindividuation is composed of at least two factors, Self-Awareness and Altered Experience, which may be differentiated into more refined subcomponents that reliably distinguish individuating from deindividuating situations.

Neither of the two alternative forms of a Deindividuation × Race or Deindividuation × Insult interaction effect were obtained. It was surprising that deindividuation did not interact with the race variable. Donnerstein and Donnerstein (1976) found that race interacted with anonymity, potential retaliation, and censure, all of which were manipulated in the individuation–deindividuation independent variable. These interactions were produced by whites' fear of high levels of retaliation from blacks. In the present study, however, whites reported the same level of expected counteraggression from blacks ($M = 6.3$ on a 10-point scale) and whites ($M = 6.4$).[2] Thus, the state of deindividuation neither made group members hyperresponsive to immediate stimuli nor oblivious to them. A similar finding had been obtained for high- versus low-aggressive models (Prentice-Dunn and Rogers, 1980). The models had the same effect in the individuating and deindividuating conditions. One of the next steps in constructing a theory of deindividuation will be to reexamine and refine general propositions that a state of deindividuation will render a person more (or less) responsive to environmental cues. An important conceptual and empirical task is to identify the specific classes of environmental stimuli to which a state of deindividuation would make a group member more or less responsive.

Interracial Aggression

The Race × Insult interaction effect (see Figure 19.1) indicated that nonangered white subjects aggressed more against same-race victims than against different-race victims; in contrast, angered white subjects aggressed more against different-race victims than same-race victims. The interaction effect was stronger in the present study than in Baron's (1979). A key difference between the two experiments was whether the aggressors were acting alone or in a group. Clearly, the interaction emerges more strongly within a group setting. As Miller and Dollard (1941) noted, "People in a crowd behave as they would otherwise, only more so" (p. 218).

If whites were not insulted, they engaged in reverse discrimination (cf. Dutton, 1976) and behaved consistently with current norms. Similarly, noninsulted blacks act consistently with current norms for their race (Wilson and Rogers, 1975). When emotionally aroused, however, the whites' behavior was consistent with the older, traditional pattern of discrimination against blacks. When emotionally aroused, blacks also act consistently with their traditional pattern (Wilson and Rogers, 1975). This finding is but one specific manifestation of Sargent's (1948) thesis that anger can lead to several overt responses, one of which is regression to more primitive behaviors.

When emotionally aroused, both blacks (Wilson and Rogers, 1975) and whites (present study) regress to chronologically earlier modes of appropriate racial be-

---

[2] The authors wish to thank Ed Donnerstein for suggesting this interpretation.

havior. This pattern of interracial aggression could aptly be labeled *regressive racism*. Although racism may be defined to include beliefs of inferiority of a different race, we prefer to adopt the definition of racism as simply the differential treatment of people based solely on their race (P. Katz, 1976). Regressive racism differs in at least two important ways from the two other major forms of individual racism. First, both the dominative and aversive racists (cf. Kovel, 1970) firmly believe in the inferiority of blacks; they never embraced the new egalitarian norms. Thus, they would not exhibit reverse discrimination. Second, the concepts of dominative and aversive racism have been applied almost exclusively to whites. Regressive racism, on the other hand, may be found in the members of any race. To a large extent, black and white Americans take a Janus-faced view of interracial encounters, one face looking forward but the other face focusing grimly on the past.

Katz and his colleagues (e.g., Katz and Glass, 1979) could interpret the Race × Insult interaction effect as evidence of ambivalence-amplification. Although the data from our program of interracial aggression may be consistent with either the regressive racism or ambivalence-amplification interpretations, there are three major problems with the latter analysis of our data. First, ambivalence-amplification assumes that an individual cannot cope with mixed feelings: that a different-race person's behavior that threatens our positive or negative attitudes *must* be resolved by enhancing behavior that will confirm one aspect of our feelings and suppress the other. There is no direct evidence for this assumption. It is more intuitively compelling to believe that people can act on a mixture of positive and negative feelings toward other people. Second, there is no evidence to support the proposed mediating process, whether it is labeled threat to self-esteem, guilt, distress over inequity, or dissonance. Several of Katz's studies (especially Katz, Glass, Lucido, and Farber, 1979) have tested the ambivalence-amplification hypothesis that ambivalence heightens guilt and that the ensuing behavior reduces guilt. Despite the great difficulties of measuring this mediator, Katz's studies have found no supporting evidence. Furthermore, the present study included an item assessing guilt feelings in the postexperiment questionnaire. The data analyses indicated that the independent variables did not differentially affect guilt, nor did the subjects' behavior reduce guilt. Third, it is very difficult for the ambivalence-amplification hypothesis to account for blacks' aggressive behavior (Wilson and Rogers, 1975). In contrast, the concept of regressive racism can account for the behavior of both whites and blacks.

Additional support for and an extension of the concept of regressive racism may be found in the research from the Donnersteins' program. An interaction effect highly similar to the one displayed in Figure 1 has been reported between the race-of-victim variable and (a) anonymity-to-victim (Donnerstein et al., 1972), (b) threatened retaliation (Donnerstein et al., 1972), and (c) potential censure (Donnerstein and Donnerstein, 1973). Thus, in Figure 19.1, if the two points on the ab-

abcissa labeled "no-insult" and "insult" were to be relabeled (a) "nonanonymous" and "anonymous," (b) "retaliation" and "no-retaliation," or (c) "potential censure" and "no censure," the same interaction would be present.

Crosby, Bromley, and Saxe (1980) interpreted the Donnersteins' findings as revealing that antiblack hostility is actually pervasive, but subtle and covert. The Donnersteins reported that whites were very hostile toward blacks in those conditions in which their negative behavior could not be detected or punished. These latter conditions (anonymity, no retaliation, no censure) are well-established *disinhibitors* of aggression that increase aggressive responding. The present study disclosed that a well-established *instigator* of aggression, anger arousal, produced the same effect as the disinhibitors. Perhaps regressive racism is revealed not only by emotional arousal, but also by a host of variables that disinhibit or instigate aggression. Moreover, regressive racism may be revealed whenever unprejudiced values are not fully internalized (one case of incomplete internalization occurs when the unprejudiced values are relatively new). Regressive racism may be a sufficiently inclusive concept to give coherence to a variety of interracial behavior. Interpreting the Donnersteins' data in terms of regression racism is speculative, and perhaps this extension of the concept goes too far. The concept of regressive racism, however, accounts quite well for anger-mediated interracial aggression.

Both the blacks and whites in our studies and the whites in the Donnersteins' studies have been highly sensitive to the race of their victim. Sadly, neither whites nor blacks accorded the other race egalitarian treatment. Awaiting us, black and white together, is the fire next time; it is smoldering in the regressive racist.

References

BARON, R. A. *Human aggression.* New York: Plenum Press, 1977.

BARON, R. A. Effects of victim's pain cues, victim's race, and level of prior instigation upon physical aggression. *Journal of Applied Social Psychology,* 1979, **9,** 103–114.

BRIGHAM, J., AND WRIGHTSMAN, L. *Contemporary issues in social psychology* (4th ed.). Monterey, Calif.: Brooks/Cole, in press.

BUSS, A. H. *Self-consciousness and social anxiety.* San Francisco: Freeman, 1980.

CAMPBELL, A. *White attitudes toward black people.* Ann Arbor: University of Michigan, Institute for Social Research, 1971.

CAPLAN, N. The new ghetto man: A review of recent empirical studies. *Journal of Social Issues,* 1970, **26,** 59–73.

CROSBY, F., BROMLEY, S. AND SAXE, L. Recent unobtrusive studies of black and white discrimination and prejudice: A literature review. *Psychological Bulletin,* 1980, **87,** 546–563.

DAVIS, E., AND FINE, M. The effects of the findings of the U.S. National Advisory Commission on Civil Disorders: An experimental study of attitude change. *Human Relations,* 1975, **28,** 209–227.

DIENER, E. Deindividuation: Causes and consequences. *Social Behavior and Personality,* 1977, **5**, 143–155.

DIENER, E. Deindividuation, self-awareness, and disinhibition. *Journal of Personality and Social Psychology,* 1979, **37**, 1160–1171.

DIENER, E. Deindividuation: The absence of self-awareness and self-regulation in group members. In P. Paulus (Ed.), *The psychology of group influence.* Hillsdale, N.J.: Erlbaum, 1980.

DIENER, E., LUSK, R., DeFOUR, D., AND FLAX, R. Deindividuation: Effects of group size, density, number of observers, and group member similarity on self-consciousness and disinhibited behavior. *Journal of Personality and Social Psychology,* 1980, **39**, 449–459.

DONNERSTEIN, E., AND DONNERSTEIN, M. Variables in interracial aggression: Potential ingroup censure. *Journal of Personality and Social Psychology,* 1973, **27**, 143–150.

DONNERSTEIN, E., AND DONNERSTEIN, M. Research in the control of interracial aggression. In R. Geen, and E. O'Neal (Eds.), *Perspectives on aggression.* New York: Academic Press, 1976.

DONNERSTEIN, E., DONNERSTEIN, M., SIMON, S., AND DITRICHS, R. Variables in interracial aggression: Anonymity, expected retaliation, and a riot. *Journal of Personality and Social Psychology,* 1972, **22**, 236–245.

DUTTON D. G. Tokenism, reverse discrimination, and egalitarianism in interracial behavior. *Journal of Social Issues,* 1976, **32**, 93–108.

FENIGSTEIN, A., SCHEIER, M. F., AND BUSS. A. H. Public and private self-consciousness: Assessment and theory. *Journal of Consulting and Clinical Psychology,* 1975, **43**, 522–527.

FESTINGER, L., PEPITONE, A., AND NEWCOMB, T. Some consequences of de-individuation in a group. *Journal of Abnormal and Social Psychology,* 1952, **47**, 382–389.

GRIFFIN, B. Q., AND ROGERS, R. W. Reducing interracial aggression: Inhibiting effects of victim's suffering and power to retaliate. *Journal of Psychology,* 1977, **95**, 151–157.

ICKES, W., LAYDEN, M. A., AND BARNES, R. D. Objective self-awareness and individuation: An empirical link. *Journal of Personality,* 1978, **46**, 146–161.

KATZ, I., AND GLASS, D. An ambivalence-amplification theory of behavior toward the stigmatized. In W. G. Austin and S. Worchel (Eds.), *The social psychology of intergroup relations.* Monterey, Calif.: Brooks/Cole, 1979.

KATZ, I., GLASS, D., LUCIDO, D., AND FARBER, J. Harmdoing and victim's racial or orthopedic stigma as determinants of helping behavior. *Journal of Personality,* 1979, **47**, 340–364.

KATZ, P. Racism and social science: Towards a new commitment. In P. Katz (Ed.), *Towards the elimination of racism.* New York: Pergamon Press, 1976.

KOVEL, J. *White racism: A psychological history.* New York: Pantheon, 1970.

MILLER, N., AND DOLLARD, J. *Social learning and imitation.* New Haven, Conn.: Yale University Press, 1941.

MILLS, J. A. A procedure for explaining experiments involving deception. *Personality and Social Psychology Bulletin,* 1976, **2**, 3–13.

OVERALL, J. E., SPIEGEL, D. K., AND COHEN, J. Equivalence of orthogonal and nonorthogonal analysis of variance. *Psychological Bulletin,* 1975, **82**, 182–186.

PRENTICE-DUNN, S., AND ROGERS, R. W. Effects of deindividuating situational cues and aggressive models on subjective deindividuation and aggression. *Journal of Personality and Social Psychology,* 1980, **39**, 104–113.

ROSS, J. M., VANNEMAN, R. D., AND PETTIGREW, T. F. Patterns of support for George Wallace: Implications for racial change. *Journal of Social Issues,* 1976, **32**, 69–91.

RUDE, G. *The crowd in history.* New York: Wiley, 1964.

SARGENT, S. S. Reaction to frustration—A critique and hypothesis. *Psychological Review,* 1948, **55**, 108–114.

SCHEIER, M. F., FENIGSTEIN, A., AND BUSS, A. H. Self-awareness and physical aggression. *Journal of Experimental Social Psychology,* 1974, **10**, 264–273.

Southern Commission on the Study of Lynching. *Lynchings and what they mean.* Atlanta, G.: Author, 1931.

TAYLOR, D. G., SHEATSLEY, P. B., AND GREELEY, A. M. Attitudes toward racial integration. *Scientific American,* 1978, **238**(6), 42–49.

WILSON, L., AND ROGERS, R. W. The fire this time: Effects of race of target, insult, and potential retaliation on black aggression. *Journal of Personality and Social Psychology,* 1975, **32,** 857–864.

ZIMBARDO, P. G. The human choice: Individuation, reason, and order versus deindividuation, impulse, and chaos. In W. J. Arnold and D. Levine (Eds.), *Nebraska Symposium on Motivation* (Vol. 17). Lincoln: University of Nebraska Press, 1970.

20

Naturalistic Studies of Aggressive Behavior: Aggressive Stimuli, Victim Visibility, and Horn Honking

Charles W. Turner, John F. Layton, and Lynn Stanley Simons

Three studies extended laboratory research on aggression to a naturalistic setting which involved horn honking from drivers as a measure of aggression; the studies were adapted from Doob and Gross. The results of a survey (Study 1) of 59 drivers suggested that they were frequently irritated by and aggressive toward other drivers. A second study (using a 3×2 factorial design with 92 male drivers) indicated that manipulation of a rifle in an aggressive context and victim visibility (dehumanization) both significantly influenced horn honking rates subsequent to obstruction at a signal light. A third study with 137 male drivers and 63 female drivers examined the interactive effects of a rifle, an aggressively connotated bumper sticker, and individual subject characteristics (sex and an exploratory index of self-perceived status) on horn honking. The results for three studies in naturalistic settings offer possible extensions of laboratory-based findings on aggression. The role of inhibitions in modifying the pattern of results was also discussed.

There has been considerable recent controversy about the validity of laboratory studies of aggression (Buss, Booker, and Buss, 1972; Feshback and Singer, 1971; U.S. Surgeon General's . . . Committee on Television, 1972). For example, some researchers have suggested that the commonly used measures of aggression may not have external validity. Thus, variables which affect laboratory-based responses may not influence naturally occurring aggressive responses.

One possible limitation of the laboratory setting is that subjects may sharply

Reprinted with permission from the authors and *The Journal of Personality and Social Psychology,* Vol. 31, No. 6, 1975. Copyright 1975 by the American Psychological Association.

Numerous individuals assisted in the completion of the present research and manuscript. We are especially endebted to Irwin Altman, Martin Chemers, Don Hartmann, Alex and Rhonda Klistoff, Ken Murray, Robert Van Wyngarden, and Jack White for their advice and assistance. The research was partly supported by University of Utah Research Committee Funds 1745 to Charles W. Turner.

modify their behavior if they believe that someone is carefully monitoring and evaluating their reactions. According to Turner and Simons (1974), when subjects were led to believe that the experimenter was monitoring their responses to weapons, they were less likely than nonaware subjects to shock their partners. Awareness of the experimenter's purpose apparently caused subjects to inhibit aggressive behavior. In order to reduce the subject's inhibitions, researchers have introduced a variety of deceptions to minimize beliefs that the experiment was designed to evaluate aggressive behavior. However, many subjects today may be too sophisticated for the deceptions commonly used in the laboratory. Thus, some laboratory findings could be artifactual for two reasons. First, subjects may be responding primarily to awareness of deceptions rather than the experimental treatments (Page and Scheidt, 1971; Stricker, Messick, and Jackson, 1969). Second, their primary motivation may often be to portray themselves in a favorable light to the experimenter (Rosenberg, 1969).

Since some laboratory results may be produced by experimental artifacts such as evaluation apprehension, suspicion, negativism, and sophistication, it is important that attempts be made to investigate aggressive behavior in subjects who are not aware that they are being studied. The primary purpose of the present research was to assess whether naturalistic manipulations conceptually similar to laboratory procedures can affect human aggressive responses. Laboratory researchers have attempted to manipulate arousal, aggressive stimulation (Berkowitz, 1974; Geen and O'Neal, 1969), and dehumanization (Milgram, 1965; Zimbardo, 1969). In an exploratory attempt to extend laboratory research to a naturalistic setting, a rifle in an aggressive context (aggressive stimuli) and victim visibility (dehumanization) were manipulated in the present study for obstructed (and possibly aroused) drivers at a signal light.

In order to develop an appropriate naturalistic setting to measure aggressive responses, guidelines were adopted from Webb, Campbell, Schwartz, and Sechrest's (1966) analysis of unobtrusive measures. The following criteria were adopted to reduce subjects' perceptions that they were being studied and to develop adequately sensitive and independent measures of aggression: (a) There should be relatively low inhibitions about the behavior so that the base level responding would be considerably above zero probability, (b) the response should not be likely to produce contagion effects on others' aggression, (c) the subjects's anonymity should be preserved, (d) the experimental setting should be reasonably naturalistic so that the procedures would not be an unusual imposition on the subjects or endanger them in any way, (e) the subjects should remain in the experimental setting for short periods of time such that they would not be exposed to more than one experimental treatment, and (f) the experimental treatments could be randomly assigned to subjects.

STUDY 1

Doob and Gross (1968) have offered one possible procedure for a naturalistic study of aggression. Their findings suggest that horn honking might be an aggressive reaction toward low-status drivers who prevent the flow of traffic at a signal light. Anecdotal evidence suggests that many drivers become aggressive when "frustrated" by the behavior of other drivers. Parry (1968) surveyed English drivers concerning their aggressive reactions while driving. His findings suggest that hostile reactions while driving included facial expressions, verbalizations such as swearing, hand gestures, tailgating, light flashing (high beams), and horn honking. Some drivers also reported actual fist fights or attempts to chase other drivers off the road. Parry's findings suggest that many drivers may become angry and aggressive while driving. In order to determine whether similar hostile reactions occur in Salt Lake City, a survey was conducted based on Parry's questionnaire. Thus, Study 1 was designed to determine whether there was a sufficiently high base rate of anger and aggressive responses from drivers so that experimental treatments might be expected to produce reliable results.

METHOD

Subjects and Procedure

The subjects were sampled from the population of frequent drivers in Salt Lake City. One hundred homes were randomly selected from the city address directory. When an investigator located a residence, the most frequent male or the most frequent female driver (randomly determined) was asked to complete Parry's (1968) driving survey. Acceptable subjects were located in 93 homes. The subjects were given a stamped envelope to return the questionnaire. If they did not respond within 2 weeks, they were again encouraged to complete the questionnaire. The subjects were assured of the complete confidentiality of their responses. Fifty-nine (63%) of the delivered questionnaires were returned.

RESULTS AND DISCUSSION

Twelve questions (of 77) from Parry's (1968) questionnaire were selected for analysis, and the results (see Table 20.1) are reported separately for males and females. The findings suggest that a high proportion of "frequent" drivers sometimes become angry or are irritated by the driving behaviors of other drivers. For example, 77% of males and 56% of females reported "swearing under their breaths" at other drivers, while 50% of male and 15% of female drivers reported "flashing

TABLE 20.1

Percentages of female and male respondents reporting hostile reactions to questionnaire items

Questionnaire item	Male respondents (%)	Female respondents (%)
I am easily provoked when driving.	23	18
I lose my temper when another driver does something silly.	40*	41
I have been known to flash my lights at others in anger.	50	15
I get annoyed if the traffic lights changed to red as I approached them.	23	23
I make rude signs at other motorists when I am provoked.	15	11
At times, I've felt that I could gladly kill another driver.	12	18
If someone suddenly turns without signaling, I get annoyed.	58	92
I swear out loud at other drivers.	23	41
I swear under my breath at other drivers.	77	56
I have given chase to a driver who has annoyed me.	12	4
If the driver behind has his lights shining in my mirror, I pay him back in some way.	23	12
I am usually impatient at traffic lights.	19	7

Note: The samples are based on the responses of 26 men and 27 women.

*One subject did not complete the item.

their lights in anger'' at other drivers. While overt hostile responses were not reported by a majority of drivers on every question (e.g., Questions 1, 2, 10, and 12), there does appear to be evidence that hostile reactions to other drivers are a frequent occurrence. If the verbal reports are accurate reflections of actual driving situations, then a large number of drivers might be frequently irritated by the behavior of other drivers. This anger or irritation could sometimes lead to an overt aggressive response if such a response is readily available (e.g., horn honking following an obstruction at a signal light).

STUDY 2

The present research was primarily designed to extend laboratory-based procedures for investigating possible determinants of aggressive reactions in a naturalistic setting. Hence, the procedure of Doob and Gross (1968) was adapted in order to manipulate exposure to aggressive stimuli and to attempt manipulations of inhibitions by dehumanization of the subject's "victim." The alleged victim (an experimental confederate) would potentially frustrate all subjects by obstructing them at a signal light.

Berkowitz and LePage (1967) manipulated aggressive stimulation by exposing some subjects to a pistol and a shotgun. One possible analogous field manipulation of aggressive stimuli would be to present a rifle in the gun rack of a pickup truck, especially since rifles are often carried that way in Utah. However, a high proportion (perhaps 50%) of Utah males have used rifles frequently in a "sporting" context. It is possible that weapons (a rifle or a pistol) are not always perceived as aggressive stimuli when they have been observed frequently in a nonaggressive context; for example, the rifles may be perceived as sporting equipment somewhat like a fishing pole or skis. In the present experiment, an attempt was made to vary the salience of an aggressive meaning for a rifle by pairing it with an ostensible bumper sticker having an aggressive or a nonaggressive label. This manipulation is somewhat analogous to one employed by Berkowitz and Alioto (1973). They led some subjects to believe that a filmed football game was a "grudge" match, while other subjects were encouraged to think of the game simply as a sporting event. Subjects watching an apparent grudge match were more likely to see an aggressive meaning to the players' actions and were more likely to shock a partner who had previously angered them. Berkowitz and Alioto's (1973) findings suggest that the *context* of stimulus materials may play an important role in determining whether the material is viewed with an aggressive meaning.

In an attempt to vary the subject's inhibitions about being aggressive, the mutual visibility of the victim and subject was varied. According to findings by Milgram (1965), subjects appeared to be more willing to administer shock to a fellow subject when they were both less likely to see and to be seen by the victim. In addition, Zimbardo (1969) has proposed that deindividuation of both the subject and victim (dehumanization) can increase the probability of aggressive behavior.

METHOD

Subjects

Experimental treatments were randomly assigned to 92 male drivers who served as subjects. Nine additional subjects were dropped from the sample, since they were females ($n = 4$) or male drivers of older vehicles ($n = 5$). These subjects were approximately evenly distributed across conditions. The subjects were an arbitrarily selected sample of drivers of late-model vehicles (less than 6 years of age) in a 20×20 block region of a mixed business-residential district of Salt Lake City. Only newer car drivers were employed because high-status victims seem to lead to inhibitions in honking (Doob and Gross, 1968). It is possible that older car drivers would perceive themselves as having low status relative to the victim, which could lead to inhibitions masking the effects of independent variables. The experimental treatments were run on Saturdays from 9 A.M. to 5 P.M. It was assumed that Saturdays would produce a broader based sample of drivers from the potential population of all drivers, since fewer would be working. Moreover, the influence of "rush hour" traffic conditions could be minimized by testing on Saturdays.

Experimental Design

A 3×2 between-subjects factorial design was employed to manipulate aggressive stimulation and victim visibility (dehumanization). The subject was obstructed at a signal light for 12 sec. by an older model (1964) pickup truck with a gun rack in the rear window. The aggressive stimulation variable had three levels: (a) The gun rack was left empty (control), (b) a .303-calibre military rifle was placed in the gun rack and a bumper sticker was attached to the truck in order to reduce the perceived aggressiveness of the rifle (Rifle and "Friend" bumper sticker), or (c) the rifle was paired with a bumper sticker designed to increase the perceived aggressiveness of the rifle (Rifle and "Vengeance" bumper sticker). The bumper stickers were attached to the tailgate of the truck directly in line with the subject's vision, and they could be easily removed and reattached after each trial. The bumper stickers measured approximately 4×15 inches (102×381 mm), and the words (3 inches, or 67 mm, high) were printed with broad lettering (⅜ inch, or 9.5 mm, thick), so that they could be easily read at 50 feet (15m). The words *friend* and *vengeance* were selected from the aggressive or altruistic lists of Parke, Ewall, and Slaby (1972). Ratings by 30 college students indicated that the word *vengeance* was highest (without also being rated high in anxiety) and *friend* was lowest on an aggressive-nonaggressive dimension from the words in Parke et al.'s lists.

Victim visibility (dehumanization) was manipulated by closing a curtain across the rear window of the pickup (without obstructing the view of the gun rack) in half of the conditions (low visibility) and leaving the curtain open in the other conditions (high visibility). The experimental conditions were run in blocks of six such that each condition was completed before any conditions was replicated.

Procedure

The procedure was modeled closely after Doob and Gross (1968). An experimental confederate driving a pickup truck timed his arrival at an intersection at approximately the same time that the light turned red. If a male driver of a late-model apparently privately owned vehicle came to a complete stop behind the confederate before the light changed to green, the driver confederate started the trial (if the conditions were not satisfied, the trial was aborted). When the light turned green, the driver-confederate started a stop-watch, faced straight ahead, and kept his brake lights on to avoid any indication that he might be having trouble with the pickup. At the end of 12 sec., the confederate moved forward with the traffic. Thus, the subjects were obstructed at the light for 12 sec. The first driver in line behind the confederate was always considered to be the subject. An observer was placed in an inconspicuous spot at the intersection so that the subject would be unlikely to see him. The observer rated demographic characteristics of the subject before the trial began (e.g., age and sex of subject; age of car, make of car, number of occupants, and general traffic density). Based on the observer's ratings, nine subjects were dropped from the sample, since they were either females or male drivers of older vehicles. The deleted subjects were inadvertently exposed to treatments when the driver-confederate could not see them clearly in his mirror. The driver's side view mirror was partly obstructed with tape so that his reflection would not be visible to the subjects behind him when the curtain was closed. The tape also prevented the confederate from clearly seeing the subject's vehicle, and he misjudged the sex or vehicle age of some subjects. The observer's judgments were employed to establish the vehicle age or sex of subjects. The observers started a stopwatch when the light turned green and recorded the latency and frequency of honks from the subject. The observers received at least 1 hour of pretraining in the rating procedures.

RESULTS AND DISCUSSION

Subjects' honking responses were dichotomized into those honking (scored as 1) and those not honking (scored as 0). According to Lunney (1970) and D'Agostino (1971), analysis of variance procedures may be applied to dichotomous data when sample sizes are reasonably large (for error $df = 20$) and the sample proportions (means) fall between .20 and .80. Typically, hypotheses about proportions are tested with the binomial distribution, but the binomial is closely approximated by the normal distribution for n greater than 10, especially when the "population proportion" or null hypothesis is approximately .50. Since these conditions were satisfied with the present data ($n = 15.2$; $p = .42$), the analysis of variance procedures were employed.

The rates of honking are reported in Table 20.2. Five independent planned con-

TABLE 20.2
Percentage of horn honking in the experimental conditions of Study 2

| | Aggressive stimulation | | |
| | Rifle present | | |
Victim visibility	Control	Friend bumper sticker	Vengeance bumper sticker	Mean %
Low visibility (curtain closed)	33.3_I ($n = 15$)	46.7_{II} ($n = 15$)	76.5_{III} ($n = 17$)	52.2
High visibility (curtain open)	21.4_{IV} ($n = 14$)	29.4_V ($n = 17$)	42.9_{VI} ($n = 14$)	31.2
Mean %	27.4	38.0	59.7_{III}	

Note: Cell entries refer to the percentages of subjects producing at least one honk. Numbers in parentheses refer to cell sizes. The Roman numerals specify the order of the contrast weights which were applied to means. For example, in Contrast B ($-\frac{1}{2}$, $-\frac{1}{2}$, $+1$, 0, 0, 0), the contrast weight $+1$ was applied to Condition III (rifle present/vengeance bumper sticker/low visibility).

trasts (Kirk, 1968, pp. 73–76; 178–182) were computed on the six cell means using an unweighted-means solution for the unequal sample sizes. The contrast weights for each mean are presented in order according to the cell subscripts (I–VI) which are reported in Table 19.2: Contrast A $= -\frac{1}{3}, -\frac{1}{3}, -\frac{1}{3}, +\frac{1}{3}, +\frac{1}{3}, +\frac{1}{3}$; Contrast B $= -\frac{1}{2}, -\frac{1}{2}, +1, 0, 0, 0$; Contrast C $= -1, +1, 0, 0, 0, 0$; Contrast D $= 0, 0, 0, -\frac{1}{2}, -\frac{1}{2}, +1$; Contrast E $= 0, 0, 0, -1, +1, 0$. The results of the planned contrast analysis indicated that the closed curtain significantly increased the rate of honking compared with the open curtain treatment (Contrast A: $F(1, 86) = 4.43, p < .05$). In addition, the honking rate for the rifle/vengeance condition ($\bar{X} = .765$) was significantly higher than the average ($\bar{X} = .400$) of the other two conditions ([rifle/friend + control] $\times \frac{1}{2}$) when the curtain was closed (Contrast B: $F(1, 86) = 5.98, p < .02$), but the effect was not significant when the curtain was open (Contrast D: $F(1, 86) = 1.37, p > .20$). The two remaining contrasts (C and E) which compared the rifle/friend condition to the control condition were also nonsignificant ($F < 1.0$).

The results of the present study are generally consistent with the reasoning that led to the procedures, since both victim visibility and the rifle/vengeance condition increased horn honking. Thus, the present findings tentatively suggest that dehumanization and the presence of a rifle which is perceived as an aggressive stimulus can increase the probability of aggressive responding in a naturalistic setting. The rifle did not significantly influence the rate of honking when it was in a friendly or "prosocial" context (Contrasts C and E), nor did the rifle/vengeance condition significantly influence honking when the victim was visible (Contrast D). The fact that the rifle/vengeance condition honking rate was significantly higher only when the victim was not visible may be consistent with laboratory procedures used to study aggressive behavior. That is, most researchers in the laboratory typically isolate the victim from the subject in order to lower inhibitions about giving shocks. Similarly, in the present study, reduced visibility of the victim (when the curtain was closed) might have increased the rate by reducing inhibitions.

STUDY 3

One limitation of the procedure in Study 2 was that the rifle and the vengeance bumper sticker were not independently manipulated. Thus, the findings for the rifle/vengeance condition might have been due to either object alone or to the interactive effects of both objects. In Study 3, the rifle and the vengeance bumper sticker were independently manipulated so that their hypothesized interactive effects on horn honking could be tested. The vengeance bumper sticker was designed to increase the likelihood that subjects would perceive an aggressive connotation to the rifle. Previous laboratory findings suggest that uninhibited subjects would be more aggressive when they viewed stimuli with an aggressive meaning (Berkowitz and Alioto, 1973). Since it was possible that inhibitions might mask any effects of the rifle and the vengeance bumper sticker, the dehumanization (curtain) manipulation of Study 2 was employed for all conditions in an attempt to lower inhibitions.

The work of Doob and Gross (1968) suggests that there may be strong individual differences (e.g., status and sex differences) in drivers' reactions to obstruction as a stoplight. Based, in part, on Doob and Gross's findings, male drivers of older vehicles and female drivers were not used as subjects in Study 2 because it was assumed that they would inhibit horn honking. Doob and Gross found that male subjects honked less at higher- than lower-status victims, possibly because high-status victims produced inhibitions about honking. Presumably, the lower the *subject's* self-perceived status, the higher the *victim's* status (relatively) is likely to appear. Thus, self-perceived status relative to the victim may influence willingness to honk as an aggressive response.

In the present investigation, the subjects were divided into two groups based on the age of the vehicle they were driving. This procedure was employed in an attempt to derive an exploratory measure of subjects' self-perceived status. It is possible that the older a person's car, the less likely he would be to perceive himself as higher in status than the confederate in the pickup truck. Since the vehicle age variable could reflect other differences than self-perceived status (e.g., differential likelihoods of being frustrated due to different experiences with stalled automobiles), the variable was included only as an exploratory assessment of possible status differences.[1]

Several researchers (summarized by Bardwick, 1971, chap. 7) have found different patterns of male and female aggressive behavior. Further, Doob and Gross (1968) found different horn honking reactions in men and women, since women had longer latencies for their first honks. Based on these findings, it was assumed that males and females might not respond with horn honking in the same way to the manipulations of the present study if horn honking reactions to obstruction at a signal light reflect aggressive responses. Hence, the subject's sex was recorded to permit separate comparisons of the experimental manipulations for males and females. It is possible that the effects of the rifle or the vengeance bumper sticker would be significant only with male subjects driving new vehicles, since other subjects might inhibit horn honking responses.

---

[1] Status was varied by vehicle age rather than model for several reasons. It was assumed that different models might not be consistently perceived as representing high or low status by all drivers (e.g., sports car vs. luxury sedans). Moreover, it was difficult to determine how subjects would judge older model, expensive vehicles relative to newer model, less expensive vehicles. There might be considerable inconsistency in perceptions of status for model-cost and model age variations.

METHOD

Subjects

Male ($n = 137$) and female ($n = 63$) drivers of apparently privately owned vehicles served as subjects. Subjects were selected by the same procedures used in Study 2 except that no restrictions were placed on the age of drivers' vehicles or sex of subjects. Four additional subjects were dropped from the sample due to recording errors (i.e., not recording age of vehicle or sex of subject).

Experimental Design

Each subject was exposed to one level of a weapon (rifle vs. no rifle) manipulation and one level of the bumper sticker (vengeance bumper sticker vs. no bumper sticker) manipulation in an attempt to independently manipulate perceived aggressiveness of the rifle. The status of the subjects was also classified according to a median split (approximately) on the age of the subject's vehicle (new vehicle: less than or equal to 4 years old; old vehicle: more than 4 years old).

Procedure

The procedure was identical to Study 2 except for the independent manipulation of the bumper sticker and the rifle. In addition, both male and female drivers were exposed to the treatments. Two pickup trucks (1969 models) were used to introduce the experimental conditions. These trucks were 5 to 6 years newer than the truck used in Study 2. Hence, the victim's perceived status in Study 3 might have been higher than that of the victim in Study 2.

Hidden observers started a stopwatch when the light turned green and recorded the latency and frequency of honks. The raters also recorded information about subject's age, sex, and age of vehicle. The raters received at least 1 hour of pretraining in the rating procedures. In order to assess the reliability of observer ratings, two raters were employed for two separate samples of subjects (Sample A = 62 subjects from Study 3; Sample B = 46 subjects from an unpublished study testing other hypotheses about the effects of a rifle). The percentage agreement between the raters for presence or absence of a honk was 100% in Sample A and 96% in Sample B. The reliability (r_{xx}) for rated age of subject's auto was .83 for Sample A and .78 for Sample B; most of the disagreements occurred for older autos. Reliabilities for frequency of honking were .94 in Sample A and .87 in Sample B; reliability for latency of honking was .90 for both Samples A and B. No attempt was made to record or rate duration of honks, since reliabilities of ratings would have been too low, and adequately sensitive, portable tape recorders were not available to record the honks.

RESULTS AND DISCUSSION

As in Study 2, subject's horn honking responses were dichotomized into those honking (scored as 1) and those not honking (scored as 0). According to the reasoning advanced above, it was assumed that the effects of the rifle would be most pronounced when it appeared in an aggressive context (i.e., the bumper sticker). Predictions about the interactive effect of the rifle and the bumper sticker can be tested most directly by a planned comparison procedure. The primary interaction hypothesis was tested by a contrast that compared the rifle/vengeance condition to the average of the other three rifle and vengeance combinations (Contrast A: $-\frac{1}{3}$, $-\frac{1}{3}$, $-\frac{1}{3}$, $+1$). Two additional independent contrasts were selected: The second contrast compared the rifle/no bumper sticker condition to the two no rifle conditions (Contrast B: $-\frac{1}{2}$, $-\frac{1}{2}$, $+1$, 0); the third contrast compared the two no rifle conditions (Contrast C: -1, $+1$, 0, 0). The pooled results for Contrasts B and C represent deviations from the main hypothesis. The mean rates (proportions) of subject honking which are reported separately for new- and old-vehicle male driv-

TABLE 20.3
Percentage of horn honking in the experimental conditions of Study 3

Experimental conditions			
No rifle present		Rifle present	
No bumper sticker (I)	Vengeance bumper sticker (II)	No bumper sticker (III)	Vengeance bumper sticker (IV)
Male drivers of new vehicles			
50.0 ($n = 18$)	33.3 ($n = 12$)	30.4 ($n = 23$)	65.0 ($n = 20$)
Male drivers of old vehicles			
56.2 ($n = 16$)	38.1 ($n = 21$)	46.2 ($n = 13$)	14.3 ($n = 14$)
Female drivers			
80.0 ($n = 15$)	50.0 ($n = 16$)	50.0 ($n = 16$)	50.0 ($n = 16$)

Note: Cell entries refer to the percentages of subjects producing at least one honk. Numbers in parentheses refer to cell sizes. The Roman numerals specify the order of the contrast weights which were applied to means. For example, in Contrst A $(-\frac{1}{3}, -\frac{1}{3}, -\frac{1}{3}, +1)$, the contrast weight $+1$ was applied to Condition IV (rifle present/vengeance bumper sticker).

TABLE 20.4
Summary of the contrast analysis for horn honking rates in Study 3

Source	df	MS	F
Male drivers of new vehicles			
Hypothesis (Contrast A)	1	.948	4.03*
Deviation from hypothesis (Contrast B + C)	2	.193	<1
Male drivers of old vehicles			
Hypothesis	1	1.228	5.23*
Deviation from hypothesis (Contrast B + C)	2	.128	<1
Error for males	129	.235	
Female drivers			
Hypothesis	1	.117	<1
Deviation from hypothesis (Contrast B + C)	2	.471	1.92
Error for females	59	.244	

Note: The harmonic mean for new-vehicle males was $\bar{n} = 17.21$, for old-vehicle males, $\bar{n} = 15.48$, and for females, $\bar{n} = 15.72$.

*$p < .05$.

ers and for female drivers are presented in Table 20.3. The results for the contrast analysis are summarized in Table 20.4.

New-Vehicle Male Drivers

The results for the planned comparisons indicated that the rate of honking in the rifle/vengeance condition with new-vehicle male drivers was significantly higher than the average of the other three rifle/bumper sticker conditions (Contrast A: F $(1, 129) = 4.03, p < .05$). The other three conditions did not differ significantly from each other (Contrast B + C: $F < 1.0$). It should also be noted that there is no single alternative contrast to A for new-vehicle male drivers which could be significant, maximum alternative contrast F $(1, 129) = 1.6, p > .20$. Hence, the results tentatively support the predictions leading to the present procedures. Although the rifle/vengeance condition was significantly different from the average of the other three conditions, a careful inspection of the means reported in Table 18.3 indicates that it was not significantly different from the control condition. One possible explanation for the somewhat weaker results obtained in Study 3 is based on the fact that the confederate-victims drove newer vehicles (3–4 years old) in Study 3 as compared to an 8–9-year-old vehicle in Study 2. Perhaps some subjects inhibited honking to victims in the newer trucks in Study 3 because the victims were perceived to be of relatively high status. Since Doob and Gross (1968) found evidence of inhibitory reactions toward high-status victims, perhaps the weaker findings in the present study were partly accounted for by inhibitory processes.[2]

Old-Vehicle Male Drivers

The three planned contrasts (A, B, C) were employed to assess the effects of treatments for old-vehicle male drivers. The results for Contrast A indicated that the rifle/vengeance condition produced significantly lower rates of honking than the other three conditions, F (1, 129) = 5.23, $p < .05$. The other three conditions did not differ significantly (Contrast B + C: $F < 1.0$). One possible explanation can be offered for these findings: When the old-vehicle drivers were exposed to the rifle in an aggressive context (the bumper sticker), they were more likely to perceive an aggressive meaning to the rifle and hence to their own honking responses. If they perceived their honks as potentially aggressive, they might have inhibited reactions in the presence of a higher-status victim. These interpretations are somewhat similar to those offered by Ellis, Weinir, and Miller (1971), who found that subjects produced lower levels of shock giving in the presence of a rifle and a pistol. Apparently for their subjects, the weapons produced inhibitions about being aggressive. The present findings tentatively suggest that the presence of a rifle in an aggressive context (like the bumper sticker) for some male subjects may produce inhibitions rather than stimulate more aggression.

Female Drivers

The planned contrasts (A, B, C) were also applied to the cell means for female subjects. Neither Contrast A ($F < 1.0$) nor the deviation Contrasts B + C, F (2, 59) = 1.92, $p < .20$, were significant. The differences between conditions also were not significant when female subjects were divided by the age of their vehicles (new and old). However, the results for the females must be interpreted cautiously, since there were fewer subjects in any condition; thus, the condition differences might be significant with sample sizes as large as those obtained for the male subjects. The lower frequency of female subjects resulted from the fact that most drivers were males, at least on Saturdays.

---

[2] In order to assess the robustness of the aggressive stimulation manipulation which was reflected in the comparison of the rifle/vengeance condition to the control (no rifle/no bumper sticker) condition, the results for Study 2 and Study 3 were reanalyzed. A 2 × 2 factorial analysis of variance was computed using Study 2–Study 3 as one factor and the aggressive stimulation manipulation as the second factor. The results indicated that the aggressive stimulation factor was significant, F (1, 56) = 5.78, $p < .02$, while neither the other factor (study replication) nor the interaction was significant ($F < 1.0$). The absence of a significant interaction suggests that the pattern of results was similar in the two studies for the effect of the rifle/vengeance condition versus the control condition.

GENERAL DISCUSSION

The primary reason for employing a naturalistic paradigm in the present research was to explore the possibility that laboratory procedures could be extended to a setting where subjects were unaware that they were being studied. While there are many advantages to naturalistic studies, one disadvantage results from the fact that it is difficult to obtain validity or manipulation checks from subjects to determine their perceptions of the experimental treatments. For example, it was not possible in the present research to assess directly any effects of the vengeance bumper sticker on subjects' perception of the rifle or to measure independently possible differences in inhibitions produced by the victim visibility or vehicle age variables. As a consequence, any inferences about possible mediating principles can be offered only very tentatively, since alternative interpretations can be offered for the present results. Additional research is required before any firm conclusions are warranted about the present manipulations and the dependent measures.

The results of a survey and two naturalistic experiments in the present research tentatively suggest that findings somewhat analogous to laboratory research on aggression can be produced in a naturalistic setting. For example, male subjects in Study 2 were more likely to honk at a victim when he was not visible. Similarly, Milgram (1965) found in a laboratory setting that subjects were more likely to harm a victim who was not visible. Zimbardo's (1969) construct of dehumanization provides one possible explanation for the effects of victim visibility. According to the construct, inhibitions against harming a victim are lowered in the absence of cues which "humanize" a victim. The present curtain manipulations might have "dehumanized" the victim by removing visual cues from him which might have reduced possible inhibitions against horn honking as an aggressive response.

However, there is another interpretation possible for effects of the curtain manipulation. For example, the horn honking subsequent to obstruction at the light might be interpreted better as a "signal" response rather than as aggression. Since the subjects could not see the driver-confederate when the curtain was closed, they might have thought that he was being inattentive at the light. Thus, they might have used their horns to signal that the light had changed. Anecdotal evidence suggests that drivers often honk at others to attract their attention or to warn them about some danger. This alternative interpretation of the horn honking measure cannot be dismissed. Nevertheless, the pattern of findings (including the results of the survey in Study 1) suggests that drivers may become frustrated and angry at other drivers, and this anger or frustration can lead to various hostile reactions such as light flashing, swearing, or hand gestures. Presumably, horn honking might also be perceived as an aggressive response by subjects, especially in the presence of aggressive stimuli.

Male drivers of new vehicles in Study 2 gave more honks when they were exposed to the rifle/vengeance bumper sticker condition but only if they could not see the confederate. In one sense, the findings of Study 3 replicate the results for Study 2, since the honking rate in the rifle/vengeance condition was significantly higher (Contrast A) only for male drivers of new vehicles (the confederate was not visible for any subjects in Study 3).

In the present research, the vengeance bumper sticker condition was added to the rifle manipulation in an attempt to extend laboratory studies of an aggressive context for stimuli (Berkowitz and Alioto, 1973) in a naturalistic setting. The vengeance bumper sticker context was selected in an attempt to increase the salience of the aggressive connotation for the rifle. Otherwise, it was possible that an aggressive meaning of the rifle would not be salient in the gun rack of a pickup truck (e.g., it might be viewed as sporting equipment such as a fishing pole or skis). The present results are somewhat analogous to the results of Berkowitz and Alioto (1973), who found that the context in which a football game was presented played an important role in determining what aggressive reactions followed the game. When the filmed game was characterized as a ''grudge match'' as opposed to a sporting event, subjects were more likely to perceive an aggressive meaning to the football players' actions, and they were more likely to shock a partner who had previously insulted them. Similarly, the vengeance bumper sticker might have modified horn honking responses in the presence of the rifle because its aggressive meaning was more salient.

One important finding in Study 3 was the strong individual differences in subjects' reactions. Although male drivers of new vehicles with nonvisible victims honked more when exposed to both the rifle and the vengeance bumper sticker, the higher rates did not occur for all subjects. When male drivers of new vehicles *could* see their victim, or when male drivers of old vehicles and female drivers were exposed to the rifle/vegeance condition, they did not honk more. One possible explanation for the lower honking rates is that these subjects might have inhibited horn honking responses, especially in the presence of the rifle/vengeance condition. For example, if male drivers of old vehicles perceived themselves to be of lower status than the confederate, they might have inhibited horn honking as an aggressive response due to fears of retaliation from the high-status driver in front of them. The results suggest the possibility that the presence of aggressive stimuli might lead to lower levels of aggression from many individuals due to inhibitions about engaging in aggressive behavior. Hence, there might be important limitations on the generalizability for the effects of aggressive stimuli on horn honking responses and, possibly, on other aggressive and antisocial responses.

As with the victim visibility manipulations, there are several possible explanations for the present aggressive stimulus manipulations. For example, the rifle/bumper sticker combination might have served as a classically conditioned aggressive stimulus which elicited aggression (Berkowitz and LePage, 1967); it might have served as a retrieval cue (Tulving and Thomson, 1973) to remind subjects of previous experiences with aggressive stimuli (e.g., violent portrayals in the mass media), or it might have served as a cue which changed the subject's perceptions of the aggressive meaning of their responses (Berkowitz and Alioto, 1973; Berkowitz and Turner, 1974).

Since there are alternative explanations for the present findings which cannot be dismissed, no firm conclusions can be offered about which principles best explain the results until additional research is completed. Still, the present research provides procedures which might be used to extend laboratory research to naturalistic settings where subjects do not know that they are being studied.

References

BARDWICK, J. M. *Psychology of women: a study of bio-cultural conflicts.* New York: Harper & Row, 1971.

BERKOWITZ, L. Some determinants of impulsive aggression: Role of mediated associations with reinforcements for aggression. *Psychological Review,* 1974, **84,** 165–176.

BERKOWITZ, L., AND ALIOTO, J. T. The meaning of an observed event as a determinant of its aggressive consequences. *Journal of Personality and Social Psychology,* 1973, **28,** 206–217.

BERKOWITZ, L., AND LEPAGE, A. Weapons as aggression-eliciting stimuli. *Journal of Personality and Social Psychology,* 1967, **7,** 202–207.

BERKOWITZ, L., AND TURNER, C. W. Perceived anger level, instigating agent, and aggression. In H. London and R. E. Nisbett (eds.), *Thought and feeling: cognitive alteration of feeling states.* Chicago: Aldine, 1974.

BUSS, A. H., BOOKER, A., AND BUSS, E. Firing a weapon and aggression. *Journal of Personality and Social Psychology,* 1972, **22,** 296–302.

D'AGOSTINO, R. B. A second look at analysis of variance on dichotomous data. *Journal of Educational Measurement,* 1971, **8,** 327–333.

DOOB, A. N., AND GROSS, A. E. Status of frustrator as an inhibitor of horn-honking responses, *Journal of Social Psychology,* 1968, **76,** 213–218.

ELLIS, D. P., WEINIR, P., AND MILLER, L. Does the trigger pull the finger? An experimental test of weapons as aggression eliciting stimuli. *Sociometry,* 1971, **34,** 453–465.

FESHBACK, S., AND SINGER, R. D. *Television and aggression: an experimental field study.* San Francisco: Jossey-Bass, 1971.

GEEN, R. C., AND O'NEAL, E. C. Activation of cue-elicited aggression by general arousal. *Journal of Personality and Social Psychology,* 1969, **11,** 289–292.

KIRK, R. E. *Experimental design: procedures for the behavioral sciences.* Belmont, Calif.: Brooks/Cole, 1968.

LUNNEY, G. H. Using analysis of variance with a dichotomous dependent variable: An empirical study. *Journal of Educational Measurement,* 1970, **7,** 263–269.

MILGRAM, S. Some conditions of obedience and disobedience to authority. *Human Relations,* 1965, **18,** 57–76.

PAGE, M. M., AND SCHEIDT, R. J. The elusive weapons effect: demand awareness, evaluation apprehension, and slightly sophisticated subjects. *Journal of Personality and Social Psychology,* 1971, **20,** 304–318.

PARKE, R. D., EWALL, W., AND SLABY, R. G. Hostile and helpful verbalizations as regulators of nonverbal aggression. *Journal of Personality and Social Psychology,* 1972, **23,** 243–248.

PARRY, M. *Aggression on the road.* London: Tavistock, 1968.

ROSENBERG, M. The conditions and consequences of evaluation. In R. Rosenthal and R. Rosnow (eds.), *Artifacts in behavioral research.* New York: Academic Press, 1969.

STRICKER, L. J., MESSICK, S., AND JACKSON, D. N. Evaluating deception in psychological research. *Psychological Bulletin,* 1969, **77,** 273–295.

TULVING, E., AND THOMSON, D. M. Encoding specificity and retrieval processes in episodic memory. *Psychological Review,* 1973, **80,** 352–373.

TURNER, C. W., AND SIMONS, L. S. Effects of subject sophistication and evaluation apprehension on aggressive responses to weapons. *Journal of Personality and Social Psychology,* 1974, **30,** 341–348.

United States Surgeon General's Scientific Advisory Committee on Television and Social Behavior. *Television and growing up: the impact of television violence.* Washington, D.C.: U.S. Government Printing Office, 1972.

WEBB, E. J., CAMPBELL, D. T., SCHWARTZ, R. D., AND SECHREST, L. *Unobtrusive measures: nonreactive research in the social sciences.* Chicago: Rand McNally, 1966.

ZIMBARDO, P. G. The human choice: individuation, reason, and order versus deindividuation, impulse and chaos. In W. J. Arnold and D. Levine (eds.), *Nebraska Symposium on Motivation* (vol. 17). Lincoln: University of Nebraska Press, 1969.

VI

PREJUDICE AND ATTRIBUTION

21

The Nonverbal Mediation
of Self-Fulfilling Prophecies
in Interracial Interaction

Carl O. Word, Mark P. Zanna, and Joel Cooper

Two experiments were designed to demonstrate the existence of a self-fulfilling prophecy mediated by nonverbal behavior in an interracial interaction. The results of Experiment 1, which employed naive, white job interviewers and trained white and black job applicants, demonstrated that black applicants received (a) less immediacy, (b) higher rates of speech errors, and (c) shorter amounts of interview time. Experiment 2 employed naive, white applicants and trained white interviewers. In this experiment subject-applicants received behaviors that approximated those given either the black or white applicants in Experiment 1. The main results indicated that subjects treated like the blacks of Experiment 1 were judged to perform less adequately and to be more nervous in the interview situation than subjects treated like the whites. The former subjects also reciprocated with less proximate positions and rated the interviewers as being less adequate and friendly. The implications of these findings for black unemployment were discussed.

Sociologist Robert Merton (1957), by suggesting that an originally false definition of a situation can influence the believer to act in such a way as to bring about that situation, is generally credited with focusing attention on the phenomenon of the self-fulfilling prophecy. The present investigation is concerned with such a phenomenon in face-to-face, dyadic interactions. In this context it is hypothesized that one person's attitudes and expectations about the other person may influence the believer's actions, which in turn, may induce the other person to behave in a way that confirms the original false definition. Interpersonally, this phenomenon has been documented in schools, with teachers' expectations influencing students' per-

---

Reprinted with permission of the authors and *Journal of Experimental Social Psychology, 10*. Copyright 1974 by Academic Press, Inc.

This research was supported by N.I.H. Biomedical Research Grants #5 S05 FR07057-04 and #5 S05 RR07057-07.

formances, and in psychology laboratories, with experimenters' expectations influencing subjects' responses (cf. Rosenthal, 1971).

In the present study attention will be directed toward (1) possible nonverbal mediators of this effect, and (2) the reciprocal performances of the interactants. The focus, in addition, will be on the interaction of black and white Americans with a view toward examining the employment outcomes of black job applicants interviewed by whites.

ATTITUDES AND IMMEDIACY

Mehrabian (1968) has recently reported a series of studies linking attitudes toward a target person and the concomitant nonverbal behavior directed toward that person. The results of these studies have consistently found that closer interpersonal distances, more eye contact, more direct shoulder orientation, and more forward lean are a consequence of more positive attitudes toward an addressee. Mehrabian (1969) has considered such nonverbal behaviors in terms of "immediacy" and has defined immediacy "as the extent to which communication behaviors enhance closeness to and nonverbal interaction with another . . . greater immediacy is due to increasing degrees of physical proximity and/or increasing perceptual availability of the communicator to the addressee" (p. 203).

A related series of studies has been conducted by Kleck and his associates (Kleck, 1968; Kleck, Buck, Goller, London, Pfeiffer and Vukcevic, 1968; Kleck, Ono and Hastorf, 1966) pursuing Goffman's (1963) observation that normals tend to avoid stigmatized persons. They have begun to document what might be called a nonverbal stigma effect. For example, normal interactants were found to terminate interviews sooner (Kleck *et al.*, 1966) and to exhibit greater motoric inhibition (Kleck, 1968) with a handicapped person (i.e., leg amputee), and to employ greater interaction distances with an epileptic stranger (Kleck *et al.*, 1968). This set of studies, then, also suggests that those persons who possess a personal characteristic which is discrediting in the eyes of others are treated with less immediate behaviors. In addition to such discrediting characteristics as a physical disability or a criminal record, Goffman (1963) includes blackness in a white society as a stigmatizing trait.

Thus, a body of data suggests that (1) attitudes toward an individual are linked with nonverbal behavior emitted toward that individual, and (2) positive attitudes lead to more immediate nonverbal behaviors. Two questions that now arise are concerned with whether such behaviors are (1) decoded or understood by the target and (2) reciprocated.

DECODING AND RECIPROCATING IMMEDIACY

Recent studies suggest that such evaluative, nonverbal behaviors are both decoded and reciprocated. Mehrabian (1967) found friendliness ratings of an interviewer varied as a function of the physical interaction distance, and the immediacy of head and body positions given subjects. Eye contact has been extensively investigated. Both Kleck and Nuessle (1968) and Jones and Cooper (1971) found that a high degree of eye contact produced higher evaluations of the communicator and produced more positive evaluations on the part of the subjects than did low eye contact.

Since individuals apparently are able to decode affective components of communications from variations in immediacy behavior, it seems reasonable to expect they would reciprocate such variations. This proposition also has received support. Rosenfeld (1967), for example, found that subjects treated to more smiles and positive head nods did reciprocate with more of each.

Thus individuals apparently decode less immediacy as indicating less friendly behavior and reciprocate with less friendly (i.e., less immediate) behavior of their own. Since individuals seldom are able to monitor their own nonverbal behaviors, they are more likely to attribute the reciprocated immediacy, not to their own, original nonverbal behavior, but instead to some disposition inherent in their cointeractant (cf. Jones and Nisbett, 1971). With this nonverbal reciprocation, then, a self-fulfilling prophecy is born.

WHITE-BLACK INTERACTION IN A JOB INTERVIEW SETTING

So far we have been concerned with describing possible mechanisms of interpersonal, self-fulfilling prophecies. The discussion now turns to consider such a process in black-white, dyadic interactions. It has been demonstrated time and again that white Americans have generalized, negative evaluations (e.g., stereotypes) of black Americans. This has been shown most recently in our own subject population by Darley, Lewis, and Glucksberg (1972). Such negative evaluations, of course, represent the kind of attitudes that can initiate an interpersonal, self-fulfilling prophecy. The general hypothesis that the present study sought to investigate, therefore, was that whites interacting with blacks will emit nonverbal behaviors corresponding to negative evaluations and that blacks, in turn, will reciprocate with less immediate behaviors. If the context in which the interaction occurs involves a job interview, with the white interviewing the black, such reciprocated behavior may be interpreted as less adequate performance, thus confirming, in part, the interviewer's original attitude.

These general expectations are operationalized by two subhypotheses: First, black, as compared to white, job applicants will receive less immediate nonverbal communications from white job interviewers; second, recipients of less immediate nonverbal communications, whether black or white, will reciprocate these communications and be judged to perform less adequately in the job interview situation than recipients of more positive nonverbal communications. The first hypothesis was tested in Experiment 1, which employed naive, white job interviewers and trained white and black job applicants; the second in Experiment 2, which used naive, white job applicants and trained white job interviewers who were instructed to emit either immediate or nonimmediate cues.

EXPERIMENT 1

METHOD

Overview

In the context of a study on group decision-making white subjects, as representatives of a team in competition with other teams, interviewed both white and black job applicants. The applicants were trained to respond similarly in both the verbal and nonverbal channels. The interview situation itself was arranged to give the subject-interviewers the opportunity to treat their applicants differentially without the knowledge (1) that their own behavior was being monitored, or (2) that race of the applicants was the experimental variable.

Subjects (Interviewers) and Confederates (Applicants and Team Members)

Subject-interviewers were 15 white, Princeton males recruited to participate in a study of group decision-making conducted by Career Services and the Psychology Department. They were informed that the study would last approximately one hour and a half and that they would be paid $2.00 and possibly $5.00 more. One of the subjects was eliminated when he indicated that he was aware of the purpose of the study before the debriefing period. No other subject volunteered this sort of information after intensive probing, leaving an *n* of 14.

Confederate-applicants were two black and three white high school student volunteers referred by their high school counselor. Each was told that the study was concerned with cognitive functioning and that the experimenter was interested in finding out how subjects made up their minds when forced to choose between nearly identical job applicants. All confederates in both experiments were naive with respect to the hypotheses. Intensive probing following the experiment indicated that they did not become aware. The three confederates who served as the subject's "team members" and the experimenter were male Princeton volunteers.

Procedure

Upon arrival the subjects entered a room containing two confederate team members, who introduced themselves and acted friendly. Another confederate entered and acted friendly, as well. Then the experimenter entered, handed out written instructions and answered any questions.

The instructions informed subjects that the four people in the room constituted a team; that they were to compete with four other teams in planning a marketing campaign; and that they needed to select another member from four high school applicants. In order to increase incentive and concern, an additional $5.00 was promised to the team which performed best in the competition. Using a supposedly random draw, the subject was chosen to interview the applicants. He was then handed a list of 15 questions which was to serve as the interview material, told he had 45 minutes to interview all four high school students and taken to the interview room where the first confederate-applicant was already seated.

In order to measure the physical distance that the interviewer placed himself from the applicant, the experimenter upon entering the interview room, feigned to discover that there was no chair for the interviewer. Subjects were then asked to wheel in a chair from an adjoining room.

Subjects were led to believe that there would be four interviews so that the race variable would be less apparent to them. In addition, to eliminate any special effect that might occur in the first and last interview, an a priori decision was made not to analyze the data from the first "warm-up" interview and not to have a fourth interview. The "warm-up" job candidate was always white. Half the subjects then interviewed a black followed by a white applicant; the other half interviewed a white then a black candidate. After completion of the third interview, subjects were told that the fourth applicant had called to cancel his appointment. After the third interview, subjects were paid and debriefed.

Applicant Performance

Confederate-applicants were trained to act in a standard way to all interviewers. First, they devised answers to the 15 questions such that their answers, though not identical, would represent equally qualifying answers. Confederates then rehearsed these answers until two judges rated their performances to be equal. Confederates were also trained to seat themselves, shoulders parallel to the backs of their chairs ($10°$ from vertical) and to make eye contact with the interviewer 50% of the time. A code was devised to signal confederates during their interviews if they deviated from the pose or began to reciprocate the gestures or head nods given them.

Dependent Measures

Immediacy behaviors. Following Mehrabian (1968, 1969), four indices of psychological immediacy were assessed: (1) Physical Distance between interviewer and interviewee, measured in inches; (2) Forward Lean, scored in 10° units, with zero representing the vertical position and positive scores representing the torso leaning toward the confederate; (3) Eye Contact, defined as the proportion of time the subject looked directly at the confederate's eyes; and (4) Shoulder Orientation, scored in units of 10° with zero representing the subject's shoulders parallel to those of the confederate and positive scores indicating a shift in either direction. Two judges,[1] placed behind one-way mirrors, scored the immediacy behaviors.

More distance and shoulder angle represent less immediate behaviors while more foreward lean and more eye contact represent more immediate behaviors. An index of total immediacy was constructed by summing the four measures, standardized, and weighted according to the regression equation beta weights established by Mehrabian (1969). Final scores of this index represent $(-.6)$ distance $+ (.3)$ forward lean $+ (.3)$ eye contact $+ (-.1)$ shoulder orientation. Positive scores represent more immediate performances.

Related Behaviors. Two related behaviors, which indicate differential evaluations of the applicants (cf. Mehrabian, 1969), were also assessed: (1) Interview length indicates the amount of time from the point the subject entered the interview room until he announced the interview was over, in minutes. This measure was taken by the experimenter. (2) Speech Error Rate, scored by two additional judges from audiotapes, represents the sum of (a) sentence changes, (b) repetitions, (c) stutters, (d) sentence incompletions, and (e) intruding, incoherent sounds divided by the length of the interview and averaged over the two judges. Higher scores represent more speech errors per minute.

RESULTS

Reliabilities and Order Effects

Reliabilities, obtained by correlating the judges' ratings, ranged from .60 to .90 (see Table 21.1). Preliminary analyses also indicated that there were no effects for the order in which confederate-applicants appeared, so that the results are based on data collapsed across this variable.

---

[1] All judges employed in the present research were Princeton undergraduates. Each worked independently and was naive concerning the hypothesis under investigation. Intensive probing indicated that they did not become aware of the hypothesis.

TABLE 21.1
Mean interviewer behavior as a function of race of job applicant; Experiment 1

Behavior	Relia-bility	Blacks	Whites	t^b	p
Total immediacy[a]	—	−.11	.38	2.79	<.02
Distance	.90	62.29 inches	58.43 inches	2.36	<.05
Forward lean	.68	−8.76 degrees	−6.12 degrees	1.09	n.s.
Eye contact	.80	62.71%	61.46%	<1	n.s.
Shoulder orientation	.60	22.46 degrees	23.08 degrees	<1	n.s.
Related behaviors					
Interview length	—	9.42 min.	12.77 min.	3.22	<.01
Speech error rate	.88	3.54 errors/min.	2.37 errors/min.	2.43	<.05

[a]See text for weighing formula, from Mehrabian (1969).
[b]t test for correlated samples was employed.

Immediacy Behaviors

The results, presented in Table 21.1, indicate that, overall, black job candidates received less immediate behaviors than white applicants ($t = 2.79$; $df = 13$; $p < .02$). On the average, blacks received a negative total immediacy score; whites received a positive one. This overall difference is primarily due to the fact that the white interviewers physically placed themselves further from black than white applicants ($t = 2.36$; $df = 13$; $p < .05$). None of the other indices of immediacy showed reliable differences when considered separately.

Related Behaviors

The results for interview length and speech error rate are also presented in Table 21.1. Here it can be seen that blacks also received less immediate behaviors. White interviewers spent 25% less time ($t = 3.22$; $df = 13$; $p < .01$) and had higher rates of speech errors ($t = 2.43$; $df = 13$; $p < .05$) with black as compared to white job candidates.

The results of the first experiment provide support for the hypothesis that black, as compared to white, job applicants receive less immediate nonverbal communications from white job interviewers. Indirectly the results also provide support for the conceptualization of blackness as a stigmatizing trait. The differences in time (evidenced by 12 of 14 interviewers), in total immediacy (evidenced by 10 of 14 interviewers), and in speech error rate (evidenced by 11 of 14 interviewers) argues for an extension of the stigma effect obtained by Kleck and his associates to include black Americans.

EXPERIMENT 2

METHOD

Overview

A second experiment was conducted to ascertain what effect the differences black and white applicants received in Experiment 1 would have on an applicant's job interview performance. In the context of training job interviewers, subject-applicants were interviewed by confederate-interviewers under one of two conditions. In the Immediate condition, as compared to the Nonimmediate condition, interviewers (1) sat closer to the applicant, (2) made fewer speech errors per minute, and (3) actually took longer to give their interviews. The main dependent measures were concerned with the interview performance of the applicant, both in terms of its judged adequacy for obtaining the job and in terms of its reciprocation of immediacy behaviors.

Subjects (Job Applicants) and Confederates (Interviewers)

Thirty white male Princeton University students were recruited ostensibly to help Career Services train interviewers for an upcoming summer job operation. No subjects were eliminated from the study, leaving an *n* of 15 in each condition. The two confederate-interviewers were also white male Princeton students.

Procedure

Upon arrival each subject was given an instruction sheet which informed him that Career Services had contracted with the Psychology Department to train Princeton juniors and seniors in the techniques of job interviewing and that one of the techniques chosen included videotaping interviewers with job applicants for feedback purposes. The subject was then asked to simulate a job applicant, to be honest, and to really compete for the job, so as to give the interviewer real, life-like practice. To make the simulation more meaningful, subjects were also informed that the applicant chosen from five interviewed that evening would receive an additional $1.50.

Subjects were taken to the interview room and asked to be seated in a large swivel chair, while the Experimenter turned on the camera. The confederate-interviewer then entered, and assumed either an immediate or nonimmediate position which will be described in more detail below. Exactly five minutes into the interviewing in both conditions, a guise was developed whose result was that the experimenter had to reclaim the chair in which the subject was sitting. The subject was then asked to take a folding chair leaning against the wall and to continue the interview. The distance from the interviewer which the subject placed his new chair was one of the study's dependent measures designed to assess reciprocated immediacy.

When the interview ended, the experimenter took the subject to another room where a second investigator, blind as to the condition of the subject, administered self-report scales and answered any questions. Subject was then paid and debriefed.

Immediacy Manipulation

As in the Kleck and Nuessle (1968) and the Jones and Cooper (1971) studies, systematic nonverbal variations were introduced by specifically training confederates. Two confederate-interviewers alternated in the two conditions. In the Immediate condition, confederates sat at a chair on the side of a table. In the Nonimmediate condition, confederates sat fully behind the table. The difference in distance from the subject's chair was about four inches, representing the mean difference in distance white interviewers gave black and white applicants in Experiment 1.[2]

In addition, the confederate-interviewers in the Immediate condition were trained to behave as precisely as possible like the subject-interviewers in Experiment 1 had acted toward white applicants. In the Nonimmediate condition, interviewers were trained to act as subject-interviewers had acted toward Blacks in Experiment 1. The factors used to simulate the immediacy behaviors found in the first experiment were speech error rate, length of interview and, as has been previously mentioned, physical distance. Eye contact, shoulder orientation and forward lean did not show significant differences in Experiment 1 and thus were held constant in Experiment 2 (with the levels set at 50% eye contact, 0° shoulder orientation and 20° forward lean).

---

[2] By having the interviewer sit either behind or at the side of the table, the impact of the four inch difference in distance was intentionally maximized in terms of psychological immediacy.

Dependent Measures

Three classes of dependent variables were collected: (1) judges' ratings of interview performance; (2) judges' ratings of reciprocated immediacy behaviors; and (3) subjects' ratings of their post-interview mood state and attitudes toward the interviewer.

Applicant performance.   Applicant interview performance and demeanor were rated by a panel of two judges from videotapes of the interviews. The videotapes were recorded at such an angle that judges viewed only the applicant, not the confederate-interviewer. The judges were merely instructed about the type of job subjects were applying for, and were asked to rate (1) the overall adequacy of each subject's performance and (2) each subject's composure on five (0–4) point scales. High scores, averaged over the judges, represent more adequate and more calm, relaxed performances, respectively.

Reciprocated immediacy behaviors.   Two additional judges, placed behind one-way mirrors as in Experiment 1, recorded subjects' forward lean, eye contact, and shoulder orientation in accordance with the procedures established by Mehrabian (1969). Distance was directly measured after each interview, and represents the distance, in inches, from the middle of the interviewer-confederate's chair to the middle of the subject's chair, after the interruption. Speech errors were scored by another panel of two judges from audiotapes of the interviews, also according to Mehrabian's (1969) procedures. High scores represent more speech errors per minute.

Applicant mood and attitude toward the interviewer.   After the interview, subjects filled out a series of questionnaires designed to assess their mood state and their attitudes toward the interviewer. Following Jones and Cooper (1971), subjects' moods were expected to vary as a function of immediacy conditions. The mood scale adapted from that study was employed. It consisted of six polar adjectives (e.g., happy-sad) separated by seven-point scales. Subjects were asked to respond to each pair according to "the way you feel about yourself."

Two measures of subjects' attitudes toward the interviewer were collected. First, subjects were asked to rate the friendliness of the interviewer on an 11-point scale, with zero representing an "unfriendly" and 10 representing a "friendly" interviewer, respectively. Second, in order to assess subjects' attitudes concerning the adequacy of the interviewer as an individual, they were asked to check the six adjectives best describing their interviewer from a list of 16 drawn from Gough's Adjective Checklist. Final scores represent the number of positive adjectives chosen minus the number of negative adjectives checked.

TABLE 21.2
Mean applicant responses under two conditions of interviewer immediacy; Experiment 2

Response	Relia-bility	Nonimmediate	Immediate	F	P
Applicant performance					
Rated performance	.66	1.44	2.22	7.96	<.01
Rated demeanor	.86	1.62	3.02	16.46	<.001
Immediacy behaviors					
Distance	—	72.73 inches	56.93 inches	9.19	<.01
Speech error rate	.74	5.01 errors/min.	3.33 errors/min.	3.40	<.10
Self reported mood and attitudes					
Mood	—	3.77	5.97	1.34	n.s.
Interviewer friendliness	—	4.33	6.60	22.91	<.001
Interviewer adequacy	—	−1.07	1.53	8.64	<.01

RESULTS

Reliabilities and Interviewer Effects

Reliabilities, obtained by correlating judges' ratings, ranged from .66 to .86 (see Table 21.2). Preliminary analyses also indicated that there were no effects for interviewers, so that the results presented are based on data collapsed across this variable.

Applicant Performance

It was predicted from an analysis of the communicative functions of nonimmediacy that applicants would be adversely affected by the receipt of nonimmediate communications. Those effects were expected to manifest themselves in less adequate job-interview performances.

Subjects in the two conditions were rated by two judges from video-tapes. The main dependent measure, applicant adequacy for the job, showed striking differences as a function of immediacy conditions (see Table 21.2). Subjects in the Nonimmediate condition were judged significantly less adequate for the job ($F = 7.96$; $df = 1/28$; $p < .01$). Subjects in the Nonimmediate condition were also judged to be reliably less calm and composed ($F = 16.96$; $df = 1/28$; $p < .001$).

Reciprocated Immediacy Behaviors

Following Rosenfeld (1967) among others, it was expected that subjects en-countering less immediate communications would reciprocate with less immediate behaviors of their own. This expectation was supported by both the measures of physical distance and speech error rate (see Table 21.2).

Subjects in the Immediate condition, on the average, placed their chairs eight inches closer to the interviewer after their initial chair was removed; subjects in the Nonimmediate conditions placed their chairs four inches further away from their interviewer. The mean difference between the two groups was highly signif-icant ($F = 9.19$; $df = 1/28$; $p < .01$).

As in Experiment 1 mean comparisons for the forward lean, eye contact, and shoulder orientation measures of immediacy did not reach significance. The com-bination of these measures, using the weighting formula devised by Mehrabian (1969), however, was reliably different (means of $-.29$ and $.29$ in the Nonim-mediate and Immediate conditions, respectively; $F = 5.44$; $df = 1/28$; $p < .05$).

The rate at which subjects made speech errors also tended to be reciprocated with subjects in the Nonimmediate condition exhibiting a higher rate than subjects in the Immediate condition ($F = 3.40$; $df = 1/28$; $p < .10$).

Applicant Mood and Attitude Toward the Interviewer

It was expected that subjects receiving less immediate (i.e., less positive) com-munication would (1) feel less positively after their interviews, and (2) hold less positive attitudes toward the interviewer himself. These expectations were only partially supported (see Table 21.2). Although subjects in the nonimmediate con-dition reported less positive moods than subjects in the Immediate condition, this difference was not statistically reliable.

Subjects in the less immediate condition did, however, rate their interviewers to be less friendly ($F = 22.91$; $df = 1/28$; $p < .001$) and less adequate overall ($F = 8.64$; $df = 1/28$; $p < .01$) than subjects in the more immediate condition.

DISCUSSION

Results from the two experiments provide clear support for the two subhy-potheses, and offer inferential evidence for the general notion that self-fulfilling prophecies can and do occur in interracial interactions.

The results of Experiment 1 indicated that black applicants were, in fact, treated to less immediacy than their white counterparts. Goffman's (1963) conception of blackness as a stigmatizing trait in Anglo-American society is, thus, given exper-imental support—insofar as that classification predicts avoidance behaviors in interactions with normals. These results may also be viewed as extending the stigma effect documented by Kleck and his associates with handicapped persons.

That the differential treatment black and white applicants received in Experiment 1 can influence the performance and attitudes of job candidates was clearly demonstrated in Experiment 2. In that experiment those applicants, treated similarly to the way Blacks were treated in Experiment 1, performed less well, reciprocated less immediacy, and found their interviewers to be less adequate. Taken together the two experiments provide evidence for the assertion that nonverbal, immediacy cues mediate, in part, the performance of an applicant in a job interview situation. Further, the experiments suggests that the model of a self-fulfilling prophecy, mediated by nonverbal cues, (1) is applicable to this setting, and (2) can account, in part, for the less adequate performances of black applicants (cf. Sattler, 1970).

Social scientists have often tended to focus their attention for such phenomena as unemployment in black communities on the dispositions of the disinherited. Such an approach has been termed "victim analysis" for its preoccupation with the wounds, defects and personalities of the victimized as an explanation for social problems (Ryan, 1971). The present results suggest that analyses of black-white interactions, particularly in the area of job-seeking Blacks in white society, might profit if it were assumed that the "problem" of black performance resides not entirely within the Blacks, but rather within the interaction setting itself.

References

DARLEY, J. M., LEWIS, L. D., AND GLUCKSBERG, S. Stereotype persistence and change among college students: one more time. Unpublished Manuscript, Princeton University, 1972.

GOFFMAN, E. *Stigma: notes on the management of spoiled identity.* Englewood Cliffs, New Jersey: Prentice-Hall, 1963.

JONES, R. E., AND COOPER, J. Mediation of experimenter effects. *Journal of Personality and Social Psychology,* 1971, **20,** 70–74.

JONES, E. E., AND NISBETT, R. E. The actor and the observer: divergent perceptions of the causes of behavior. In E. E. Jones, D. E. Kanouse, H. H. Kelley, R. E. Nisbett, S. Valins and B. Weiner (Eds.), *Attribution: perceiving the causes of behavior.* New York: General Learning Press, 1971.

KLECK, R. E. Physical stigma and nonverbal cues emitted in face-to-face interactions. *Human Relations,* 1968, **21,** 19–28.

KLECK, R., BUCK, P. L., GOLLER, W. L., LONDON, R. S., PFEIFFER, J. R., AND VUKCEVIC, D. P. Effects of stigmatizing conditions on the use of personal space. *Psychological Reports,* 1968, **23,** 111–118.

KLECK, R. E., AND NUESSLE, W. Congruence between indicative and communicative functions of eye contact in interpersonal relations. *British Journal of Social and Clinical Psychology,* 1968, **7,** 241–246.

KLECK, R. E., ONO, H., AND HASTORF, A. H. The effects of physical deviance upon face-to-face interaction. *Human Relations,* 1966, **19,** 425–436.

MEHRABIAN, A. Orientation behvaiors and nonverbal attitude communication. *Journal of Communication,* 1967, **17,** 324–332.

MEHRABIAN, A. Inference of attitudes from the posture, orientation, and distance of a communicator. *Journal of Consulting and Clinical Psychology,* 1968, **32,** 296–308.

MEHRABIAN, A. Some referents and measures of nonverbal behavior. *Behavior Research Methods and Instrumentation,* 1969, **1,** 203–207.

MERTON, R.K. *Social theory and social structure.* New York: Free Press, 1957.

ROSENFELD, H. M. Nonverbal reciprocation of approval: an experimental analysis. *Journal of Experimental and Social Psychology,* 1967, **3,** 102–111.

ROSENTHAL, R. Teacher expectations. In G. S. Lesser (Ed.), *Psychology and the educational process.* Glenview, Illinois: Scott, Foresman, 1971.

RYAN, W. *Blaming the victim.* New York: Pantheon, 1971.

SATTLER, J. Racial "experimenter effects" in experimentation, testing, interviewing and psychotherapy. *Psychological Bulletin,* 1970, **73,** 136–160.

22

Another Put-Down of Women?
Perceived Attractiveness as a Function
of Support for the Feminist Movement

Philip A. Goldberg, Marc Gottesdiener,
and Paul R. Abramson

Photographs were taken of 30 young women whose attitudes toward the feminist movement identified them as supporters or nonsupporters of the feminist movement. Ratings of the photographs for physical attractiveness yielded no differences between the two groups. However, subjects asked to identify the women who supported the women's liberation movement significantly chose the photographs of the less attractive women. Both male and female subjects responded this way, irrespective of their own professed attitudes toward the feminist movement. The findings are related to work done in the areas of stereotyping and sexism.

It is a matter of easy observation to note that the feminist movement has aroused strong feelings, negative and positive. It is also a matter of easy observation to note that physical attractiveness is an important element in interpersonal attractiveness (cf. Berscheid and Walster, 1972; Jourard and Secord, 1955).

Imputing unfavorable characteristics to people whose attitudes we do not like very much is simply one well-evidenced halo effect. An amusing example of this phenomenon as it applies to physical characteristics is provided by Kassarjian (1963). In his study, Kassarjian found that during the 1960 presidential election campaign, partisans overestimated the height of their candidate and underestimated the height of the opposing candidate.

Reprinted with permission from the authors and *The Journal of Personality and Social Psychology,* Vol. 32, No. 1, 1975. Copyright 1975 by the American Psychological Association.

This research was supported in part from a National Science Foundation Grant from the University of Connecticut Computer Center gj-9 and the Connecticut College Psychology Department Research Fund.

The authors are indebted to Michael Morgan, Cleary Smith, and Phil Biscuti, who served as the photographers, and to Linda C. Abramson, who assisted in data preparation and analysis.

This present study attempts to investigate the stereotype view of the physical attractiveness of young women identified as supporters of the women's liberation movement. The specific question that guided this search were: (a) How does the perception of attractiveness affect the attribution of feminist attitudes; (b) How well does this attribution correspond to reality; (c) To what extent are these judgments influenced by the person's own feminist attitudes; and (d) What differences, if any, are there in the judgments of men and women?

STUDY I

METHOD

Subjects

The subjects were 40 male and 29 female introductory psychology students from the University of Connecticut who volunteered as part of their course requirement.

Procedure

Thirty female undergraduates were haphazardly selected from the campus of Connecticut College to participate in a study about women's liberation. These young women filled out a questionnaire ascertaining their attitudes toward the women's liberation movement and they were photographed. The questionnaire assessed whether they strongly supported the movement, supported the movement with little reservation, supported the movement but had serious reservations about it, did not support the movement, or strongly opposed the movement. The photographs were portraits, including in addition to the face only the shoulders of the female.

The 30, 7×5 inch (17.8×12.7 cm), black and white photographs were encased in plastic and bound into a booklet. Two complete booklets were used. Each photograph was numbered, and the number assigned was recorded on the women's movement questionnaire which corresponded to the female in the photograph.

The 69 introductory psychology students from the University of Connecticut judged the attractiveness of these photographs. The 69 subjects were individually administered a booklet and rated the attractiveness of the 30 photographs on a 5-point Likert scale. The raters were instructed to keep in mind that, in the general propulation, a score of 5 (extremely good looking) would be obtained by 8% of the population, a score of 4 (better than average looking) by about 17%, a score of 3 (average) by about 50%, a score of 2 (somewhat below average) by about 17%, and a score of 1 (considerably below average) by about 8% of the population. These rating procedures were identical to those used by Murstein (1972) in a study of physical attractiveness and marital choice.

RESULTS

The results indicated that there were no significant sex differences in the ratings of any of the photographs. In addition, the data illustrated that the male and female mean ratings of attractiveness of the overall photographs were highly correlated ($r = .93$). Since there were no pronounced sex differences in the photo ratings, the scores of all the raters were pooled.

The mean attractiveness rating of all of the photographs was 2.81, with a standard deviation of .34. The ratings ranged from a mean of 3.59 ($SD = .87$) to a mean of 1.85 ($SD = .79$).

To investigate whether there were any real differences existing in attractiveness between women who support the women's liberation movement and those who do not, the 30 photographed females were split at the median into high and low support for the women's liberation movement groups. (The actual data indicated that 15 females strongly supported the movement, 14 had serious reservations about it, and 1 female did not support it at all.) Using the mean attractiveness ratings of these females, the results indicated that there were no significant differences in attractiveness ($t = 1.24$) between the high ($M = 2.69$, $SD = .50$) and low ($M = 2.92$, $SD = .49$) women's liberation movement support groups.

STUDY 2

METHOD

Subjects

The subjects were 41 male and 41 female introductory psychology students at the University of Connecticut who volunteered as part of their course requirement.

Procedure

The subjects were individually administered a booklet containing the 30 photographs described in the Procedure section of Study 1. They were informed that 15 of the women very strongly supported the women's liberation movement and 15 did not. The subjects were instructed to look at the photographs and put the numbers of the women who they believed supported the women's liberation movement in a column marked *Support* and the numbers of the women who did not support the women's liberation movement in a column marked *Not Support*. It was stressed that they make sure that there were 15 numbers in each column. In addition, each subject filled out the women's liberation questionnaire.

RESULTS

The mean attractiveness ratings obtained in Study 1 for each of the numbered photographs were substituted for the numbers which the subjects assigned to either supporting or not supporting the women's liberation movement columns. A one-way analysis of variance was performed for the difference between the mean attractiveness ratings of the perceived support versus perceived not support photographed women. The data strongly indicated that the women perceived as supporting the women's liberation movement ($M = 1.75$, $SD = .12$) were rated significantly less attractive than the women perceived as not supporting the women's liberation movement ($M = 2.86$, $SD = .11$; $F = 33.36$, $p < .001$).

The data were also analyzed separately by the sex of subject. Females rated those women they perceived as supporting the women's liberation movement ($M = 2.77$, $SD = .10$) as significantly less attractive than the women they perceived as not supporting the movement ($M = 2.85$, $SD = .10$; $F = 11.65$, $p < .001$). Males also rated the women they perceived as supporting the women's liberation movement ($M = 2.74$, $SD = .13$) as significantly less attractive than the women they perceived as not supporting the movement ($M = 2.87$, $SD = .13$; $F = 21.73$; $p < .001$).

The present findings also indicate that the subjects' own attitudes toward the women's liberation movement were uncorrelated with their mean attractiveness ratings. Female subjects' attitudes toward the women's liberation movement were uncorrelated with their mean attractiveness rating of women whom they perceived as not supporting the movement ($r = .07$) and their mean attractiveness rating of women whom they perceived as supporting the movement ($r = -.11$). Similar findings were obtained for male subjects. Their attitudes toward the women's liberation movement were uncorrelated with their mean attractiveness rating of women they perceived as not supporting the movement ($r = .19$) and their mean attractiveness rating of women they perceived as supporting the movement ($r = -.20$).

GENERAL DISCUSSION

The general results seem clear: Women who support the feminist movement are believed to be less attractive physically than their sisters who do not support the movement, though the evidence is that there is no true difference in attractiveness between the two groups.

The variables that did not yield differences are, perhaps, the most interesting. *Both* men and women responded similarly in associating unattractiveness with support for the women's liberation movement. This finding is, perhaps, consistent with the findings of Goldberg (1968) and Pheterson, Kiesler, and Goldberg (1971) who found that women shared in the general cultural prejudice against women.

Do the results represent further evidence of prejudice against women? One could argue that the data suggest that the subjects accepted a widely shared cultural stereotype, which is not, however, a prejudice. Or, that the derogation involved was directed against specific kinds of women from which one could not reasonably infer a more general misogyny.

Perhaps, we understand that expression of concern for "law and order" often cloak racist attitudes and that opposition to "zionists" often means opposition to Jews generally. So too, it may be suggested that the put-down of feminists might very well serve as a cover for a more generally misogynous attitude.

The finding that one's personal attitudes toward the women's liberation movement were uncorrelated with the categorization of supporters of the movement was unexpected, yet it is consistent with previous research on stereotyping with ethnic groups. Bettleheim and Janowitz's (1964) data, which were based on extensive interviews, found no significant relationship between personal stereotypes made about blacks or Jews and an individual's own attitudes toward these two groups. Similar results were obtained by Karlins, Coffman, and Walters (1969) with college students' stereotypes of 10 national and ethnic groups. They conclude that individuals possess stereotypes even though they may not often express them or feel affected by them.

Fishbein (1967) argues that it is a misconception to assume that an individual's attitude toward an object is a major determinant of his behavior with respect to that object. He summarizes research which indicates that attitudes were not related to behavior in any consistent fashion, but appear to be consequences or determinants of beliefs or behavioral intentions, rather than predictors of overt behavior.

The potential for trouble inherent in the ethnic stereotype is well known (cf. Allport, 1954). Ethnic prejudice and prejudice toward women are not identical phenomena, but there is an increasing literature attesting to uncomfortable parallels. The results from this study may provide one more example for the pervasive social put-down to which women are subject.

References

ALLPORT, G. W. *The nature of prejudice.* Cambridge, Mass.: Addison-Wesley, 1954.
BERSCHEID, E., AND WALSTER, E. Beauty and the best. *Psychology Today,* March 1972, 42–46.

BETTLEHEIM, B., AND JANOWITZ, M. *Social change and prejudice.* New York: Free Press, 1964.

FISHBEIN, M. A consideration of beliefs and their role in attitude measurement. In M. Fishbein (ed.), *Readings in attitude theory and measurement.* New York: Wiley, 1967.

GOLDBERG, P. A. Are women prejudiced against women? *Trans-action,* 1968, **5,** 28–30.

JOURARD, S. M., AND SECORD, P. F. Body cathexis and personality. *British Journal of Psychology,* 1955, **46,** 130–138.

KARLINS, M., COFFMAN, T. L., AND WALTERS, G. On the fading of social stereotypes: Studies in three generations of college students. *Journal of Personality and Social Psychology,* 1969, **13,** 1–16.

KASSARJIAN, H. H. Voting intentions and political perception. *Journal of Psychology, 1963,* **56,** 85–88.

MURSTEIN, B. I. Physical attractiveness and marital choice. *Journal of Personality and Social Psychology,* 1972, **22,** 8–12.

PHETERSON, G. I., KIESLER, S. B., AND GOLDBERG. P. A. Evaluation of the performance of women as a function of their sex, achievement, and personal history. *Journal of Personality and Social Psychology,* 1971, **19,** 114–118.

23

The Ultimate Attribution Error: Extending Allport's Cognitive Analysis of Prejudice

Thomas F. Pettigrew

Allport's *The Nature of Prejudice* is a social psychological classic. Its delineation of the components and principles of prejudice remains modern, especially its handling of cognitive factors. The volume's cognitive contentions are outlined, and then extended with an application from attribution theory. An "ultimate attribution error" is proposed: (1) when prejudiced people perceive what they regard as a negative act by an outgroup member, they will more than others attribute it dispositionally, often as genetically determined, in comparison to the same act by an outgroup member; (2) when prejudiced people perceive what they regard as a positive act by an outgroup member, they will more than others attribute it in comparison to the same act by an ingroup member to one or more of the following: (a) "the exceptional case," (b) luck or special advantage; (c) high motivation and effort, and (d) manipulable situational context. Predictions are advanced as to which of these responses will be adopted and under which conditions the phenomenon will be magnified. A brief review of relevant research is also provided.

Gordon Allport was a modest man. Yet in his reserved, even shy, manner, he was justly proud of *The Nature of Prejudice* (Allport, 1954). His place in psychology as a personality theorist had long been established with the publication of *Personality: A Psychological Interpretation* (Allport, 1937). But it was his book on intergroup prejudice that most directly expressed his deepest concerns and values, that translated his more abstract work into concrete ideas for reform and social change.

Were he alive today, Allport would undoubtedly be honored and pleased by the

Reprinted with the permission of the authors and Division 8 of the American Psychological Association and of Sage Publications, Inc., from *Personality and Social Psychology Bulletin,* Vol. 5, No. 4, 1979, pp. 461–476. Brief portions of this chapter are drawn from two other publications (Pettigrew, 1978, 1979). The author wishes to express his appreciation for valuable help from John Jemmont, Joel Johnson, Ellen Langer, Shelley Taylor, and Joachim Winkler.

Bulletin's joint observance of the silver anniversaries of the "separate is inherently unequal" public school decision of the U.S. Supreme Court and the publication of *The Nature of Prejudice*. For while the Supreme Court's ruling was widely publicized, the issuance of his book was obscure. The cloth-bound, unabridged version that appeared in 1954 attracted relatively little attention and only modest sales. It was not until the paperback edition appeared four years later, one-fifth shorter and selling in drug stores and airports, that the full impact of the volume began to be felt (Allport, 1958). Sales multiplied and continue brisk to this day, making *The Nature of Prejudice* one of the most widely read social psychological books both inside and outside of the discipline.

Allport would also be pleased by the judgements of the book's intellectual content now being rendered a generation later. He realized that the specific examples tellingly applied throughout the book would become dated. And two of his students, Professor Bernard Kramer, of the University of Massachusetts at Boston, and I, are currently revising and updating it. Like Gray's *Anatomy*, Allport's *The Nature of Preducice* may continue through multiple editions as the field's standard reference.

Though the examples age quickly, Allport hoped that the volume's basic outline would stand the test of time. "The content will have to change," he often remarked, "but I think the book's contribution is its table of contents." Indeed, its table of contents has organized the scholarly study of prejudice. *The Nature of Prejudice* delineated the area of study, set up its basic categories and problems, and cast it in a broad, eclectic framework that remains today. The book continues to be cited as the definitive theoretical statement of the field, and it remains unchallenged throughout social science as *the* book on prejudice.

Both Smith (1978) and Aronson (1978) note that the value perspective of the volume marks its twenty-five years; Smith calls it "something of a period piece," Aronson "a time capsule of what it was like in 1954." But both admire its ageless quality of scholarship. Writes Smith:

> Allport's compendious book still invites reading and defies summary. What seemed wise and judicious in 1954 mostly still seems so today. . . . His pervasive fair-mindedness, his democratic values, and his concern for evidence continues to set a model for humane, problem-focused social science . . . [F]rom Allport, we can still get wise guidance in our attempts to give more human substance to our democratic aspirations [Smith, 1978, pp. 31–32.]

Aronson concurs:

> Gordon Allport's book was a harbinger and a reflection of the thinking that went into the Supreme Court decision [Brown *v.* Board of Education, 1954]. *The Nature of Prejudice* is a remarkable mixture of careful scholarship and humane values. Allport marshalled an impressive array of data and organized these data clearly and passionately. The book has influenced an entire generation of social psychologists, and deservedly so. . . . What is modern about the book is Allport's perspective; he carefully chose among existing theories and data to come up with a brilliant and

accurate statement of the eclectic causes and possible cures of prejudice. . . . This is a tribute to the wisdom, scholarship, and judgment of a graceful mind. Allport avoided the twin pitfalls of championing one position to the exclusion of all others, or of giving each position equal status. [Aronson, 1978. p. 92.]

The modern ring of Allport's principles is, perhaps, best illustrated in his sophisticated handling of the cognitive components of prejudice. Allport was a closet Gestaltist. He remained dubious about the Gestalt theorists' nativistic view of perception (Allport and Pettigrew, 1957). But during his year of study in Germany following his Harvard doctorate, he developed a deep respect for and lasting interest in Gestalt theory. And this clearly guided his writing on the cognitive aspects of prejudice three decades later.

In the early 1950s, it was fashionable to think of ethnic stereotypes as aberrant cognitive distortions of "prejudiced personalities." The dominant influence of *The Authoritarian Personality* (Adorno et al., 1950) furthered this conception. Allport broke sharply from this view. He insisted that the cognitive correlates of prejudice were natural extensions of normal processes. Ten of the volume's 31 chapers include a discussion of cognitive factors (2, 8, 10, 11, 12, 13, 18, 19, 25, and 27), for cognitive factors are central to his approach. Prejudice for Allport involves both affective and cognitive components; he defined it as "an antipathy based upon a faulty and inflexible generalization" (Allport, 1954, p. 9).

Allport emphasized his fresh view in the title of the second chapter, "The normality of prejudgment." And it begins with a rhetorical question: "Why do human beings slip so easily into ethnic prejudice? They do so because the two essential ingredients . . . —*erroneous generalization* and *hostility*—are natural and common capacities of the human mind" (Allport, 1954, p. 17). Fifteen years prior to Tajfel's (1969, 1970) important research on the point, Allport maintained that the mere "separation of human groups" was enough to trigger the psychological processes that lead to intergroup prejudice. And the primary process was held to be that of categorization, an essential, "least effort" means of handling sensory overload upon which "orderly living depends" (Allport, 1954, p. 20).

The emphasis upon categorization continues throughout *The Nature of Prejudice*. In chapter 8, visibility and strangeness are shown to be critical in much the same manner that cognitive social psychologists today demonstrate the critical role of perceptual salience. Visible differences, runs the argument, imply real differences. And attitudes symbolically "condense" around visible cues.

In chapter 10, Allport discusses cognitive processes generally. Here the then "New Look in Perception," centered around such figures as Bruner and Postman (1948), influenced Allport's treatment. Selection and accentuation receive particular attention. Yet today's emphasis upon attribution theory is anticipated. Citing Heider's (1944) famous paper on phenomenal causality, Allport (1954, pp. 169–170, 177) maintains that "cause and effect thinking" is especially crucial, and that the human tendency "to regard *causation* as something *people* are responsible for . . . predisposes us to prejudice." It is this recognition of "the fundamental attribution error'' (Ross, 1977) to which we shall return in a suggested extension of Allport's cognitive analysis.

So armed, the reader of *The Nature of Prejudice* is then led to see how categorization is centrally involved in a variety of related phenomena: emotionally toned intergroup labels ("nouns that cut slices" was Allport's [1954, p. 178] early statement of labeling theory); stereotypes ("highly *available* . . . exaggerated belief[s] associated with a category" [Allport, 1954, p. 191, italics added]); the phenomenological approach to prejudice; the development of prejudice in children and in later life; and the cognitive correlates of prejudiced and tolerant personalities (through "dichotomization," " the need for definiteness," and "tolerance for ambiguity" [Allport, 1954, pp. 400–403, 438]).

To be sure, there are aspects of Allport's cognitive analysis that are dated or confusing. He stresses Zipf's (1949) "principle of least effort" as the primary motivational component underlying human cognition. Though here again, old wine may be returning in new bottles in the recent interest in "mindlessness" (Langer et al., 1978; Langer and Newman, 1979). But Harding (private communication) points out that three further emphases in *The Nature of Prejudice* are in need of correction. First, Allport is too expansive in his use of the concept of "category." He employs it to describe personal values as well as using it as synonymous with the process of generalization, and these expansions lead him into making the difficult distinction between rational and irrational categories. Second, the focus on the irreversibility of stereotypes as the key criterion of their rationality now seems misplaced in light of recent advances in research on stereotyping. Finally, Allport's assumption that the intergroup antipathy involved in prejudice is *based upon* a faulty generalization seems unduly restrictive; the causal order can and does flow from hate to stereotype as well.

Yet, taken as a whole, Allport's cognitive analysis of prejudice seems surprisingly current a generation later. And this fact is all the more remarkable given the exciting advances made in this area in recent years by cognitive social psychologists. Certainly, Allport's Gestalt leanings helped him to project the future trends in the field, for it has been Gestalt influence via Fritz Heider that has shaped the field's progress in recent years.

THE ULTIMATE ATTRIBUTION ERROR

Building on attribution theory, we can expand Allport's analysis by proposing a systematic patterning of intergroup misattributions shaped in part by prejudice. The proposal is an extension of "the fundamental attribution error" noted by Heider (1958) and explicated by Ross (1977). It refers to observers' consistent underestimations of situational pressures and overestimations of actors' personal dispositions on their behavior. This "error" occurs over a wide range of situations, has extensive social implications, and can be easily demonstrated in the laboratory.

Thus, Jones and Harris (1967) showed that listeners inferred a "correspondence" between communicators' private opinions and their anti-Castro remarks, even though the listeners were well aware that the remarks were made only to obey the explicit instructions of the experimenter. In another straightforward study (Ross, Amabile, and Steinmetz, 1977), subjects played a quiz game with the assigned roles of "questioner" and "contestant." Though the game allowed the "questioners" the enormous advantage of generating all of the questions from their own personal store of knowledge, later ratings of general knowledge were higher for the "questioners." This dispositional attribution was made not only by uninvolved observers but by the disadvantaged "contestants" themselves.

Note the three common elements of the fundamental attribution error illustrated in these experiments: (1) powerful situation forces (the experimenter's instructions and the quiz game format) are minimized; (2) internal, dispositional characteristics of the salient person (the communicator and the questioner) are causally magnified; and (3) role requirements (of being an experimental subject or quiz contestant) are not fully adjusted for in the final attribution.

There are systematic exceptions to the phenomenon. Actors often attribute their own behavior to situational causes when there are salient extrinsic rewards (Lepper and Greene, 1975), few choices open, and few similarities with past behavior (Monson and Snyder, 1977). And we shall maintain that observers often employ external, situational attributions to explain "away" positive behavior by members of disliked outgroups. In any event, it should be noted that the typical attribution investigation to date maximizes those conditions likely to elicit situational attributions from actors and dispositional attributions from observers. Yet the "real world" seems more likely to elicit the reverse attributional pattern, correctly or in error, when actors are performing familiar acts in situations under their control.

There are other qualifications. Dispositional and situational causal attributions do not form a neat, one-dimensional continuum: That is, a dispositional attribution is not necessarily the opposite of a situational attribution. In addition, there appears to be a positivity bias for intimate others, such that you grant them the benefit of the doubt by attributing positive actions to dispositional causes and negative actions to situational causes (Taylor and Koivumaki, 1976).

Granting members of a disliked outgroup the benefit of the doubt, however, may not be so common. Taylor and Koivumaki (1976, p. 408) suggest that "a person who is disliked or hated may well be viewed as responsible for bad behaviors and not responsible for good ones. In other words, we may find a corresponding 'negativity' effect for disliked others." It is this possibility of a "negativity" effect extended to the intergroup level that forms the basis of the proposed ultimate attribution error.

The proposal follows, too, from Heider's (1958) writings on the attribution process. For example, Heider believed that negative self-attribution might be avoided in order to protect one's self-esteem. Here we propose that a stereotyped view of the outgroup needs to be protected from a positive outgroup evaluation. If the outgroup member is seen as performing a negative act consistent with our negative view, the fundamental error of dispositional attributions will be enhanced. And often when race and ethnicity are involved, these attributions will take the form of believing the actions to be a result of immutable, genetic characteristics of the derogated group in general—the bedrock assumption of racist doctrine.

But the problematic instance arises when we perceive the outgroup member is "out of role"—that is, the outgrouper is performing a positive act inconsistent with our negative view of the group. The most primitive defense, of course, is simply to deny the act altogether or to reevaluate it as potentially negative. An ambitious act becomes "pushy"; an intelligent act becomes "cunning." But here we are concerned with the causal attributions made when the outgroup behavior is perceived as unquestionably positive in direct violation of the stereotype held.

Two aspects of this process may trigger what Heider (1958) and Kelley (1967) call "egocentric assumptions." The positive behavior is by definition positive in the terms of the prejudiced perceivers themselves; thus, it fits their conceptions of behavior that they might well expect of themselves. Here a familiar point of attribution theory is relevant. Perceivers possess the historical data about themselves performing similar acts, but little or no such data about outgroup members performing them. They have generally attributed such positive behavior by themselves as dispositionally caused, as further evidence of their being decent, upstanding human beings. Yet it is precisely this mode of explanation that is now in conflict with their established conceptions of the disliked outgroup. Moreover, if the perceivers are themselves the direct beneficiaries of the outgroup member's positive act, the attributional issue is made still more problematic. Phrased more generically, the more personally involving the outgrouper's positive act is, the more difficult it becomes to explain "away" the stereotype-challenging behavior.

Balance theory suggests that one way out of the dilemma is to change our views of the entire outgroup. And the intergroup contact research literature (Pettigrew, 1971), inspired by Allport's (1954, chapter 16) carefully qualified but often misinterpreted contact hypothesis, has established that this possibility does in fact occur under specified situational conditions. But this alternative typically requires the perception of repeated positive acts by the outgroup and is usually strongest among the initially least prejudiced. Here we concentrate our attention upon what happens when one positive act by an outgroup member is perceived by more prejudiced individuals.

In the interaction under discussion, the perceiver does not possess cause-and-effect information over time. And it is precisely this type of attributional situation that invokes Kelley's (1972) discounting effect. Discounting the importance of a particular dispositional cause in the presence of other "plausible" causes arises with greater force when the outgroup member performs a positive rather than negative act. In contrast to the contention of Jones and Davis (1965) that in-role acts are more discounted and therefore less "confident" than out-of-role acts, we are here proposing that intergroup perception may often reverse this pattern. In other words, the same antisocial behavior that would qualify *within* a social group as out-of-role will frequently be seen as in-role *across* social groups if it matches hostile stereotypes and expectations. Likewise, prosocial acts by the disliked out-group will be regarded as out-of-role and made problematic by a range of plausible causes.

This same reasoning has already been applied and verified for low-status actors in the classical research by Thibaut and Riecken (1955) and its replication by Ring (1964). Negative acts by low-status actors in Ring's work were apparently *less* discounted by subjects than the same acts by high-status actors. And both studies showed that positive acts were *more* dispositionally attributed for high-status actors and *more* discounted and situationally attributed for low-status actors. In a comment pertinent to the present argument, Kelley (1972, p. 9) notes that greater uncertainty would be expected for the positively behaving low-status actor "inasmuch as both internal and external reasons are plausible"—that is, a lower-status person could have performed the positive act because of personal qualities and/or because of the greater power of the attributor. We are applying in this paper a similar logic across racial and ethnic groups of varying status differentials, though later research may show that the ultimate attribution error is strongest in the causal conclusions of higher-status groups.

A further suggestion by Kelley (1972) allows us to organize systematically the attributional possibilities. He advances the interesting idea that "attributional processes are closely linked to the effective exercise of control" (Kelley, 1972, p. 23). In ambiguous situations, he predicts that there will be a bias toward attributing cause to factors that are potentially controllable by the perceiver. But in situations with potentially important consequences, there will be a bias toward causes that may be less controllable yet could link to the vital consequences. Crossing the perceived degree and locus of control, Table 23.1 generates four possible attributional directions for resolving the explanatory problem raised by the perception of a positive act by a member of a disliked group.

The abscissa is the familiar, if oversimplified, internal–external locus of control dimension (Rotter, 1966; Phares, 1976). The ordinate follows Kelley's suggestion

TABLE 23.1
Classification Scheme for "Explaining Away" Positive Behavior by a Member of a Disliked Outgroup

		Perceived Locus of Control of Act	
		Internal	External
Perceived Degree of Controllability of Act	Low	A. The Exceptional Case	B. Luck or Special Advantage
	High	C. High Motivation and Effort	D. Manipulable Situational Context

and specifies two levels of perceived control of the act by the attributor. If Kelley is correct, there will be a bias toward regarding the problematic act as having been or potentially becoming influenced by the attributor. Consequently, alternatives C and D should be more frequently employed than alternatives A and B. But, as Kelley emphasized and both the Thibaut–Riecken and Ring studies demonstrated, there are frequent instances in which the attributor still chooses the low-control alternatives. For example, if long-term future involvement with the outgroup member is anticipated (an "important consequence"), a high internal evaluation of the person as quite exceptional might well be preferred. Likewise, the perception of minority group members successfully acquiring a sought-after-goal can easily be regarded as the result of luck or an unfair special advantage—external attributions beyond the attributor's control. The furor over and distorted views of affirmative action programs in recent years illustrate cell B. More fundamentally, low-status perceivers who are attributing cause to positive actions of high-status outgroup members are likely to adopt the two low-control alternative explanations. Let us briefly consider each of Table 23.1's four alternatives.

The Exceptional Case

This alternative can be derived from a consideration of cognitive heuristics (Tversky and Kahneman, 1974). Negative intergroup stereotypes offer striking examples of the operation of heuristics. They are readily *available* images that act as displacing *anchors* from which to judge outgroup behavior and with which to match outgroup behavior for its *representatives*. The anchoring effect underscores once again the initial tendency not to perceive the act as positive, but rather to assimilate it toward the negative stereotype. But if perceived, the pro-social act by an outgroup member appears to be odd, even deviant. Similar instances cannot be called up readily in the bigot's mind. And, as it does not match the stereotype, the actor and the behavior seem *un*representative. This line of analysis suggests that one mode of resolution for the prejudiced perceiver will be to exclude this particular actor from the disliked outgroup. This resolution can even lead to generous, if often patronizing, exaggeration of the positive qualities of this exceptional person in order to differentiate this "good" individual from the "bad" outgroup in a fashion not unlike that found for solo-role minorities in the laboratory (Taylor, in press) and in the field (Kanter, 1977). Popular expressions capture the phenomenon: "She is really the exception that proves the rule"; "He's really different; he's bright and hard-working, not like other Chicanos." Allport (1954) observed this common response, and described it as "fence mending" our stereotypes.

Luck or Special Advantage

Table 23.1 classification scheme resembles the scheme proposed by Weiner et al. (1972) for attributions of achievement behavior. They derived their scheme by crossing locus of control with the perceived degree of stability. Their four attributional modes—ability, effort, task difficulty, and luck—only roughly coincided with those described here, for the goals of the two schemes are different. Weiner and his colleagues sought to understand attributions of achievement behavior viewed across successive points in time. Here we attempt to categorize attributions for a wide range of positive behaviors at one point in time. Thus, rather than the controllability of the cause, Weiner et al. focused upon the time-linked degree of stability of the attributed cause—relatively stable ability and task difficulty attributions versus the more unstable dimensions of effort and luck.

The luck attribution, then, assumes a somewhat different meaning in the present context. The positive outgroup act can be seen as beyond the control of either the attributor or the actor and therefore of little significance: "He's dumb like the rest of his group, but he won anyway of sheer luck." With no means of inferring luck from any past variable pattern of outcomes (save, perhaps, for occasional perceptions of behavior by *other* members of the outgroup), the use of this attribution is less "rationally" established than in the research reported by Weiner et al. (1972).

More common, then, is the attribution of special advantage. The actor is seen as having behaved positively and having achieved a stereotype-breaking result because the actor had the benefit of a special advantage conferred by virtue of the outgroup status. Typically, the special advantage is regarded as discriminatory and accompanied by a sense of resentment and fraternal deprivation (Runciman, 1966; Vanneman and Pettigrew, 1972). Black Americans have traditionally explained away positive behavior and outcomes of white Americans in this manner. But the generality of the phenomenon is suggested by the recent vehemance of many whites, including many who label themselves "liberals," against affirmative action programs for minorities.

High Motivation and Effort

Individual members of a disliked outgroup are often seen by more benign bigots as "overcoming" through great personal effort the handicaps involved with belonging to an unfortunate people. The traditional phrase for such "compensatory" behavior in American race relations was "a credit to his race." As indicated in Table 23.1, the distinction between this mode and the exceptional case attribution involves the amount of control that is seen as exerted over the cause. Outgroup members who work hard at being antistereotypical in their behavior are not seen as intrinsically exceptional, since they are perceived to be responding positively to aspects of the interaction under some control of others. They are not viewed as true exceptions, for they would return to their "true," stereotypical state were it not for their keen motivation. But both exceptional and striving outgroup members are important exemplars for prejudiced individuals to point to as "proof" that discrimination and other situational factors are not responsible for negative behaviors and outcomes of the outgroup: "They made it, didn't they? So there must be something personally wrong with the rest of them."

Manipulable Situational Context

Most, but by no means all, structural factors of interaction are out of our immediate control. Those factors that are within our power to manipulate are thus highly valued. The symbolic interactionist wing of social psychology, in particular Goffman (1959, 1969, 1971), has vividly demonstrated the many forms such situational manipulations can assume. An outgroup member's positive act can, therefore, be seen not as a function of effort but as a consequence of situational factors at least partly influenced by others: "What could the cheap Scot do but pay the whole check once everybody stopped talking and looked at him?" Here we return to the role point of Jones and Davis (1965). This attribution often arises when the situationally-defined role is regarded as more powerful and salient than the group role.

Three Further Points

Two comparative frames of reference are involved with these predictions. Tajfel's (1970) skillful work suggests that virtually all human beings are subject to patterned differences in their perceptions of ingroup versus outgroup behavior. Thus, the ultimate attribution error is likely to characterize the attributions of most human beings, not just those of prejudiced individuals. So the predictions advanced in this paper are relative in two ways: (1) The perceptions of acts by outgroup members will tend to show the predicted trends relative to perceptions of the same acts by ingroup members; and (2) prejudiced individuals will tend to show these phenomena more sharply than others.

These predictions also assume reasonably high salience of group membership. Adherence to religious doctrine (Charters and Newcomb, 1958) and even tolerance for pain (Buss and Portnoy, 1967; Lambert et al., 1960) have been enhanced by making group membership salient. Doise and Sinclair (1973) demonstrated how favorability ratings of both ingroups and outgroups can be altered by varying degrees of group salience. And McKillip, DiMiceli, and Luebke (1977) showed that under conditions of high salience of sex roles men made greater use of the male competence stereotype and women of the female warmth stereotype. Similarly, the ultimate attribution error is most likely to occur when perceivers are conscious of both their and the actor's group memberships.

Finally, not all group relations, fortunately, are hostile and marked by negative stereotypes. The fascinating research on East Africa by Brewer and Campbell (1976) underlines this fact and suggests that there is no inexorable strain toward boundary convergence and Sumner's (1906) full-blown version of extreme ethnocentrism. The intensity of the ultimate attribution error should therefore vary considerably across contrasting intergroup situations. As a first approximattion, we propose two related hypotheses. The ultimate attribution error will be greatest when the groups involved have histories of intense conflict and possess especially negative stereotypes of each other. It will also be greatest when racial and ethnic differences co-vary with national and socioeconomic differences; or, more strongly phrased, the more bounded the two groups, the greater the ultimate attribution error is likely to be.

The Formal Predictions

Based on this discussion, we propose that: *Across-group perceptions are more likely than within-group perceptions, especially for prejudiced individuals, to include the following:*

1. *For acts perceived as negative (antisocial or undesirable), behavior will be attributed to personal, dispositional causes. Often these internal causes will be seen as innate characteristics, and role requirements will be overlooked.*

2. *For acts perceived as positive (prosocial or desirable), behavior will be attributed to any one or the combination of the following: A. to the exceptional, even exaggerated, special case individual who is contrasted with his/her group; B. to luck or special advantage and often seen as unfair; C. to high motivation and effort; and/or D. to manipulable situational context.* (See Table 23.1.)

3. *For acts perceived as positive, the most probable causal attributions are:*

a. *2A and 2B when the consequences of the attribution are deemed potentially important;*

b. *2C and 2D when the short-term control of the behavior is valued;*

c. *2A and 2B when the attributor is lower status than the actor;*

d. *2C and 2D when the attributor is higher status than the actor;*

e. *2A and 2C when the behavior is culturally regarded as generally dispositionally determined;*

f. *2B and 2D when the behavior is culturally regarded as generally situationally determined;*

g. *2A when the outgroup member is a solo or token participant and thus separated from the outgroup, highly salient, and less threatening;*

h. *2B when successful and valued outcomes are likely to result from the behavior;*

i. *2C when the behavior is culturally regarded as difficult; and*

j. *2D when a situationally defined role is seen as more powerful and salient than the membership role.*

The attributional tendencies of 1, 2, and 3 above will be enhanced:

4. *When there exists high salience for perceivers of both their own and the actor's group memberships;*

5. *When perceivers are highly involved in the actor's behavior (e.g., when they are the target). Extremely high involvement will often lead to either attitude change toward the entire outgroup and/or to multiple attributions of cause; and*

6. *When the groups represented in the interaction:*

a. *have had histories of intense conflict and possess especially negative stereotypes of each other or*

b. *have their racial and ethnic differences covary with national and socioeconomic differences; further, the more bounded the groups, the greater the ultimate attribution error.*

INITIAL EVIDENCE

While the ultimate attribution error evolves directly from the theoretical writings of Heider, Kelley, Weiner, Ross, Tversky, Kahneman, Campbell, and others, its direct empirical base is thin. The research on the influence of status upon causal attributions by Thibaut and Riecken (1955) and Ring (1964) is importantly, even if only indirectly, relevant, and it shaped the predictions. Indirectly relevant, too, is the ingenious work on group discrimination of Tajfel. He and his coworkers have found strong experimental evidence of a "generic norm of outgroup behavior" among English school boys "divided into groups defined by flimsy and unimportant criteria" (Tajfel, 1970). Familiarity with the experimental setting only increases the group discrimination (Tajfel and Billig, 1974). The ultimate attribution error attempts to describe part of the cognitive mediating process of this "norm of outgroup behavior."

The few directly relevant studies support the present contentions. Duncan (1976), for example, showed 100 white undergraduates a videotape depicting one person (either black or white) ambiguously shoving another (either black or white). His subjects tended to attribute the shove to personal, dispositional causes when the harm-doer was black, but to situational causes when the harm-doer was white. In addition, the shove was labeled as more violent when it had been administered by a black. These findings support prediction 1 above, though they depend heavily upon Duncan's having developed videotapes that portray truly equivalent degrees of aggression. In any event, Allport and Postman (1947) investigated the operation of this violent stereotype of blacks three decades ago in their famous research on rumor transmission. One of their pictures employed to initiate a rumor chain showed a white man holding a razor while arguing with a black man. In over half of their all-white experimental groups, the final report indicated that the black in the picture (instead of the white) held the razor. The black was sometimes said to be "brandishing it wildly" and "threatening" the white.

Taylor and Jaggi (1974) report the most complete test to date. They conducted their study with 30 Hindu office clerks in southern India. First, the subjects rated the concepts "Muslim" and "Hindu" on twelve evaluative traits. Their responses indicate sharply different stereotypes of the ingroup and the outgroup. Hindus were seen as significantly more generous, hospitable, kind, friendly, sociable, sincere, and honest, while Muslims were regarded as more rude and more often cheaters.

Taylor and Jaggi next gave their subjects short descriptions of either a Moslim or a Hindu behaving positively or negatively in one of four contexts involving the subject: a shopkeeper being either generous or cheating; a teacher either praising or scolding the subject as a student; an actor either supplying appropriate help or ignoring the slightly injured subject; and a householder either sheltering or ignoring the subject when caught in the rain. For each of the sixteen descriptions (outgroup or ingroup × positive or negative behavior × four situations), the subjects chose the major reason for the actor's behavior from a list of four or five, one reflecting an internal attribution and the remainder external attributions. The results are summarized in Table 23.2.

These unequivocal results are consistent with the ultimate attribution error predictions of this paper. Dispositional attributions are employed largely for the ingroup's positive behavior and the outgroup's negative behavior. Such attributions for these two types of acts are more frequent in *all* comparisons with negative ingroup and positive outgroup acts. These findings support predictions 1 and 2 and provide at least inferential support for prediction 6. The relatively bounded Muslim and Hindu groups have experienced a history of intense conflict on the Indian sub-continent, including the 1971 Pakistani-Indian war that took place not long before this research was conducted. Consequently, the Taylor and Jaggi investi-

TABLE 23.2
Internal Attributions of Hindu Subjects

Situation	Hindu Ingroup Actor		Muslim Outgroup Actor	
	Positive Behavior	Negative Behavior	Positive Behavior	Negative Behavior
Shopkeeper	43%	3%	10%	40%
Teacher	43%	3%	10%	23%
Help to Injured	67%	3%	10%	33%
Householder	80%	0%	20%	33%

Source: Adapted from Taylor and Jaggi (1974, Table 2).

gation focuses upon a situation where the ultimate attribution error would be expected to operate with special force.

The Taylor-Jaggi research design can be faulted, however. By first eliciting ratings on evaluative traits, the study probably made the cross-group stereotypes especially salient for their subjects; and consistent with prediction 4, the responses on the second task were thereby heightened. Fortunately, then, this work has been partially replicated in the United States without this design problem. Wang and McKillip (1978) utilized the accident responsibility approach to determine the effects of nationality on judging an automobile accident. Thirty foreign Chinese college students (from Taiwan, Hong Kong, and Singapore), 30 American college students, and 30 adult residents of Carbondale, Illinois, read one of the three versions of an automobile accident: Chinese driver and American victim, American driver and Chinese victim, and no national identifications at all. All subjects then assigned a fine (from zero to fifty dollars) to either the driver or the victim, rated the driver and victim on seven evaluative traits, and answered a 13-item measure of American and Chinese ethnocentrism.

Wang and McKillip's findings strongly support prediction 1. The three groups did not differ in their assessments of the control accident without national identification; each tended to find the driver culpable and rated the victim more favorably. But the three groups differed in degree of ethnocentrism, with the Chinese students and town residents significantly more ethnocentric than the American students. Accordingly, the Chinese students and the town residents revealed large nationality biases in their judgments, but the American students did not. Thus, the Chinese subjects tended to place responsibility for the accident on either the American victim or driver and rate them negatively in comparison to the Chinese involved in the accidents. Resident subjects yielded precisely the opposite results. And the American college subjects revealed no responsibility differences by nationality and only slight trait differences in favor of the American targets.

A third relevant study by Banks, McQuarter, and Pryor (unpublished) had black and white high school students evaluate identical achievement performances of black and white targets in terms of ability and effort. Consistent with predictions 2 and 3d and 3e combined, the white subjects placed greater emphasis upon the role of effort in evaluating blacks than in evaluating whites. Yet equal emphasis was given to the importance of ability for both races. Black subjects did not reveal the effect. Banks and his colleagues explain this failure to replicate with the black subjects in terms of differential racial association and familarity of their two subject groups. The students came from a high school that had only a token enrollment of less than ten percent blacks. Hence, the black subjects were more familiar with their white peers than the white subjects were with them, and this differential familiarity may have influenced the findings. Independent measures of racial prejudice in the two groups were not gathered, but the present approach would lead one to predict that the black subjects were less anti-white than the white subjects were anti-black. But if Banks et al. (unpublished) are correct in their interaction and familiarity explanation, it would be consistent with prediction 6 concerning the effects of convergent boundaries.

Other studies, too, suggest that lower-status groups do not show the same effects as upper-status groups. Doise and Sinclair (1973) found that lower-status Swiss boys often gave *more* favorable outgroup than ingroup ratings even after group status had been made salient through interaction with upper-status, college-preparatory boys. By contrast, the college-bound Swiss boys typically rated their group differentially more favorable as group status became more salient. Similarly, studies on sex biases among American undergraduates often find effects that resemble the ultimate attribution error among men but not among women. On a male-oriented task, women are regarded by both sexes as succeeding relative to men less because of ability than of luck (Deaux and Emswiller, 1974). The reverse did not hold true on a female-oriented task.

Do these findings across race, social class, and sex indicate that the ultimate attribution error is limited to high-status, dominant groups? This possibility arose earlier in our discussion of the Thibaut-Riecken (1955) and Ring (1964) experiments. Future work directed to this precise point is required. But for now it seems likely that the predicted trends act in similar ways across status groups (save for the differences specified in predictions 3c and 3d). Three related explanations for these results are tenable. First, the lower-status groups may simply not be as prejudiced as the higher-status groups. Only the Doise and Sinclair (1973) study employed an independent measure of prejudice, and they found slightly more *favorable* ratings were assigned the outgroup in an individual setting where group status was not salient. Moreover, Doise and Sinclair (1973, p. 153) discerned a sharp split within their lower-status group between those who gave higher or gave lower ratings to the higher-status outgroup. This observation suggests that identification with the more powerful outsiders and/or acceptance of their more restricted role among some lower-status members may block and even reverse the operation of the predicted trends. A second explanation, then, is that "group consciousness" may be critical. Consistent with this possibility, Banks et al. (unpublished) found that their black subjects revealed no effect, while the Swiss lower-status boys and American undergraduate women, presumably with their "group consciousness" less raised, revealed a derogation of their own group. Finally, it may well be that social class differences among Swiss adolescents and sex differences among American college students simply do not evoke the depth of group identification and differentiation needed to trigger the ultimate attribution error.

We conclude, then, that tests of the ultimate attribution error require independent measures of prejudice and stereotyping. This requirement raises particular problems for experimental research that attempts to investigate black-white intergroup perceptions with American college students. As repeated studies have demonstrated in recent years, racial climates on college campuses have changed sharply since Allport (1954) wrote *The Nature of Prejudice*. Blatant measures of racial, ethnic, and religious prejudice once served researchers well. Now unobtrusive, subtle measures are required to detect the avoidance patterns and other more complex forms of interracial interaction that characterize today's college life. A range of such indicators have been ingeniously devised by social psychologists, such as differential helping behavior (Gaertner, 1973, 1975, 1976), differential aggressive behavior (Donnerstein and Donnerstein, 1976, 1978), and a variety of other nonverbal measures (Weitz, 1972; Word et al., 1974). The nonverbal measures particularly commend themselves for tests of the ultimate attribution error, since voice judgments of, for example, taped speech, seating patterns, and interaction distance are easily and naturally included in such research designs.

In addition to the need for independent measures of prejudice, two other requirements of such tests of the present predictions are suggested by the limitations of the relevant studies discussed here. First, the full range of resolutions outlined in Table 23.1 to the conflict created by perceiving positive outgroup acts needs to be addressed. Second, the subjects in most of these experiments were removed from the behavior itself. Having behavior described to you and actually seeing it and being involved in it yourself are, of course, radically different phenomenal experiences. The fact that these results generally support the predictions is, perhaps, made more compelling by this distant stance. Nonetheless, direct tests of the ultimate attribution error require face-to-face interaction of the perceiver and actor.

A FINAL WORD

The cognitive analysis of prejudice has traditionally centered upon the concept of stereotype. Allport (1954) expanded this analysis with his emphasis upon categorization and the biases of normal cognitive processing. Recent advances in cognitive social psychology allow further expansion of this approach. One direction for this expansion involves systematic intergroup misattributions, an example of which has been outlined in this paper. We hope that the ultimate attribution error and other applications of recent work to the prejudice domain will begin to receive greater attention within social psychology so as to allow specification of the linkages between these mediating cognitions and intergroup behavior.

References

ADORNO, T. W., FRENKEL-BRUNSWIK, E., LEVINSON, D. J., AND SANFORD, R. N. *The authoritarian personality.* New York: Harper & Row, 1950.

ALLPORT, G. W. *Personality: A psychological interpretation.* New York: Holt, Rinehart & Winston, 1937.

ALLPORT, G. W. *The nature of prejudice.* Reading, Mass.: Addison-Wesley, 1954.

ALLPORT, G. W. *The nature of prejudice.* Garden City, N.Y.: Doubleday Anchor, 1958.

ALLPORT, G. W., AND PETTIGREW, T. F. Cultural influence on the perception of movement: The trapezoidal illusion among Zulus. *Journal of Abnormal and Social Psychology,* 1957, **55,** 104–113.

ALLPORT, G. W., AND POSTMAN, L. *The psychology of rumor.* New York: Holt, Rinehart & Winston, 1947.

ARONSON, E. Reconsiderations: *The nature of prejudice. Human Nature,* July 1978, **1,** 92–94, 96.

BANKS, W. C., McQUARTER, G. V., AND PRYOR J. In consideration of a cognitive-attributional basis for stereotypy. Unpublished manuscript. Princeton University, 1977.

BREWER, M. B., AND CAMPBELL, D. T. *Ethnocentrism and intergroup attitudes: East African evidence.* Beverly Hills: Sage, 1977.

BRUNER, J. S., AND POSTMAN, L. An approach to social perception. In W. Dennis (ed.), *Current trends in social psychology.* Pittsburgh, Pa.: University of Pittsburgh Press, 1948.

BUSS, A. H., AND PORTNOY, N. W. Pain tolerance and group identification. *Journal of Personality and Social Psychology,* 1967, **6,** 106–108.

CHARTERS, W. W., AND NEWCOMB, T. M. Some attitudinal effects of experimentally increased salience of a membership group. In G. E. Swanson, T. M. Newcomb, and E. L. Hartley (eds.), *Readings in social psychology* (revised ed.). New York: Holt, Rinehart & Winston, 1952.

DEAUX, K. AND EMSWILLER, T. Explanation of successful performance on sex-linked tasks: What is skill for the male is luck for the female. *Journal of Personality and Social Psychology,* 1974, **29,** 80–85.

DONNERSTEIN, M., AND DONNERSTEIN, E. Variables in interracial aggression. *Journal of Social Psychology,* 1976, **100,** 111–121.

DONNERSTEIN, M., AND DONNERSTEIN, E. Direct and vicarious censure in the control of interracial aggression. *Journal of Personality,* 1978, **48,** 162–175.

DOISE, W., AND SINCLAIR, A. The categorization process in intergroup relations. *European Journal of Social Psychology,* 1973, **3,** 145–157.

DUNCAN, B. L. Differential social perception and attribution of intergroup violence: Testing the lower limits of stereotyping of blacks. *Journal of Personality and Social Psychology,* 1976, **34,** 590–598.

GAERTNER, S. Helping behavior and racial discrimination among liberals and conservatives. *Journal of Personality and Social Psychology,* 1973, **25,** 335–341.

GAERTNER, S. The role of racial attitudes in helping behavior. *Journal of Social Psychology,* 1975, **97,** 95–101.

GAERTNER, S. Nonreactive measures in racial attitude research: A focus on "liberals." In P. A. Katz (ed.), *Towards the elimination of racism.* Elmsford, N.Y.: Pergamon, 1976.

GOFFMAN, E. *The presentation of self in everyday life.* New York: Doubleday Anchor, 1959.

GOFFMAN, E. *Strategic interaction.* Philadelphia: University of Pennsylvania Press, 1969.

GOFFMAN, E. *Relations in public.* New York: Harper & Row, 1971.

HARDING, J. Suggestions for revision of Allport's *The nature of prejudice.* Private communication, 1977.

HEIDER, F. Social perception and phenomenal causality. *Psychological Review,* 1944, **51,** 358–374.

HEIDER, R. *The psychology of interpersonal relations.* New York: Wiley, 1958.

JONES, E. E., AND DAVIS, K. E. From acts to dispositions: The attribution process in person perception. In L. Berkowitz (ed.), *Advances in experimental social psychology.* Vol. 2. New York: Academic Press, 1965.

JONES, E. E., AND HARRIS, V. A. The attribution of attitudes. *Journal of Experimental Social Psychology,* 1967, **3,** 1–24.

KANTER, R. M. Some effects of proportions on group life: Skewed sex ratios and responses to token women. *American Journal of Sociology,* 1977, **82,** 965–990.

KELLEY, H. H. Attribution theory in social psychology. In D. Levine (ed.), *Nebraska symposium on motivation, 1967.* Vol. 15. Lincoln: University of Nebraska Press, 1967.

KELLEY, H. H. Attribution in social interaction. In E. E. Jones, D. E. Kanouse, H. H. Kelley, R. E. Nisbett, S. Valins, and B. Weiner, *Attribution: Perceiving the causes of behavior.* Morristown, N.J.: General Learning Press, 1972.

LAMBERT, W. E., LIEBERMAN, E., AND POSER, E. G. The effect of increased salience of membership group on pain tolerance. *Journal of Personality,* 1960, **28,** 350–357.

LANGER, E. J., BLANK, A., AND CHANOWITZ, B. The mindlessness of ostensibly thoughtful action. *Journal of Personality and Social Psychology,* 1978, **36,** 635–642.

LANGER, E. J., AND NEWMAN, H. M. The role of mindlessness in a typical social psychological experiment. *Personality and Social Psychology Bulletin,* 1979, **5,** 295–306.

LEPPER, M. R., AND GREENE, D. Turning play into work: Effects of adult surveillance and extrinsic rewards on children's intrinsic motivation. *Journal of Personality and Social Psychology,* 1975, **31,** 479–486.

McKILLIP, J., DiMICELI, A. J., AND LUEBKE, J. Group salience and stereotyping. *Social Behavior and Personality,* 1977, **5,** 81–85.

MONSON, T. C., AND SNYDER, M. Actors, observers, and the attribution process. *Journal of Experimental Social Psychology,* 1977, **13,** 89–111.

PETTIGREW, T. F. *Racially separate or together?* New York: McGrawHill, 1971.

PETTIGREW, T. F. Three issues in ethnicity: Boundaries, deprivations, and perceptions. In J. M. Yinger and S. J. Cutler (eds.), *Major social issues: A multidisciplinary view.* New York: Free Press, 1978.

PETTIGREW, T. F. Foreword. In G. W. Allport, *The nature of prejudice.* Reading, Mass.: Addison-Wesley, 1979.

PHARES, E. J. *Locus of control in personality.* Morristown, N.J.: General Learning Press, 1976.

RING, K. Some determinants of interpersonal attraction in hierarchical relationships: A motivational analysis. *Journal of Personality,* 1964, **32,** 651–665.

ROSS, L. D. The intuitive psychologist and his shortcomings: Distortions in the attribution process. In L. Berkowitz (ed.), *Advances in experimental social psychology.* Vol. 10. New York: Academic Press, 1977.

ROSS, L. D., AMABILE, T. M., AND STEINMETZ, J. L. Social roles, social control, and biases in social-perception processes. *Journal of Personality and Social Psychology,* 1977, **35,** 485–494.

ROTTER, J. B. Generalized expectancies for internal versus external control of reinforcement. *Psychological Monographs,* 1966, **80**(1), Whole no. 609.

RUNCIMAN, W. G. *Relaive deprivation and social justice.* London: Routledge & Kegan Paul, 1966.

SMITH, M. B. The psychology of prejudice. *New York University Education Quarterly,* 1978, **9**(2), 29–32.

SUMNER, W. G. *Folkways.* New York: Ginn, 1906.

TAJFEL, H. Cognitive aspects of prejudice. *Journal of Social Issues,* 1969, **25,** 79–97.

TAJFEL, H. Experiments in intergroup discrimination. *Scientific American,* 1970, **223**(5), 96–102.

TAJFEL, H., AND BILLIG, M. Familiarity and categorization in intergroup behavior. *Journal of Experimental Social Psychology,* 1974, **10,** 159–170.

TAYLOR, D. M., AND JAGGI, V. Ethnocentrism and causal attribution in a south Indian context. *Journal of Cross-Cultural Psychology,* 1974, **5,** 162–171.

TAYLOR, S. E. The token in the small group: Research findings and theoretical implications. In J. Sweeney (ed.), *Psychology and politics.* New Haven, Conn.: Yale University Press, in press.

TAYLOR, S. E., AND KOIVUMAKI, J. H. The perception of self and others: Acquaintanceship, affect, and actor–observer differences. *Journal of Personality and Social Psychology,* 1976, **33,** 403–408.

THIBAUT, J. W., AND RIECKEN, H. Some determinants and consequences of the perception of social causality. *Journal of Personality,* 1955, **25,** 115–129.

TVERSKY, A., AND KAHNEMAN, D. Judgment under uncertainty: Heuristics and biases. *Science,* 1974, **185,** 1124–1131.

VANNEMAN, R. D., AND PETTIGREW, T. F. Race and relative deprivation in the urban United States. *Race,* 1972, **13,** 461–486.

WANG, G., AND McKILLIP, J. Ethnic identification and judgments of an accident. *Personality and Social Psychology Bulletin,* 1978, **4,** 296–299.

WEINER, B., FRIEZE, I., KUKLA, A., REED, L., REST, S., AND ROSENBAUM, R. M. Perceiving the causes of success and failure. In E. E. Jones, D. E. Kanouse, H. H. Kelley, R. E. Nisbett, S. Valins, and B. Weiner, *Attribution: Perceiving the causes of behavior.* Morristown, N.J.: General Learning Press, 1972.

WEITZ, S. Attitude, voice and behavior: A repressed affect model of interracial interaction. *Journal of Personality and Social Psychology,* 1972, **24,** 14–21.

WORD, C. O., ZANNA, M. P., AND COOPER, J. The nonverbal mediation of self-fulfilling prophecies in interracial interaction. *Journal of Experimental Social Psychology,* 1974, **10,** 109–120.

ZIPF, G. K. *Human behavior and the principle of least effort.* Reading, Mass.: Addison-Wesley, 1949.

24

Experiments in Group Conflict

Muzafer Sherif

What are the conditions which lead to harmony or friction between groups of people? Here the question
is approached by means of controlled situations in a boys' summer camp.

Conflict between groups—whether between boys' gangs, social classes, "races"
or nations—has no simple cause, nor is mankind yet in sight of a cure. It is often
rooted deep in personal, social, economic, religious and historical forces. Nevertheless, it is possible to identify certain general factors which have a crucial influence on the attitude of any group toward others. Social scientists have long sought
to bring these factors to light by studying what might be called the "natural history" of groups and group relations. Intergroup conflict and harmony is not a subject that lends itself easily to laboratory experiments. But in recent years there has
been a beginning of attempts to investigate the problem under controlled yet life-
like conditions, and I shall report here the results of a program of experimental
studies of groups which I started in 1948. Among the persons working with me
were Marvin B. Sussman, Robert Huntington, O. J. Harvey, B. Jack White, William R. Hood and Carolyn W. Sherif. The experiments were conducted in 1949,
1953 and 1954; this article gives a composite of the findings.

We wanted to conduct our study with groups of the informal type, where group
organization and attitudes would evolve naturally and spontaneously, without formal direction or external pressures. For this purpose we conceived that an isolated
summer camp would make a good experimental setting, and that decision led us
to choose as subjects boys about eleven or twelve years old, who would find camping
natural and fascinating. Since our aim was to study the development of group relations among these boys under carefully controlled conditions, with as little in-

terference as possible from personal neuroses, background influences or prior experiences, we selected normal boys of homogeneous background who did not know
one another before they came to the camp.

They were picked by a long and thorough procedure. We interviewed each boy's
family, teachers and school officials, studied his school and medical records, obtained his scores on personality tests and observed him in his classes and at play
with his schoolmates. With all this information we were able to assure ourselves
that the boys chosen were of like kind and background: all were healthy, socially
well-adjusted, somewhat above average in intelligence and from stable, white,
Protestant, middle-class homes.

None of the boys was aware that he was part of an experiment on group relations. The investigators appeared as a regular camp staff—camp directors, counselors and so on. The boys met one another for the first time in buses that took
them to the camp, and so far as they knew it was a normal summer of camping.
To keep the situation as lifelike as possible, we conducted all our experiments
within the framework of regular camp activities and games. We set up projects
which were so interesting and attractive that the boys plunged into them enthusi-
astically without suspecting that they might be test situations. Unobtrusively we
made records of their behavior, even using "candid" cameras and microphones
when feasible.

We began by observing how the boys became a coherent group. The first of our
camps was conducted in the hills of northern Connecticut in the summer of 1949.

FIGURE 24.1
Members of one group of boys raid the bunkhouse of another group during the first experiment of the
author and his associates, performed at a summer camp in Connecticut. The rivalry of the groups
was intensified by the artificial separation of their goals. (Photograph by Muzafer Sherif.)

When the boys arrived, they were all housed at first in one large bunkhouse. As was to be expected, they quickly formed particular friendships and chose buddies. We had deliberately put all the boys together in this expectation, because we wanted to see what would happen later after the boys were separated into different groups. Our object was to reduce the factor of personal attraction in the formation of groups. In a few days we divided the boys into two groups and put them in different cabins. Before doing so, we asked each boy informally who his best friends were, and then took pains to place the "best friends" in different groups as far as possible. (The pain of separation was assuaged by allowing each group to go at once on a hike and campout.)

As everyone knows, a group of strangers brought together in some common activity soon acquires an informal and spontaneous kind of organization. It comes to look upon some members as leaders, divides up duties, adopts unwritten norms of behavior, develops an *esprit de corps*. Our boys followed this pattern as they shared a series of experiences. In each group the boys pooled their efforts, organized duties and divided up tasks in work and play. Different individuals assumed different responsibilities. One boy excelled in cooking. Another led in athletics. Others, though not outstanding in any one skill, could be counted on to pitch in and do their level best in anything the group attempted. One or two seemed to disrupt activities, to start teasing at the wrong moment or offer useless suggestions. A few boys consistently had good suggestions and showed ability to coordinate the efforts of others in carrying them through. Within a few days one person had proved himself more resourceful and skillful than the rest. Thus, rather quickly, a leader and lieutenants emerged. Some boys sifted toward the bottom of the heap, while others jockeyed for higher positions.

We watched these developments closely and rated the boys' relative positions in the group, not only on the basis of our own observations but also by informal sounding of the boys' opinions as to who got things started, who got things done, who could be counted on to support group activities.

As the group became an organization, the boys coined nicknames. The big, blond, hardy leader of one group was dubbed "Baby Face" by his admiring followers. A boy with a rather long head became "Lemon Head." Each group developed its own jargon, special jokes, secrets and special ways of performing tasks. One group, after killing a snake near a place where it had gone to swim, named the place "Moccasin Creek" and thereafter preferred this swimming hole to any other, though there were better ones nearby.

Wayward members who failed to do things "right" or who did not contribute their bit to the common effort found themselves receiving the "silent treatment,"

FIGURE 24.2
Members of both groups collaborate in common enterprises during the second experiment, performed at a summer camp in Oklahoma. At the top, the boys of the two groups prepare a meal. In the middle, the two groups surround a water tank while trying to solve a water-shortage problem. At the bottom, the members of one group entertain the other. (Photographs by Muzafer Sherif.)

ridicule or even threats. Each group selected symbols and a name, and they had these put on their caps and T-shirts. The 1954 camp was conducted in Oklahoma, near a famous hideaway of Jesse James called Robber's Cave. The two groups of boys at this camp named themselves the Rattlers and the Eagles.

Our conclusions on every phase of the study were based on a variety of observations, rather than on any single method. For example, we devised a game to test the boys' evaluations of one another. Before an important baseball game, we set up a target board for the boys to throw at, on the pretense of making practice for the game more interesting. There were no marks on the front of the board for the boys to judge objectively how close the ball came to a bull's-eye, but, unknown to them, the board was wired to flashing lights behind so that an observer could see exactly where the ball hit. We found that the boys consistently overestimated the performances by the most highly regarded members of their group and underestimated the scores of those of low social standing.

The attitudes of group members were even more dramatically illustrated during a cook-out in the woods. The staff supplied the boys with unprepared food and let them cook it themselves. One boy promptly started to build a fire, asking for help in getting wood. Another attacked the raw hamburger to make patties. Others prepared a place to put buns, relishes and the like. Two mixed soft drinks from flavoring and sugar. One boy who stood around without helping was told by others to "get to it." Shortly the fire was blazing and the cook had hamburgers sizzling. Two boys distributed them as rapidly as they became edible. Soon it was time for the watermelon. A low-ranking member of the group took a knife and started toward the melon. Some of the boys protested. The most highly regarded boy in the group took over the knife, saying, "You guys who yell the loudest get yours last."

When the two groups in the camp had developed group organization and spirit, we proceeded to the experimental studies of intergroup relations. The groups had had no previous encounters; indeed, in the 1954 camp at Robber's Cave the two groups came in separate buses and were kept apart while each acquired a group feeling.

Our working hypothesis was that when two groups have conflicting aims—i.e., when one can achieve its ends only at the expense of the other—their members will become hostile to each other even though the groups are composed of normal well-adjusted individuals. There is a corollary to this assumption which we shall consider later. To produce friction between the groups of boys we arranged a tournament of games: baseball, touch football, a tug-of-war, a treasure hunt and so on. The tournament started in a spirit of good sportsmanship. But as it progressed good feeling soon evaporated. The members of each group began to call their rivals "stinkers," "sneaks" and "cheaters." They refused to have anything more to do with individuals in the opposing group. The boys in the 1949 camp turned against buddies whom they had chosen as "best friends" when they first arrived at the camp. A large proportion of the boys in each group gave negative ratings to all the boys in the other. The rival groups made threatening posters and planned

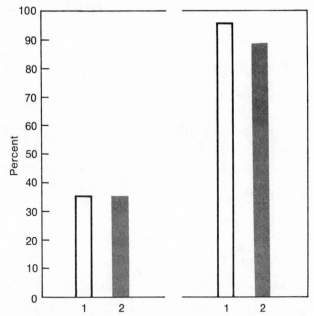

FIGURE 24.3
Friendship choices of campers for others in their own cabin are shown for Red Devils *(white)* and Bulldogs. *(gray).* At first, a low percentage of friendships were in the cabin group *(left).* After five days, most friendship choices were within the group *(right).*

raids, collecting secret hoards of green apples for ammunition. In the Robber's Cave camp the Eagles, after a defeat in a tournament game, burned a banner left behind by the Rattlers; the next morning the Rattlers seized the Eagles' flag when they arrived on the athletic field. From that time on name-calling scuffles and raids were the rule of the day.

Within each group, of course, solidarity increased. There were changes: one group deposed its leader because he could not "take it" in the contests with the adversary; another group overnight made something of a hero of a big boy who had previously been regarded as a bully. But morale and cooperativeness within the group became stronger. It is noteworthy that this heightening of cooperativeness and generally democratic behavior did not carry over to the group's relations with other groups.

We now turned to the other side of the problem: How can two groups in conflict be brought into harmony? We first undertook to test the theory that pleasant social contacts between members of conflicting groups will reduce friction between them. In the 1954 camp we brought the hostile Rattlers and Eagles together for social events: going to the movies, eating in the same dining room and so on. But far from reducing conflict, these situations only served as opportunities for the rival groups to berate and attack each other. In the dining-hall line they shoved each other aside, and the group that lost the contest for the head of the line shouted

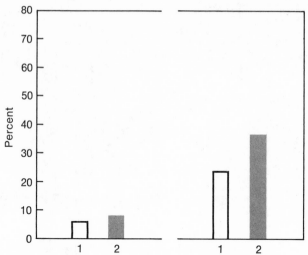

FIGURE 24.4
During conflict between the two groups in the Robber's Cave
experiment, there were few friendships between cabins *(left)*.
After cooperation toward common goals had restored good
feelings, the number of friendships between groups rose
significantly *(right)*.

"Ladies first!" at the winner. They threw paper, food and vile names at each other
at the tables. An Eagle bumped by a Rattler was admonished by his fellow Eagles
to brush "the dirt" off his clothes.

We then returned to the corollary of our assumption about the creation of conflict. Just as competition generates friction, working in a common endeavor should
promote harmony. It seemed to us, considering group relations in the everyday
world, that where harmony between groups is established, the most decisive factor
is the existence of "superordinate" goals which have a compelling appeal for both
but which neither could achieve without the other. To test this hypothesis experimentally, we created a series of urgent, and natural, situations which challenged
our boys.

One was a breakdown in the water supply. Water came to our camp in pipes
from a tank about a mile away. We arranged to interrupt it and then called the
boys together to inform them of the crisis. Both groups promptly volunteered to
search the water line for the trouble. They worked together harmoniously, and before the end of the afternoon they had located and corrected the difficulty.

A similar opportunity offered itself when the boys requested a movie. We told
them that the camp could not afford to rent one. The two groups then got together,
figured out how much each group would have to contribute, chose the film by a
vote and enjoyed the showing together.

One day the two groups went on an outing at a lake some distance away. A
large truck was to go to town for food. But when everyone was hungry and ready
to eat, it developed that the truck would not start (we had taken care of that). The

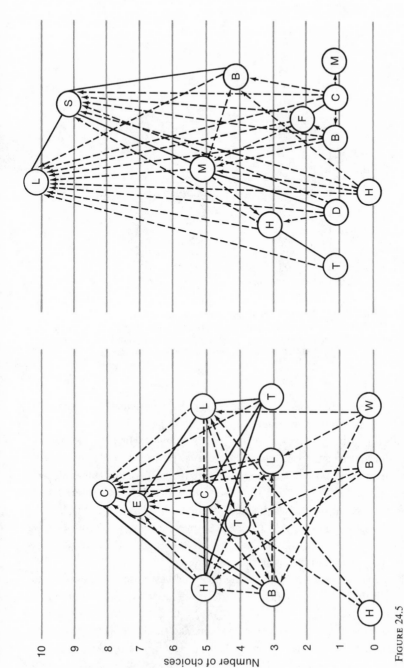

Number of choices

FIGURE 24.5
Sociograms represent patterns of friendship choice within the fully developed groups. One-way friendships are indicated by broken arrows; reciprocated friendships, by solid lines. Leaders were among those highest in the popularity scale. Bulldogs (*left*) had a close-knit organization with good group spirit. Low-ranking members participated less in the life of the group but were not rejected. Red Devils (*right*) lost the tournament of games between the groups. They had less group unity and were sharply stratified.

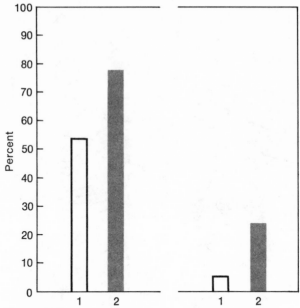

FIGURE 24.6
Negative ratings of each group by the other were common during the period of conflict *(left)* but decreased when harmony was restored *(right)*. The graphs show percentage who thought *all* (rather than *some* or *none*) of the other group were cheaters, sneaks, and so forth.

boys got a rope—the same rope they had used in their acrimonious tug-of-war—and all pulled together to start the truck.

These joint efforts did not immediately dispel hostility. At first the groups returned to the old bickering and name-calling as soon as the job in hand was finished. But gradually the series of cooperative acts reduced friction and conflict. The members of the two groups began to feel more friendly to each other. For example, a Rattler whom the Eagles disliked for his sharp tongue and skill in defeating them became a "good egg." The boys stopped shoving in the meal line. They no longer called each other names, and sat together at the table. New friendships developed between individuals in the two groups.

In the end the groups were actively seeking opportunities to mingle, to entertain and "treat" each other. They decided to hold a joint campfire. They took turns presenting skits and songs. Members of both groups requested that they go home together on the same bus, rather than on the separate buses in which they had come. On the way the bus stopped for refreshments. One group still had five dollars which they had won as a prize in a contest. They decided to spend this sum on refreshments. On their own initiative they invited their former rivals to be their guests for malted milks.

Our interviews with the boys confirmed this change. From choosing their "best friends'' almost exclusively in their own group, many of them shifted to listing boys in the other group as best friends (see Fig. 22.4). They were glad to have a second chance to rate boys in the other group, some of them remarking that they had changed their minds since the first rating made after the tournament. Indeed they had. The new ratings were largely favorable (see Fig. 22.6).

Efforts to reduce friction and prejudice between groups in our society have usually followed rather different methods. Much attention has been given to bringing members of hostile groups together socially, to communicating accurate and favorable information about one group to the other, and to bringing the leaders of groups together to enlist their influence. But as everyone knows, such measures sometimes reduce intergroup tensions and sometimes do not. Social contacts, as our experiments demonstrated, may only serve as occasions for intensifying conflict. Favorable information about a disliked group may be ignored or reinterpreted to fit stereotyped notions about the group. Leaders cannot act without regard for the prevailing temper in their own groups.

What our limited experiments have shown is that the possibilities for achieving harmony are greatly enhanced when groups are brought together to work toward common ends. Then favorable information about a disliked group is seen in a new light, and leaders are in a position to take bolder steps toward cooperation. In short, hostility gives way when groups pull together to achieve overriding goals which are real and compelling to all concerned.

Reference

SHERIF, MUZAFER, AND SHERIF, CAROLYN W. *Groups in harmony and tension.* Harper & Brothers, 1953.

25

Jigsaw Groups and the Desegregated Classroom: In Pursuit of Common Goals

Elliot Aronson and Diane Bridgeman

The desegregated classroom has not produced many of the positive results initially expected by social scientists some 25 years ago. It is argued that one of the major reasons for this failure is the overemphasis on competitiveness at the expense of interdependence in the classroom. In short, students in most classrooms very rarely cooperate with each other in pursuit of common goals. In this article, we describe a program of research in which elementary school students are "forced" to spend part of their classroom time mastering material in an interdependent structure. The results indicate that such structured interdependence increases the self-esteem, the morale, the interpersonal attraction, and the empathy of students across ethnic and racial divisions, and also improves the academic performance of minority students without hampering the performance of the ethnic majority.

There were high hopes when the Supreme Court outlawed school segregation a quarter of a century ago. If black and white children could share classrooms and become friends, it was thought that perhaps they could develop relatively free of racial prejudice and some of the problems that accompany prejudice. The case that brought about the court's landmark decision was that of *Brown* v. *Board of Education;* the decision reversed the 1896 ruling *(Plessy* v. *Ferguson)* that held that it was permissible to segregate racially, as long as equal facilities were provided for both races. In the *Brown* case, the court held that psychologically there could be no such thing as "separate but equal." The mere fact of separation implied to the minority group in question that its members were inferior to those of the majority.

The Brown decision was not only a humane interpretation of the Constitution, it was also the beginning of a profound and exciting social experiment. As Ste-

Reprinted with permission of the authors and Division 8 of the American Psychological Association and of Sage Publications, Inc., from *Personality and Social Psychology Bulletin*, Vol. 5, No. 4, 1979, pp. 438–466.

phan (1978) has recently pointed out, the testimony of social psychologists in the *Brown* case, as well as in previous similar cases in state supreme courts, suggested strongly that desegregation would not only reduce prejudice but also increase the self-esteem of minority groups and improve their academic performance. Of course the social psychologists who testified never meant to imply that such benefits would accrue automatically. Certain preconditions would have to be met. These preconditions were most articulately stated by Allport in his classic, *The Nature of Prejudice,* published the same year as the Supreme Court decision:

> Prejudice . . . may be reduced by equal status contact between majority and minority groups in the pursuit of common goals. The effect is greatly enhanced if this contact is sanctioned by institutional supports (i.e., by law, custom or local atmosphere), and provided it is of a sort that leads to the perception of common interests and common humanity between members of the two groups [Allport, 1954, p. 281].

THE EFFECTS OF DESEGREGATION

A quarter of a century after desegregation was begun, an assessment of its effectiveness is not encouraging. One of the most careful and thoroughgoing longitudinal studies of desegregation was the Riverside project conducted by Gerard and Miller (1975). They found that long after the schools were desegregated, black, white, and Mexican-American children tended not to integrate but to hang together in their own ethnic clusters. Moreover, anxiety increased and remained high long after desegregation occurred. These trends are echoed in several other studies. Indeed, the most careful, scholarly reviews of the research show few if any benefits (see St. John, 1975; Stephan, 1978). For example, according to Stephan's review, there is no single study that shows a significant increase in the self-esteem of minority children following desegregation; in fact, in fully 25 percent of the studies, desegregation is followed by a significant decrease in the self-esteem of young minority children. Moreover, Stephan reports that desegregation reduced the prejudice of whites toward blacks in only 13 percent of the school systems studied. The prejudice of blacks toward whites *increased* in about as many cases as it decreased. Similarly, studies of the effects of desegregation on the academic performance of minority children present a mixed and highly variable picture.

What went wrong? Let us return to Allport's prediction: Equal status contact in pursuit of common goals, sanctioned by authority, will produce beneficial effects. We will look at each of these three factors separately.

Sanction by Authority

In some school districts there was clear acceptance and enforcement of the ruling by responsible authority. In others the acceptance was not as clear. In still others (especially in the early years) local authorities were in open defiance of the law. Pettigrew (1961) has shown that desegregation proceeded more smoothly and with less violence in those localities where local authorities sanctioned integration. But such variables as self-esteem and the reduction of prejudice do not necessarily change for the better even where authority clearly sanctions desegregation. While sanction by authority may be necessary, it is clearly not a sufficient condition.

Equal Status Contact

The definition of equal status is a trifle slippery. In the case of school desegregation, we would claim that there is equal status on the grounds that all children in the fifth grade (for example) have the same "occupational" status; that is, they are all fifth grade students. On the other hand, if the teacher is prejudiced against blacks, he may treat them less fairly than he treats whites, thus lowering their perceived status in the classroom (see Gerard and Miller, 1975). Moreover, if, because of an inferior education (prior to desegregation) or because of language difficulties, black or Mexican-American students perform poorly in the classroom, this could also lower their status among their peers. An interesting complication was introduced by Cohen (1972). While Allport (1954) predicted that positive interactions will result if cooperative equal status is achieved, expectation theory, as developed by Cohen, holds that even in such an environment biased expectations by both whites and blacks may lead to sustained white dominance. Cohen reasoned that both of these groups accepted the premise that the majority group's competence results in dominance and superior achievement. She suggested that alternatives be created to reverse these often unconscious expectations. According to Cohen, at least a temporary exchange of majority and minority roles is therefore required as a prelude to equal status. In one study (Cohen and Roper, 1972), black children were instructed in building radios and in how to teach this skill to others. Then a group of white children and the newly trained black children viewed a film of themselves building the radios. This was followed by some of the black children teaching the whites how to construct radios while others taught a black administrator. Then all the children came together in small groups. Equal status interactions were found in the groups where black children had taught whites how to construct the radios. The other group, however, demonstrated the usual white dominance. We will return to this point in a moment.

In Pursuit of Common Goals

In the typical American classroom, children are almost never engaged in the pursuit of common goals. During the past several years, we and our colleagues have systematically observed scores of elementary school classrooms and have found that, in the vast majority of these cases, the process of education is highly competitive. Children vie with one another for good grades and the respect of the teacher. This occurs not only during the quizzes and exams but also in the informal give and take of the classroom, where children typically learn to raise their hands (often frantically) in response to questions from the teacher, groan when someone else is called upon, and revel in the failure of their classmates. This pervasive competitive atmosphere unwittingly leads the children to view one another as foes to be heckled and vanquished. In a newly desegregated school, all other things being equal, this atmosphere could exacerbate whatever prejudice existed prior to desegregation.

A dramatic example of dysfunctional competition was demonstrated by Sherif et al. (1961) in the classic "Robber's Cave" experiment. In this field experiment, the investigators encouraged intergroup competition between two teams of boys at a summer camp; this created fertile ground for anger and hostility even in previously benign, noncompetitive circumstances—like watching a movie. Positive relations between the groups were ultimately achieved only after both groups were required to work cooperatively to solve a common problem.

It is our contention that the competitive process interacts with "equal status contact." That is to say, whatever differences in ability that existed between minority children and white children prior to desegregation are emphasized by the competitive structure of the learning environment; furthermore, since segregated school facilities are rarely equal, minority children frequently enter the newly desegregated school at a distinct disadvantage, which is made more salient by the competitive atmosphere.

It was this reasoning that led Aronson and his colleagues (1975, 1978a) to develop the hypothesis that interdependent learning environments would establish the conditions necessary for the increase in self-esteem and performance and the decrease in prejudice that were expected to occur as a function of desegregation. Toward this end they developed a highly structured method of interdependent learning and systematically tested its effects in a number of elementary school classrooms. The aim of this research program was not merely to compare the effects of cooperation and competition in a classroom setting. This had been ably demonstrated by other investigators dating as early as Deutsch's (1949) experiment. Rather, the intent was to devise a cooperative classroom structure that could be utilized easily by classroom teachers on a long-term sustained basis and to evaluate the effects of this intervention via a well-controlled series of field experiments. In short, this project is an action research program aimed at developing and evaluating a classroom atmosphere that can be sustained by the classroom teachers long after the researchers have packed up their questionnaires and returned to the more cozy environment of the social psychological laboratory.

The method is described in detail elsewhere (Aronson et al., 1978a). Briefly, students are placed in six-person learning groups. The day's lesson is divided into six paragraphs such that each student has one segment of the written material.[1] Each student has a unique and vital part of the information, which, like the pieces of a jigsaw puzzle, must be put together before any of the students can learn the whole picture. The individual must learn his own section and teach it to the other members of the group. The reader will note that in this method each child spends part of her time in the role of expert. Thus, the method incorporates Cohen's findings (previously discussed) within the context of an equal status contact situation.

Working with this "jigsaw" technique, children gradually learn that the old competitive behavior is no longer appropriate. Rather, in order to learn all of the material (and thus perform well on a quiz), each child must begin to listen to the others, ask appropriate questions, and in other ways contribute to the group. The process makes it possible for children to pay attention to one another and begin to appreciate each other as potentially valuable resources. It is important to emphasize that the motivation of the students is not necessarily altruistic; rather, it is primarily self-interest, which, in this case, happens also to produce outcomes that are beneficial to others.

EXPERIMENTS IN THE CLASSROOM

Systematic research in the classroom has produced consistently positive results. The first experiment to investigate the effects of the jigsaw technique was conducted by Blaney et al. (1977). The schools in Austin, Texas, had recently been desegregated, producing a great deal of tension and even some interracial skirmishes throughout the school system. In this tense atmosphere, the jigsaw technique was introduced in 10 fifth grade classrooms in seven elementary schools. Three classes from among the same schools were also used as controls. The control classes were taught by teachers who, while using traditional techniques, were rated very highly by their peers. The experimental classes met in jigsaw groups for about forty-five minutes a day, three days a week for six weeks. The curriculum was basically the same for the experimental and control classes. Students in the jigsaw groups showed significant increases in their liking for their groupmates both within and across ethnic boundaries. Moreover, children in jigsaw groups showed a significantly greater increase in self-esteem than children in the control classrooms. This was true for Anglo children as well as for ethnic minorities. Anglos and blacks showed greater liking for school when taught in the jigsaw classrooms than when studying in traditional classrooms. (The Mexican-American students showed a tendency to like school *less* in the jigsaw classes; this will be discussed shortly.)

---

[1] The method works best with discrete, continuous written material (like social studies), but it has been successfully utilized with mathematics and language arts as well.

These results were essentially replicated by Geffner (1978) in Watsonville, California—a community consisting of approximately 50 percent Anglos and 50 percent Mexican-Americans. As a control for the possibility of a Hawthorne effect, Geffner compared the behavior of children in classrooms using the jigsaw and other cooperative learning techniques with that of children in highly innovative (but not interdependent) classroom environments as well as with traditional classrooms. Geffner found consistent and significant gains within classrooms using jigsaw and other cooperative learning techniques. Specifically, children in these classes showed increases in self-esteem as well as increases in liking for school. Negative ethnic stereotypes were also diminished. That is, children increased their positive general attitudes toward their own ethnic group as well as toward members of other ethnic groups to a far greater extent than did children in traditional and innovative classrooms.

Changes in academic performance were assessed in an experiment by Lucker, et al. (1977). The subjects were 303 fifth and sixth grade students from five elementary schools in Austin, Texas. Six classrooms were taught in the jigsaw manner, while five classrooms were taught traditionally by competent teachers. For two weeks children were taught a unit on colonial America taken from a fifth grade textbook. All children were then given the same standardized test. The results showed that Anglo students performed just as well in jigsaw classes as they did in traditional classes ($\overline{X} = 66.6$ and 67.3 respectively); minority children performed significantly better in jigsaw classes than in traditional classes ($\overline{X} = 56.6$ and 49.7 respectively).[2] The difference for minority students was highly significant. Only two weeks of jigsaw activity succeeded in narrowing the performance gap between Anglos and minorities from more than 17 percentage points to about 10 percentage points. Interestingly enough, the jigsaw method apparently does *not* work a special hardship on high ability students: Students in the highest quartile in reading ability benefited just as much as students in the lowest quartile.

UNDERLYING MECHANISMS

Increased Participation

We have seen that learning in a small interdependent group leads to greater interpersonal attraction, self-esteem and liking for school, more positive inter-ethnic and intra-ethnic perceptions, and, for ethnic minorities, an improvement in academic performance. We think that some of our findings are due to more active involvement in the learning process under conditions of reduced anxiety. In jigsaw, children are required to participate. This increase in participation should enhance interest, which would result in an improvement in performance as well as

---

[2] The mean scores have been converted to percentage of correct answers.

an increased liking for school—all other things being equal. But all other things are sometimes not equal. For example, in the study by Blaney et al. (1977), there was some indication from our observation of the groups that many of the Mexican-American children were experiencing some anxiety as a result of being required to participate more actively. This seemed to be due to the fact that these children had difficulty with the English language, which produced some embarrassment in working with a group dominated by Anglos. In a traditional classroom, it is relatively easy to "become invisible" by remaining quiet and refusing to volunteer. Not so in jigsaw. This observation was confirmed by the data on liking for school. Blaney et al. found that Anglos and blacks in jigsaw classrooms liked school better than those in the traditional classrooms, while for Mexican-Americans the reverse was true. This anxiety could be reduced if Mexican-American children were in a situation in which it was not embarrassing to be more articulate in Spanish than in English. Thus, Geffner (1978), working in a situation in which both the residential and school population was approximately 50 percent Spanish-speaking, found that Mexican-American children (like Anglos and blacks) increased their liking for school to a greater extent in the cooperative groups than in traditional classrooms.

Increases in Empathic Role-taking

Only a small subset of our results is attributable to increases in active participation in and of itself. We believe that people working together in an interdependent fashion increase their ability to take one another's perspective. For example, suppose that Jane and Carlos are in a jigsaw group. Carlos is reporting and Jane is having difficulty following him. She doesn't quite understand because his style of presentation is different from what she is accustomed to. Not only must she pay close attention, but, in addition, she must find a way to ask questions that Carlos will understand and that will elicit the additional information she needs. In order to accomplish this, she must get to know Carlos, put herself in his shoes, empathize.

Bridgeman (1977) tested this notion. She reasoned that taking one another's perspective is required and practiced in jigsaw learning. Accordingly, the more experience students have with the jigsaw process, the greater will their role-taking abilities become. In her experiment, Bridgeman administered a revised version of Chandler's (1973) role-taking cartoon series to 120 fifth grade students. Roughly half of the students spent eight weeks in a jigsaw learning environment while the others were taught in either traditional or innovative small group classrooms. Each of the cartoons in the Chandler test depicts a central character caught up in a chain of psychological cause and effect, such that the character's subsequent behavior is shaped by and fully comprehensible only in terms of the preceding events. In one of the sequences, for example, a boy who has been saddened by seeing his father off at the airport begins to cry when he later receives a gift of a toy airplane similar to the one that had carried his father away. Midway into each sequence, a second character is introduced in the role of a late-arriving bystander who witnesses the resultant behaviors of the principal character but is not privy to the causal events. Thus, the subject is in a privileged position relative to the story character, whose role the subject is later asked to assume. The cartoon series measures the degree to which the subject is able to set aside facts known only to him or herself and adopt a perspective measurably different from his or her own. For example, while the subject knows why the child in the above sequence cries when he receives the toy airplane, the mailman who delivered the toy is not privy to this knowledge. What happens when the subject is asked to take the mailman's perspective?

After eight weeks, students in the jigsaw classrooms were better able to put themselves in the bystander's place than students in the control classrooms. For example, when the mailman delivered the toy airplane to the little boy, students in the control classrooms tended to assume that the mailman knew the boy would cry; that is, they behaved as if they believed that the mailman knew that the boy's father had recently left town on an airplane—simply because they (the subjects) had this information. On the other hand, students who had participated in a jigsaw group were much more successful at taking the mailman's role—realizing that he could not possibly understand why the boy would cry upon receiving a toy airplane.

Attributions for Success and Failure

Working together in the pursuit of common goals changes the "observer's" attributional patterns. There is some evidence to support the notion that cooperation increases the tendency for individuals to make the same kind of attributions for success and failure to their partners as they do to themselves. In an experiment by Stephan et al. (1978), it was found (as it has been in several experiments by others) that when an individual succeeds at a task, he tends to attribute his success dispositionally (e.g. skill), but when he fails he tends to make a situational attribution (e.g., luck). Stephan et al. went on to demonstrate that individuals engaged in an *interdependent* task make the same kinds of attributions to their partner's performance as they do to their own. This was not the case in competitive interactions.

Effects of Dependent Variables on One Another

It is reasonable to assume that the various consequences of interdependent learning become antecedents for one another. Just as low self-esteem can work to inhibit a child from performing well, anything that increases self-esteem is likely to produce an increase in performance among underachievers. Conversely, as Franks and Marolla (1976) have indicated, increases in performance should bring about increases in self-esteem. Similarly, being treated with increased attention and respect by one's peers (as almost inevitably happens in jigsaw groups) is another important antecedent of self-esteem, according to Franks and Marolla. There is ample evidence for a two-way causal connection between performance and self-esteem (see Covington and Beery, 1976; Purkey, 1970).

OTHER COOPERATIVE TECHNIQUES

In recent years a few research teams utilizing rather different techniques for structuring cooperative behavior have produced an array of data consistent with those resulting from the jigsaw technique. For example, Cook and his colleagues (1978) have shown that interracial cooperative groups in the laboratory underwent a significant improvement in attitudes about people of other races. In subsequent field experiments, Cook and his colleagues found that interdependent groups produced more improved attitudes toward members of previously disliked racial groups than was present in noninterdependent groups. It should be noted, however, that no evidence for generalization was found; that is, the positive change was limited to the specific members of the interdependent group and did not extend to the racial group as a whole.

Johnson and Johnson (1975) have developed the "Learning Together" model, which is a general and varied approach to interdependent classroom learning. Basically, Johnson and Johnson have found evidence for greater cross-ethnic friendship ratings, greater self-esteem, and higher motivation in their cooperative groups than in control conditions. They have also found increases in academic performance.

In a different vein, Slavin (1978) and DeVries, Edwards, and Slavin (1978) have developed two highly structured techniques that combine within-group cooperation with across-group competition. These techniques, "Teams Games and Tournaments" (TGT) and "Student Teams Achievement Divisions" (STAD), have consistently produced beneficial results in lower class, multi-racial classrooms. Basically, in TGT and STAD, children form heterogeneous five-person teams; each member of a team is given a reasonably good opportunity to do well by dint of the fact that she competes against a member of a different team with similar skills to her own. Her individual performance contributes to her team's score. The results are in the same ball park as jigsaw: Children participating in TGT and STAD groups show a greater increase in sociometric, cross-racial friendship choices and more observed cross-racial interactions than control conditions. They also show more satisfaction with school than the controls do. Similarly, TGT and STAD produce greater learning effectiveness among racial minorities than do the control groups.

It is interesting to note that the basic results of TGT and STAD are similar to those of the jigsaw technique in spite of one major difference in procedure: While the jigsaw technique makes an overt attempt to minimize competition, TGT and STAD actually promote competitiveness and utilize it across teams—within the context of intrateam cooperation. We believe that this difference is more apparent than real. In most classrooms where jigsaw has been utilized, the students are in jigsaw groups for less than two hours per day. The rest of the class time is spent in a myriad of process activities, many of which are competitive in nature. Thus, what seems important in both techniques is that *some* specific time is structured around cooperativeness. Whether the beneficial results are produced *in spite* of a surrounding atmosphere of competitiveness or *because* of it is the task of future research to determine.

CONCLUSIONS

We are not suggesting that jigsaw learning or any other cooperative method constitutes the solution to our interethnic problems. What we have shown is that beneficial effects occur as a result of structuring the social psychological aspects of classroom learning so that children spend at least a portion of their time in pursuit of common goals. These effects are in accordance with predictions made by social scientists in their testimony favoring desegregating schools some 25 years ago. It is important to emphasize the fact that the jigsaw method has proved effective even if it is employed for as little as 20 percent of a child's time in the classroom. Moreover, other techniques have produced beneficial results even when interdependent learning was purposely accompanied by competitive activities. Thus, the data do not indicate the desirability of either placing a serious limit on classroom competition or interfering with individually guided education. Interdependent learning can and does coexist easily with almost any other method used by teachers in the classroom.

References

ARONSON, E., STEPHEN, C., SIKES, J., BLANEY, N., AND SNAPP, M. *The Jigsaw Classroom.* Beverly Hills: Sage Publications, 1978. (a)

ARONSON, E., BRIDGEMAN, D. L., AND GEFFNER, R. The effects of a cooperative classroom structure on students' behavior and attitudes. In D. Bar-Tal and L. Saxe (eds.), *Social Psychology of Education: Theory and Research.* Washington, D.C.: Hemisphere, 1978. (b)

BLANEY, N. T., STEPHAN, C., ROSENFIELD, D., ARONSON, E., AND SIKES, J. Interdependence in the classroom: A field study. *Journal of Educational Psychology,* 1977, **69,** 139–146.

BRIDGEMAN, D. L. Enhanced role taking through cooperative interdependence: A field study. *Child Development,* 1981, **52,** 1231–1238.

CHANDLER, M. J. Egocentrism and antisocial behavior: The assessment and training of social perspective-taking skills. *Developmental Psychology,* 1973, **9,** 326–332.

COHEN, E. Interracial interaction disability. *Human Relations,* 1972, **25**(1), 9–24.

COHEN, E., AND ROPER, S. Modification of interracial interaction disability: An application of status characteristics theory. *American Sociological Review,* 1972, **6,** 643–657.

COOK, S. W. Interpersonal and attitudinal outcomes in cooperating interracial groups. *Journal of Research and Development in Education,* 1978.

COVINGTON, M. V., AND BEERY, R. G. *Self-Worth and School Learning.* New York: Holt, Rinehart & Winston, 1976.

DEUTSCH, M. An experimental study of the effects of cooperation and competition upon group process. *Human Relations,* 1949, **2,** 199–231.

DEVRIES, D. L., EDWARDS, K. J., AND SLAVIN, R. E. Bi-racial learning teams and race relations in the classroom: Four field experiments on Teams–Games–Tournament. *Journal of Educational Psychology,* 1978.

FRANKS, D. D., AND MAROLLA, J. Efficacious action and social approval as interacting dimensions of self-esteem: A tentative formulation through construct validation. *Sociometry,* 1976, **39,** 324–341.

GEFFNER, R. A. The effects of interdependent learning on self-esteem, inter-ethnic relations, and intra-ethnic attitudes of elementary school children: A field experiment. Unpublished Doctoral Thesis, University of California, Santa Cruz, 1978.

GERARD, H., AND MILLER, N. *School Desegregation.* New York: Plenum, 1975.

JOHNSON, D. W., AND JOHNSON, R. T. *Learning Together and Alone.* Englewood Cliffs, N.J.: Prentice-Hall, 1975.

LUCKER, G. W., ROSENFIELD, D., SIKES, J., AND ARONSON, E. Performance in the interdependent classroom: A field study. *American Educational Research Journal,* 1977, **13,** 115–123.

PETTIGREW, T. Social psychology and desegregation research. *American Psychologist,* 1961, **15,** 61–71.

PURKEY, W. W. *Self-Concept and School Achievement.* Englewood Cliffs, N.J.: Prentice-Hall, 1970.

SHERIF, M., HARVEY, O. J., WHITE, J., HOOD, W., AND SHERIF, C. *Intergroup Conflict and Cooperation: The Robber's Cave Experiment.* Norman, Okla.: University of Oklahoma Institute of Intergroup Relations, 1961.

SLAVIN, R. E. Student teams and achievement divisions. *Journal of Research and Development in Education,* in press.

STEPHAN, C., PRESSER, N. R., KENNEDY, J. C. AND ARONSON, E. Attributions to success and failure in cooperative, competitive and interdependent interactions. *European Journal of Social Psychology,* 1978, **8,** 269–274.

STEPHAN, W. G. School desegregation: An evaluation of predictions made in *Brown* v. *Board of Education. Psychological Bulletin,* 1978, **85,** 217–238.

ST. JOHN, N. *School Desegregation: Outcomes for Children.* New York: John Wiley and Sons, 1975.

VII

ATTRACTION: WHY PEOPLE LIKE EACH OTHER

26

"Playing Hard to Get": Understanding an Elusive Phenomenon

Elaine Walster (Hatfield), G. William Walster, Jane Piliavin, and Lynn Schmidt

According to folklore, the woman who is hard to get is a more desirable catch than the woman who is too eager for an alliance. Five experiments were conducted to demonstrate that individuals value hard-to-get dates more than easy-to-get ones. All five experiments failed. In Experiment VI, we finally gained an understanding of this elusive phenomenon. We proposed that two components contribute to a woman's desirability: *(a)* how hard the woman is for the subject to get and *(b)* how hard she is for other men to get. We predicted that the selectively hard-to-get woman (i.e., a woman who is easy for the subject to get but hard for all other men to get) would be preferred to either a uniformly hard-to-get woman, a uniformly easy-to-get woman, or a woman about which the subject has no information. This hypothesis received strong support. The reason for the popularity of the selective woman was evident. Men ascribe to her all of the assets of uniformly hard-to-get and the uniformly easy-to-get women and none of their liabilities.

According to folklore, the woman who is hard to get is a more desirable catch than is the woman who is overly eager for alliance. Socrates, Ovid, Terence, the *Kama Sutra,* and Dear Abby all agree that the person whose affection is easily won is unlikely to inspire passion in another. Ovid, for example, argued:

> Fool, if you feel no need to guard your girl for her own sake, see that you guard her for mine, so I may want her the more. Easy things nobody wants, but what is forbidden is tempting. . . . Anyone who can love the wife of an indolent cuckold, I should suppose, would steal buckets of sand from the shore. [pp. 65–66.]

When we first began our investigation, we accepted cultural lore. We assumed that men would prefer a hard-to-get woman. Thus, we began our research by in-

This research was supported in part by National Science Foundation Grants GS 2932 and GS 30822X and in part by National Institute for Mental Health Grant MH 16661.

Reprinted with permission from the authors and *The Journal of Personality and Social Psychology,* Vol. 26, No. 1, 1973. Copyright 1973 by the American Psychological Association.

terviewing college men as to why they preferred hard-to-get women. Predictably, the men responded to experimenter demands. They explained that they preferred hard-to-get women because the elusive woman is almost inevitably a valuable woman. They pointed out that a woman can only afford to be "choosy" if she is popular—and a woman is popular for some reason. When a woman is hard to get, it is usually a tip-off that she is especially pretty, has a good personality, is sexy, etc. Men also were intrigued by the challenge that the elusive woman offered. One can spend a great deal of time fantasizing about what it would be like to date such a woman. Since the hard-to-get woman's desirability is well recognized, a man can gain prestige if he is seen with her.

An easy-to-get woman, on the other hand, spells trouble. She is probably desperate for a date. She is probably the kind of woman who will make too many demands on a person; she might want to get serious right away. Even worse, she might have a "disease."

In brief, nearly all interviewees agreed with our hypothesis that a hard-to-get woman is a valuable woman, and they could supply abundant justification for their prejudice. A few isolated men refused to cooperate. These dissenters noted that an elusive woman is not always more desirable than an available woman. Sometimes the hard-to-get woman is not only hard to get—she is *impossible* to get, because she is misanthropic and cold. Sometimes a woman is easy to get because she is a friendly, outgoing woman who boosts one's ego and insures that dates are "no hassle." We ignored the testimony of these deviant types.

We then conducted five experiments designed to demonstrate that an individual values a hard-to-get date more highly than an easy-to-get date. All five experiments failed.

Theoretical Rationale

Let us first review the theoretical rationale underlying these experiments.

In Walster, Walster, and Berscheid (1971) we argued that if playing hard to get does increase one's desirability, several psychological theories could account for this phenomenon:

1. Dissonance theory predicts that if a person must expend great energy to attain a goal, one is unusually appreciative of the goal (see Aronson and Mills, 1959; Gerard and Mathewson, 1966; Zimbardo, 1965). The hard-to-get date requires a suitor to expend more effort in her pursuit than he would normally expend. One way for the suitor to justify such unusual effort is by aggrandizing her.

2. According to learning theory, an elusive person should have two distinct advantages: (*a*) Frustration may increase drive—by waiting until the suitor has achieved a high sexual drive state, heightening his drive level by intro-

ducing momentary frustration, and then finally rewarding him, the hard-to-get woman can maximize the impact of the sexual reward she provides (see Kimball, 1961, for evidence that frustration does energize behavior and does increase the impact of appropriate rewards). (*b*) Elusiveness and value may be associated—individuals may have discovered through frequent experience that there is more competition for socially desirable dates than for undesirable partners. Thus, being "hard to get" comes to be associated with "value." As a consequence, the conditional stimulus (CS) of being hard to get generates a fractional antedating goal response and a fractional goal response, which leads to the conditioned response of liking.

3. In an extension of Schachterian theory, Walster (1971) argued that two components are necessary before an individual can experience passionate love; (*a*) He must be physiologically aroused; and (*b*) the setting must make it appropriate for him to conclude that his aroused feelings are due to love. On both counts, the person who plays hard to get might be expected to generate unusual passion. Frustration should increase the suitor's physiological arousal, and the association of "elusiveness" with "value" should increase the probability that the suitor will label his reaction to the other as "love."

From the preceding discussion, it is evident that several conceptually distinct variables may account for the hard-to-get phenomenon. In spite of the fact that we can suggest a plethora of reasons as to why playing hard-to-get strategy might be an effective strategy, all five studies failed to provide any support for the contention that an elusive woman is a desirable woman. Two experiments failed to demonstrate that outside observers perceive a hard-to-get individual as especially "valuable." Three experiments failed to demonstrate that a suitor perceives a hard-to-get date as especially valuable.

Walster, Walster, and Berscheid (1971) conducted two experiments to test the hypothesis that teenagers would deduce that a hard-to-get boy or girl was more socially desirable than was a teenager whose affection could be easily obtained. In these experiments high school juniors and seniors were told that we were interested in finding out what kind of first impression various teenagers made on others. They were shown pictures and biographies of a couple. They were told how romantically interested the stimulus person (a boy or girl) was in his partner after they had met only four times. The stimulus person was said to have liked the partner "extremely much," to have provided no information to us, or to have liked the partner "not particularly much." The teenagers were then asked how socially desirable both teenagers seemed (i.e., how likable, how physically attractive, etc.). Walster, Walster, and Berscheid, of course, predicted that the more romantic interest the stimulus person expressed in a slight acquaintance, the less socially desirable that stimulus person would appear to an outside observer. The results were diametrically opposed to those predicted. The more romantic interest the stimulus person expressed in an acquaintance, the *more* socially desirable teenagers judged him to be. Restraint does not appear to buy respect. Instead, it appears that "All the world *does* love a lover."

Lyons, Walster, and Walster (1971) conducted a field study and a laboratory experiment in an attempt to demonstrate that men prefer a date who plays hard to get. Both experiments were conducted in the context of a computer matching service. Experiment IIII was a field experiment. Women who signed up for the computer matching program were contacted and hired as experimenters. They were then given precise instructions as to how to respond when their computer match called them for a date. Half of the time they were told to pause and think for 3 seconds before accepting the date. (These women were labeled" "hard to get.") Half of the time they were told to accept the date immediately. (These women are labeled "easy to get.") The data indicated that elusiveness had no impact on the man's liking for his computer date.

Experiment IV was a laboratory experiment. In this experiment, Lyons et al. hypothesized that the knowledge that a woman is elusive gives one indirect evidence that she is socially desirable. Such indirect evidence should have the biggest impact when a man has no way of acquiring *direct* evidence about a coed's value or when he has little confidence in his own ability to assess value. When direct evidence is available, and the man possesses supreme confidence in his ability to make correct judgments, information about a woman's elusiveness should have little impact on a man's reaction to her. Lyons et al. thus predicted that when men lacked direct evidence as to a woman's desirability, a man's self-esteem and the woman's elusiveness should interact in determining his respect and liking for her. Lyons et al. measured males' self-esteem via Rosenberg's (1965) measure of self-esteem. Rosenfeld's (1964) measure of fear of rejection, and Berger's (1952) measure of self-acceptance.

The dating counselor then told subjects that the computer had assigned them a date. They were asked to telephone her from the office phone, invite her out, and then report their first impression of her. Presumably the pair would then go out on a date and eventually give us further information about how successful our computer matching techniques had been. Actually, all men were assigned a confederate as a date. Half of the time the woman played hard to get. When the man asked her out she replied:

> Mmm [slight pause] No, I've got a date then. It seems like I signed up for that Date Match thing a long time ago and I've met more people since then—I'm really pretty busy all this week.

She paused again. If the subject suggested another time, the confederate hesitated only slightly, then accepted. If he did not suggest another time, the confederate would take the initiative of suggesting: "How about some time next week—or just meeting for coffee in the Union some afternoon?" And again, she accepted the next invitation. Half of the time, in the easy-to-get condition, the confederate eagerly accepted the man's offer of a date.

Lyons et al. predicted that since men in this blind date setting lacked direct evidence as to a woman's desirability, low-self-esteem men should be more receptive to the hard-to-get woman than were high-self-esteem men. Although Lyons et al.'s manipulation checks indicate that their manipulations were successful and their self-esteem measure was reliable, their hypothesis was not confirmed. Elusiveness had no impact on liking, regardless of subject's self-esteem level.

Did we give up our hypothesis? Heavens no. After all, it had only been disconfirmed four times.

By Experiment V, we had decided that perhaps the hard-to-get hypothesis must be tested in a sexual setting. After all, the first theorist who advised a woman to play hard to get was Socrates; his pupil was Theodota, a prostitute. He advised:

> They will appreciate your favors most highly if you wait till they ask for them. The sweetest meats, you see, if served before they are wanted seem sour, and to those who had enough they are positively nauseating; but even poor fare is very welcome when offered to a hungry man. [Theodota inquired] And how can I make them hungry for my fare? [Socrates' reply] Why, in the first place, you must not offer it to them when they have had enough—but prompt them by behaving as a model of Propriety, by a show of reluctance to yield, and by holding back until they are as keen as can be; and then the same gifts are much more to the recipient than when they're offered before they are desired [see Xenophon, p. 48.]

Walster, Walster, and Lambert (1971) thus proposed that a prostitute who states that she is selective in her choice of customers will be held in higher regard than will be the prostitute who admits that she is completely unselective in her choice of partners.

In this experiment, a prostitute served as the experimenter. When the customer arrived, she mixed a drink for him; then she delivered the experimental manipulation. Half of the time, in the hard-to-get condition, she stated, "Just because I see you this time it doesn't mean that you can have my phone number or see me again. I'm going to start school soon, so I won't have much time, so I'll only be able to see the people that I like the best." Half of the time, in the easy-to-get condition, she did not communicate this information. From this point on, the prostitute and the customer interacted in conventional ways.

The client's liking for the prostitute was determined in two ways: First, the prostitute estimated how much the client had seemed to like her. (Questions asked were, for example, How much did he seem to like you? Did he make arrangements to return? How much did he pay you?) Second, the experimenter recorded how many times within the next 30 days the client arranged to have sexual relations with her.

Once again we failed to confirm the hard-to-get hypothesis. If anything, those clients who were told that the prostitute did not take just anyone were *less* likely to call back and liked the prostitute less than did other clients.

At this point, we ruefully decided that we had been on the wrong track. We decided that perhaps all those practitioners who advise women to play hard to get are wrong. Or perhaps it is only under very special circumstances that it will benefit one to play hard to get.

Thus, we began again. We reinterviewed students—this time with an open mind. This time we asked men to tell us about the advantages *and* disadvantages of hard-to-get *and* easy-to-get women. This time replies were more informative. According to reports, choosing between a hard-to-get women and an easy-to-get woman was like choosing between Scylla and Charybdis—each woman was uniquely desirable and uniquely frightening.

Although the elusive woman was likely to be a popular prestige date, she presented certain problems. Since she was not particularly enthusiastic about you, she might stand you up or humiliate you in front of your friends. She was likely to be unfriendly, cold, and to possess inflexible standards.

The easy-to-get woman was certain to boost one's ego and to make a date a relaxing, enjoyable experience, but . . . Unfortunately, dating an easy woman was a risky business. Such a woman might be easy to get, but hard to get rid of. She might "get serious." Perhaps she would be so oversexed or overaffectionate in public that she would embarrass you. Your buddies might snicker when they saw you together. After all, they would know perfectly well why you were dating *her*.

The interlocking assets and difficulties envisioned when they attempted to decide which was better—a hard-to-get or an easy-to-get woman—gave us a clue as to why our previous experiments had not worked out. The assets and liabilities of the elusive and the easy dates had evidently generally balanced out. On the average, then, both types of women tended to be equally well liked. When a slight difference in liking did appear, it favored the easy-to-get woman.

It finally impinged on us that there are *two* components that are important determinants of how much a man likes a woman: *(a)* How hard or easy she is for him to get, and *(b)* how hard or easy she is for *other men* to get. So long as we were examining the desirability of women who were hard or easy for everyone to get, things balanced out. The minute we examined other possible configurations, it became evident that there is one type of woman who can transcend the limitations of the uniformly hard-to-get or the uniformly easy-to-get woman. If a woman has a reputation for being hard to get, but for some reason she is easy for the subject to get, she should be maximally appealing. Dating such a woman should insure one of great prestige; she is, after all, hard to get. Yet, since she is exceedingly available to the subject, the dating situation should be a relaxed, rewarding experience. Such a *selectively* hard-to-get woman possesses the assets of both the easy-to-get and the hard-to-get women, while avoiding all of their liabilities.

Thus, in Experiment VI, we hypothesized that a selectively hard-to-get woman (i.e., a woman who is easy for the subject to get but very hard for any other man to get) will be especially liked by her date. Women who are hard for everyone—including the subject—to get, or who are easy for everyone to get—or control women, about whom the subject had no information—will be liked a lesser amount.

METHOD

Subjects were 71 male summer students at the University of Wisconsin. They were recruited for a dating research project. This project was ostensibly designed to determine whether computer matching techniques are in fact more effective than is random matching. All participants were invited to come into the dating center in order to choose a date from a set of five potential dates.

When the subject arrived at the computer match office, he was handed folders containing background information on five women. Some of these women had supposedly been "randomly" matched with him; others had been "computer matched" with him. (He was not told which women were which.)

In reality, all five folders contained information about fictitious women. The first item in the folder was a "background questionnaire" on which the woman had presumably described herself. This questionnaire was similar to one the subject had completed when signing up for the match program. We attempted to make the five women's descriptions different enough to be believable, yet similar enough to minimize variance. Therefore, the way the five women described themselves was systematically varied. They claimed to be 18 or 19 years old; freshmen or sophomores; from a Wisconsin city, ranging in size from over 500,000 to under 50,000; 5 feet 2 inches to 5 feet 4 inches tall; Protestant, Catholic, Jewish or had no preference; graduated in the upper 10 to 50 percent of their high school class; and Caucasians who did not object to being matched with a person of another race. The women claimed to vary on a political spectrum from "left of center" through "moderate" to "near right of center"; to place little or no importance on politics and religion; and to like recent popular movies. Each woman listed four or five activities she liked to do on a first date (i.e., go to a movie, talk in a quiet place, etc.).

In addition to the background questionnaire, three of the five folders contained five "date selection forms." The experimenter explained that some of the women had already been able to come in, examine the background information of their matches, and indicate their first impression of them. Two of the subject's matches had not yet come in. Three of the women had already come in and evaluated the subject along with her four other matches. These women would have five date selection forms in their folders. The subject was shown the forms, which consisted of a scale ranging from "definitely do *not* want to date" (-10) to "definitely want to date" ($+10$). A check appeared on each scale. Presumably the check indicated how much the woman had liked a given date. (At this point, the subject was told his identification number. Since all dates were identified by numbers on the forms, this identification number enabled him to ascertain how each date had evaluated both him and her four other matches.)

The date selection forms allowed us to manipulate the elusiveness of the woman. One woman appeared to be uniformly hard to get. She indicated that though she was willing to date any of the men assigned to her, she was not enthusiastic about any of them. She rated all five of her date choices from $+1$ to $+2$, including the subject (who was rated 1.75).

One woman appeared to be uniformly easy to get. She indicated that she was enthusiastic about dating all five of the men assigned to her. She rated her desire to date all five of her date choices $+7$ to $+9$. This included the subject, who was rated 8.

One woman appeared to be easy for the subject to get but hard for anyone else to get (i.e., the selectively hard-to-get woman). She indicated minimal enthusiasm for four of her date choices, rating them from $+2$ to $+3$, and extreme enthusiasm ($+8$) for the subject.

Two women had no date selection forms in their folders (i.e., no information women).

Naturally, each woman appeared in each of the five conditions.

The experimenter asked the man to consider the folders, complete a "first impression questionnaire" for each woman, and then decide which *one* of the women he wished to date. (The subject's rating of the dates constitute our verbal measure of liking; his choice in a date constitutes our behavioral measure of liking.)

The experimenter explained that she was conducting a study of first impressions in conjunction with the dating research project. The study, she continued, was designed to learn more about how good people are at forming first impressions of others on the basis of rather limited information. She explained that filling out the forms would probably make it easier for the man to decide which one of the five women he wished to date.

The first impression questionnaire consisted of three sections:

Liking for various dates.   Two questions assessed subject's liking for each woman: "If you went out with this girl, how well do you think you would get along?"—with possible responses ranging from "get along extremely well" (5) to "not get along at all" (1)—and "What was your overall impression of the girl?"—with possible responses ranging from "extremely favorable" (7) to "extremely unfavorable" (1). Scores on these two questions were summed to form an index of expressed liking. This index enables us to compare subject's liking for each of the women.

Assets and liabilities ascribed to various dates.   We predicted that subjects would prefer the selective woman, because they would expect her to possess the good qualities of both the uniformly hard-to-get and the uniformly easy-to-get woman, while avoiding the bad qualities of both her rivals. Thus, the second section was designed to determine the extent to which subjects imputed good and bad qualities to the various dates.

This section was comprised of 10 pairs of polar opposites. Subjects were asked to rate how friendly–unfriendly, cold–warm, attractive–unattractive, easy-going–rigid, exciting–boring, shy–outgoing, fun-loving–dull, popular–unpopular, aggressive–passive, selective–nonselective each woman was. Ratings were made on a 7-point scale. The more desirable the trait ascribed to a woman, the higher the score she was given.

Liabilities attributed to easy-to-get women.   The third scale was designed to assess the extent to which subjects attributed selected negative attributes to each woman. The third scale consisted of six statements:

> She would more than likely do something to embarrass me in public.
> She probably would demand too much attention and affection from me.
> She seems like the type who would be too dependent on me.
> She might turn out to be too sexually promiscuous.
> She probably would make me feel uneasy when I'm with her in a group.
> She seems like the type who doesn't distinguish between the boys she dates. I probably would be "just another date."

Subjects were asked whether they anticipated any of the above difficulties in their relationship with each woman. They indicated their misgivings on a scale ranging from "certainly true of her" (1) to "certainly not true of her" (7).

The experimenter suggested that the subject carefully examine both the background questionnaires and the date selection forms of all potential dates in order to decide whom he wanted to date. Then she left the subject. (The experimenter was, of course, unaware of what date was in what folder.)

The experimenter did not return until the subject had completed the first impression questionnaires. Then she asked him which woman he had decided to date.

After his choice had been made, the experimenter questioned him as to what factors influenced his choice. Frequently men who chose the selectively easy-to-get woman said that "She chose me, and that made me feel really good" or "She seemed more selective than the others." The uniformly easy-to-get woman was often rejected by subjects who complained "She must be awfully hard up for a date—she really would take anyone." The uniformly hard-to-get woman was once described as a "challenge" but more often rejected as being "snotty" or "too picky."

At the end of the session, the experimenter debriefed the subject and then gave him the names of five actual dates who had been matched with him.

RESULTS

We predicted that the selectively hard-to-get woman (easy for me but hard for everyone else to get) would be liked more than women who were uniformly hard to get, uniformly easy to get, or neutral (the no information women). We had no prediction as to whether or not her three rivals would differ in attractiveness. The results strongly support our hypothesis.

TABLE 26.1
Men's Choices in a Date

Item	Selectively hard to get	Uniformly hard to get	Uniformly easy to get	No information for No. 1	No information for No. 2
Number of men choosing to date each woman	42	6	5	11	7

Dating Choices

When we examine the men's choices in dates, we see that the selective woman is far more popular than any of her rivals. (See Table 26.1.) We conducted a chi-square test to determine whether or not men's choices in dates were randomly distributed. They were not ($\chi^2 = 69.5$, $df = 4$, $p < .001$). Nearly all subjects preferred to date the selective woman. When we compare the frequency with which her four rivals (combined) are chosen, we see that the selective woman does get far more than her share of dates ($\chi^2 = 68.03$, $df = 1$, $p < .001$).

We also conducted an analysis to determine whether or not the women who are uniformly hard to get, uniformly easy to get, or whose popularity is unknown, differed in popularity. We see that they did not ($\chi^2 = 2.86$, $df = 3$).

Liking for the Various Dates

Two questions tapped the men's romantic liking for the various dates: *(a)* "If you went out with this woman, how well do you think you'd get along?"; and *(b)* "What was your overall impression of the woman?" Scores on these two indexes were summed to form an index of liking. Possible scores ranged from 2 to 12.

A contrast was then set up to test our hypothesis that the selective woman will be preferred to her rivals. The contrast that tests this hypothesis is of the form $\Gamma_1 = 4\mu$ (selectively hard to get) $- 1$ (uniformly hard to get) $- 2\mu$ (neutral). We tested the hypothesis $\Gamma_1 = 0$ against the alternative hypothesis $\Gamma_1 \neq 0$. An explanation of this basically simple procedure may be found in Hays (1963). If our hypothesis is true, the preceding contrast should be large. If our hypothesis is false, the resulting contrast should not differ significantly from 0. The data again provide strong support for the hypothesis that the selective woman is better liked than her rivals ($F = 23.92$, $df = 1/70$, $p < .001$).

Additional Data Snooping

We also conducted a second set of contrasts to determine whether the rivals (i.e., the uniformly hard-to-get woman, the uniformly easy-to-get woman, and the

TABLE 26.2
Men's Reactions to Various Dates

	Type of date			
Item	Selectively hard to get	Uniformly hard to get	Uniformly easy to get	No information
Men's liking for dates	9.41[a]	7.90	8.53	8.58
Evaluation of women's assets and liabilities				
Selective[b]	5.23	4.39	2.85	4.30
Popular[b]	4.83	4.58	4.65	4.83
Friendly[c]	5.58	5.07	5.52	5.37
Warm[c]	5.15	4.51	4.99	4.79
Easy Going[c]	4.83	4.42	4.82	4.61
Problems expected in dating	5.23[d]	4.86	4.77	4.99

[a]The higher the number, the more liking the man is expressing for the date.
[b]Traits we expected to be ascribed to the selectively hard-to-get and the uniformly hard-to-get dates.
[c]Traits we expected to be ascribed to the selectively hard-to-get and the uniformly easy-to-get dates.
[d]The higher the number the *fewer* the problems the subject anticipates in dating.

control woman) were differentially liked. Using the procedure presented by Morrison (1967) in chapter 4, the data indicate that the rivals are differentially liked ($F = 4.43$, $df = 2/69$). As Table 26.2 indicates, the uniformly hard-to-get woman seems to be liked slightly less than the easy-to-get or control woman.

In any attempt to explore data, one must account for the fact that observing the data permits the researcher to capitalize on chance. Thus, one must use simultaneous testing methods so as not to spuriously inflate the probability of attaining statistical significance. In the present situation, we are interested in comparing the means of a number of dependent measures, namely the liking for the different women in the dating situation. To perform post hoc multiple comparisons in this situation, one can use a transformation of Hotelling's t^2 statistic, which is distributed as F. The procedure is directly analogous to Scheffé's multiple-comparison procedure for independent groups, exept where one compares means of a number of dependent measures.

To make it abundantly clear that the main result is that the discriminating woman is better liked than each of the other rivals, we performed an additional post hoc analysis, pitting each of the rivals separately against the discriminating woman. In these analyses, we see that the selective woman is better liked than the woman who is uniformly easy to get ($F = 3.99$, $df = 3/68$), than the woman who is uniformly hard to get ($F = 9.47$, $df = 3/68$), and finally, than the control women ($F = 4.93$, $df = 3/68$).

Thus, it is clear that although there are slight differences in the way rivals are liked, these differences are small, relative to the overwhelming attractiveness of the selective woman.

Assets and Liabilities Attributed to Dates

We can now attempt to ascertain *why* the selective woman is more popular than her rivals. Earlier, we argued that the selectively hard-to-get woman should occupy a unique position; she should be assumed to possess all of the virtues of her rivals, but none of their flaws.

The virtues and flaws that the subject ascribed to each woman were tapped by the polar–opposite scale. Subjects evaluated each woman on 10 characteristics.

We expected that subjects would associate two assets with a uniformly hard-to-get woman: Such a woman should be perceived to be both "selective" and "popular." Unfortunately, such a woman should also be assumed to possess three liabilities—she should be perceived to be "unfriendly," "cold," and "rigid." Subjects should ascribe exactly the opposite virtues and liabilities to the easy-to-get woman: Such a woman should possess the assets of "friendliness," "warmth," and "flexibility," and the liabilities of "unpopularity" and "lack of selectivity." The selective woman was expected to possess only assets: She should be perceived to be as "selective" and "popular" as the uniformly elusive woman, and as "friendly," "warm," and "easy-going" as the uniformly easy woman. A contrast was set up to test this specific hypothesis. (Once again, see Hays for the procedure.) This contrast indicates that our hypothesis is confirmed ($F = 62.43$, $df = 1/70$). The selective woman is rated most like the uniformly hard-to-get woman on the first two positive characteristics and most like the uniformly easy-to-get woman on the last three characteristics.

For the reader's interest, the subjects' ratings of all five women's assets and liabilities are presented in Table 26.2.

Comparing the Selective and the Easy Women

Scale 3 was designed to assess whether or not subjects anticipated fewer problems when they envisioned dating the selective woman than when they envisioned dating the uniformly easy-to-get woman. On the basis of pretest interviews, we compiled a list of many of the concerns men had about easy women (e.g., "She would more than likely do something to embarrass me in public.").

We, of course, predicted that subjects would experience more problems when contemplating dating the uniformly easy woman than when contemplating dating a woman who was easy for *them* to get, but hard for anyone else to get (i.e., the selective woman).

Men were asked to say whether or not they envisioned each of the difficulties were they to date each of the women. Possible replies varied from 1 (certainly true of her) to 7 (certainly not true of her). The subjects' evaluations of each woman were summed to form an index of anticipated difficulties. Possible scores ranged from 6 to 42.

A contrast was set up to determine whether the selective woman engendered less concern than the uniformly easy-to-get woman. The data indicate that she does ($F = 17.50$, $df = 1/70$). If the reader is interested in comparing concern engendered by each woman, these data are available in Table 26.2.

The data provide clear support for our hypotheses: The selective woman is strongly preferred to any of her rivals. The reason for her popularity is evident. Men ascirbe to her all of the assets of the uniformly hard-to-get and the uniformly easy-to-get women, and none of their liabilities.

Thus, after five futile attempts to understand the "hard-to-get" phenomenon, it appears that we have finally gained an understanding of this process. It appears that a woman can intensify her desirability if she acquires a reputation for being hard-to-get and then, by her behavior, makes it clear to a selected romantic partner that she is attracted to him.

In retrospect, especially in view of the strongly supportive data, the logic underlying our predictions sounds compelling. In fact, after examining our data, a colleague who had helped design the five ill-fated experiments noted that, "That is exactly what I would have predicted" (given his economic view of man). Unfortunately, we are all better at postdiction than prediction.

References

ARONSON, E., AND MILLS, J. The effect of severity of initiation on liking for a group. *Journal of Abnormal and Social Psychology,* 1959, **67,** 31–36.

BERGER, E. M. The relation between expressed acceptance of self and expressed acceptance of others. *Journal of Abnormal and Social Psychology,* 1952, **47,** 778–782.

GERARD, H. B. AND MATHEWSON, G. C. The effects of severity of initiation and liking for a group: A replication. *Journal of Experimental Social Psychology,* 1966, **2,** 278–287.

HAYS, W. L. *Statistics for psychologists.* New York: Holt, Rinehart, 1963.

KIMBALL, G. A. *Hilgard and Marquis' conditioning and learning.* New York: Appleton-Century-Crofts, 1961.

LYONS, J., WALSTER, AND WALSTER, G. W. Playing hard-to-get: An elusive phenomenon University of Wisconsin, Madison: Author, 1971. (Mimeo)

MORRISON, D. F. *Multivariate statistical methods.* New York: McGraw-Hill, 1967.

OVID. *The art of love.* Bloomington: University of Indiana Press, 1963.

ROSENBERG, M. *Society and the adolescent self image.* Princeton, N.J.: Princeton University Press, 1965.

ROSENFELD, H. M. Social choice conceived as a level of aspiration. *Journal of Abnormal and Social Psychology,* 1964, **68,** 491–499.

WALSTER, E. Passionate love. In B. I. Murstein (Ed.), *Theories of attraction and love.* New York: Springer, 1971.

WALSTER, E., WALSTER, G. W., AND BERSCHEID, E. The efficacy of playing hard-to-get. *Journal of Experimental Education,* 1971, **39,** 73–77.

WALSTER, G. W., AND LAMBERT, P. Playing hard-to-get: A field study. University of Wisconsin, Madison: Author, 1971. (Mimeo)

XENOPHON. *Memorabilia*. London: Heinemann, 1923.

ZIMBARDO, P. G. The effect of effort and improvisation on self persuasion produced by role-playing. *Journal of Experimental Social Psychology*, 1965, **1,** 103–120.

27

The Search for a Romantic Partner: The Effects of Self-Esteem and Physical Attractiveness on Romantic Behavior

Sara B. Kiesler and Roberta L. Baral

What do people look for in a romantic partner? Observations of everyday life suggest one "common sense" assumption: that people prefer romantic partners who are attractive and socially desirable. Yet if the person himself were truly objective, he would consider what is most practical. That is, what is the probability of a successful outcome for attempts to win various alternative partners? While attainment of a very attractive partner might be very rewarding, the chances of actually "catching" such a person are likely to be somewhat lower than are the chances of winning a less attractive partner. Hence, if one is only moderately or low attractive himself, his most realistic initial choice would be based on both the wish to maximize attractiveness of a partner and the likelihood of positive outcome.

Reprinted with permission from the authors and *Personality and Social Behavior*, Kenneth J. Gergen and David Marlowe, eds. Copyright © 1970, Addison-Wesley Publishing Company, Inc., Chapter 8, pp. 155–165.

The present study was supported by United States Public Health Grant MH-131-33-01 to the senior author, and by National Science Foundation Grants GE6156 and GY832 to the Yale University Culture and Behavior Honors Major, directed by Charles A. Kiesler. The junior author and experimenter were participants in the National Science Foundation Undergraduate Research Program.

The authors would like to thank the following persons for their extensive cooperation: H. Richardson Moody, Jr., who was the experimenter; Lola Libby, who was a confederate; and Charles A. Kiesler, who gave considerable aid in the analysis and editorial phases of the study.

This hypothesis, of course, is not new. For example, "exchange theories" (Homans, 1961; Thibaut and Kelley, 1959) assume that people estimate potential "costs" and "rewards" when choosing others.

For the average sort of person, then, realistic behavior should entail his choosing a person who is not extremely attractive. Yet "average" and "below average" people do not always choose those who are actually moderate or low in favorability. A study by Walster et al. (1966) provides example. At a college dance, dates were randomly matched in order to study the relationship of physical attractiveness to romantic behavior. In general, the investigators found that the more attractive the female date, the better liked she was and the more the man said he would like to date her, *regardless* of his *own* physical attractiveness. Furthermore, the same finding held for women. The more attractive the male date, the more he was liked and wanted as a date again, regardless of the woman's own attractiveness.

The question arises, however, why the moderately and low attractive person in the Walster et al. study tended to choose the most attractive partner (instead of a more "realistic" partner). One possibility is that subjects over-evaluated their own attractiveness. First, the subjects may have assumed that the computer has paired them on the basis of similarity, thus making it easy to believe that they were fairly equal in attractiveness to the very attractive partners. Second, all the subjects were dressed up and prepared to present their best face. In such a context, it may have been easy to distort personal attributes and end up with relatively high "self-esteem." Even if subjects did not misperceive their own physical attractiveness, they could have felt that their intelligence or personality made up for any deficiency in physical attractiveness. If they did, indeed, exaggerate their own positive attributes, then their behavior *was* realistic in that both chances of success and maximization of attractiveness were considered. Each one merely ended up choosing someone of *objectively* higher attractiveness.

That people do overevaluate themselves in many situations has long been known (see Crowne and Marlowe, 1964). For example, a prior success experience should raise a person's self-esteem and feelings of worthiness (e.g., Aronson and Carlsmith, 1962). In the romantic-behavior situation, then, certain situational conditions that raise self-esteem may be of some importance in affecting what the romantic behavior will be. We have assumed that one effect will be on the person's perceptions of a potential partner. Thus, if self-esteem is high, he will perceive a highly attractive person as attainable and will choose that person.

Let us now consider what happens when self-esteem is lowered. For instance, if a person is presented with evidence that he is not overintelligent, then it is harder for him to distort his self-image in a favorable direction. Unlike the person with high self-esteem, the person with low self-esteem should perceive the very attractive partner as relatively dissimilar and "hard to get." He should then choose someone not so desirable.

We propose, then, that the self-esteem variable will influence who seems to be a realistic choice. In the present study, our hypothesis was: The *lower* the self-esteem, the lower would be the attractiveness of a chosen romantic partner. Thus we predicted that moderately attractive males with high self-esteem would tend to choose and like very attractive females, but with low self-esteem they would tend to choose only moderately attractive females.

In order to test our hypothesized interaction between self-esteem and attractiveness of the partner, an experiment was designed in which two levels of self-esteem, high and low, and two levels of physical attractiveness of a female, high and moderate, were orthogonally varied. Male subjects were first led to believe that they were doing very well or rather poorly on an intelligence test. We assumed that the former condition would lead to temporarily high self-esteem and encourage overevaluation of self, and that the latter condition would lead to a temporary lowering of self-esteem and would discourage overevaluation of self. The subjects were then exposed to a female of very high or only moderate physical attractiveness. Subsequently, observations were made of the subjects' romantic behavior.

METHOD

Subjects

Subjects were 43 male volunteers ranging in age from 19 to 37. They were recruited from institutions in the New Haven, Connecticut, area, including Southern Connecticut College, Yale Medical School, Yale College, New Haven College, and local libraries. The subjects were mostly undergraduate and graduate students, but included among them were also a few men in various occupations (e.g., high school teacher). Of the 43 subjects, 6 were excluded from the final analysis: 1 for suspicion, 1 for knowing one of the confederates previously, 2 because they were married, and 2 because they were engaged. The subjects were paid $1.50 for their participation.

Procedure

The subjects were recruited for a one-hour study on "intelligence testing." As each subject arrived he was ushered into an ordinary faculty office at Yale University by a male experimenter. The experimenter told the subject that he was perfecting and establishing norms for a new intelligence test, which had already been successfully used on "hundreds" of students. He emphasized that the test was already very accurate and reliable, and that it predicted "success in life." He then asked the subject to take the test orally, explaining that it consisted of five parts, which would be presented sequentially.

Self-Esteem Manipulation

The subjects were randomly assigned to the low or high self-esteem condition. In both conditions the "test" was the same. Subjects were first required to repeat from memory long lists of numbers; in a second part, they were asked to define words. To reduce initial variance in performance, the test was made extremely difficult. For instance, some of the words to be defined were fictitious (e.g., "sympantic").

The manipulation of high self-esteem was accomplished by conveying to the subject the impression that his performance was better than that of most subjects. At preprogrammed intervals the experimenter nodded; at others he told the subject that other subjects had had much more trouble with the questions. In addition, the number task was made somewhat easier by reading the numbers with some rhythm. Thus performance was actually a little better in the high-self-esteem condition than in the low-self-esteem condition.

In the low-self-esteem condition, the experimenter attempted to convey the impression that the subject's performance was inferior. At intervals he frowned and looked away or mentioned that other subjects had performed better. At other intervals he asked the subject if he felt "relaxed enough."

After the "second part" of the test, the experimenter suggested a break. He then gave the subject a short questionnaire, saying that it was a "Psychology Department questionnaire administered to all subjects in studies at Yale." This questionnaire was designed to check on the self-esteem manipulation.

When the subject had filled out the questionnaire, the experimenter stood up and said he "hadn't eaten all day and really needed a cup of coffee." Low-self-esteem subjects were told that "maybe a break will help you." High-self-esteem subjects were told that "since you are doing so well we have plenty of time." The experimenter asked the subject to come with him to get a cup of coffee. Just before they left the room, the experimenter "noticed" a telephone message on his desk and remarked that "he'd better take care of it soon." This sequence was performed in order to prepare the subject for the experimenter's eventual departure.

Meeting the Potential Romantic Partner

The experimenter took the subject to a small canteen in the same building and bought the coffee. As he turned to sit down, he appeared to recognize a girl seated at one of the tables and approached her. This girl was actually one of two confederates used in the study. The experimenter greeted the confederate and asked her if she was working in the building during the summer. She was then introduced to the subject as a coed at a nearby college who was doing summer work for a psychologist, and a preprogrammed conversation began.

After a minute, the experimenter excused himself, telling the subject that "now would be a good time to make that phone call." He was gone for ten minutes, during which the confederate engaged the subject in further conversation (asking him what he did, what his likes and dislikes were, etc.). The experimenter then returned and, acting displeased, told the subject that since his "fiancée really needs the car" he would have to break off the experiment. He apologized, paid the subject, and left. The confederate then continued talking with the subject (on prearranged topics), either for a half hour or until the subject attempted to leave or asked her for a date, whichever came first.

Physical Attractiveness Manipulation

The female confederate, one of two chosen because they were highly attractive, appeared in one of two attractiveness conditions, "high" or "moderate." Both confederates appeared in both conditions in random sequence. In the high-attractiveness condition, each confederate wore becoming make-up and fashionable clothing designed to enhance her initial attractiveness. In the moderate-attractiveness condition, an attempt was made to reduce somewhat the confederate's initial attractiveness. She wore heavy glasses and no makeup, and her hair was pulled back with a rubber band. Her skirt and blouse clashed and were arranged sloppily.

During the interaction the confederates attempted to act exactly the same in all encounters. They were friendly, accepting, and interested throughout. To reduce bias, they were not informed of the self-esteem condition the subjects were in.

The Dependent Variable: Romantic Behavior

After the experimenter "concluded" the experiment and during the following half hour, the confederate recorded the frequency of behavior falling into certain prescribed categories of "romantic behavior." The categories included asking for a date, asking for information that presumably would lead to a date (e.g., asking for the confederate's phone number), offering to buy a snack or coffee for her, offering her a cigarette or mint already on hand, complimenting her, and, finally ignoring her when, at the end of the prescribed time, she said she should "get back to work" (called "ignoring the first cue to leave"). At first, it was intended also to measure the amount of time the subject stayed with the confederate, but this measure turned out to be unreliable since the confederate had to stop the experiment if the subject asked her for a date. It should be noted, however, that the majority (80 percent) did not attempt to leave before the time was up.

At the end of the experimental session, the confederate completely debriefed the subject and in particular emphasized the real nature of the intelligence test. After the debriefing, the subject was asked to rate the confederate on her attractiveness, using a 100-point scale.[1]

RESULTS

Effectiveness of the Manipulations

Subjects were asked two questions to check on the self-esteem manipulation. These were: "How would you describe your present emotional feelings?" (on a scale from "very happy" to "very sad"); and "How good do you personally feel your performance is in the task required in the present study?" (on a scale from "excellent" to "poor"). Both scales were arbitrarily scored from 1 to 20, the higher the score, the lower the self-esteem. According to analyses of variance performed on these data, subjects in the high-self-esteem condition tended to feel better than did low-self-esteem subjects ($F = 3.89$, $df = 1$, 33, $p < .10$), and they felt their performance was better ($F = 13.17$, $df = 1$, 33, $p < .01$). Thus, while the generalized mood response to the manipulation was somewhat weaker than the more direct measure of self-esteem (felt adequacy of performance), the manipulation seemed to have been effective overall.

The manipulation of physical attractiveness of the confederate was also effective at the 0.01 level (although one must have reservations about our measure).

There were no systematic differences among conditions of subject attractiveness, according to ratings made by the experimenter and confederates.

To check on possible differences between the two confederates, all data on romantic behavior (described in detail below) were subjected first to three-way analyses of variance (self-esteem × attractiveness × confederate). These analyses revealed significant tendencies for one confederate to elicit more romantic behavior than the other. However, there were no interactions of confederate with any independent variable. Thus the data for both confederates were combined for the analyses presented below.

Comparisons were also made between the first half of the experiment and the last half to check on possible confederate bias as a result of learning to recognize cues distinguishing high-self-esteem subjects from low-self-esteem subjects. However, the analysis indicated that there was no difference between phases of the experiment. In addition, a check of confederate descriptions of subjects indicated that they were unable to guess above chance levels the subjects' experimental contion.

---

[1] Although this attempt to check on the attractiveness manipulation has very limited validity, giving it during the experiment would have incurred suspicion. Informal pretest observations by a group of five persons not in the experiment supports our contention that the manipulation was effective. In fact, two judges described the "moderate" confederate as "low" in attractiveness.

Romantic Behavior

We predicted an interaction such that high-self-esteem subjects would display more romantic behavior toward the highly attractive confederate, whereas low-self-esteem subjects would display more romantic behavior toward the moderately attractive confederate. To test this prediction, we first summed the frequency of romantic behavior in each behavioral category for each subject. Any subject could get a score of 0 or 1 in each category except the "asked for date" category, in which he could obtain a score of 2 if he asked for two dates. Thus a subject's romantic behavior "score" could range from 0 to 7. The mean scores for each condition are graphically presented in Figure 27.1. Because the data were distributed in skewed fashion, the scores were transformed to produce a more normal distribution $[X = \log (X + 2)]$. These data were then subjected to an analysis of variance. According to the analysis of variance, the hypothesized interaction between self-esteem and attractiveness was supported $(F = 5.34; df = 1, 33; p < .05)$. Thus high-self-esteem subjects tended to choose the highly attractive girl, while low-self-esteem subjects tended to prefer the moderately attractive girl, as predicted.

Table 27.1 presents a breakdown of the data for each romantic-behavior category. For purposes of simplification, the percentage of subjects in each condition who displayed each type of romantic behavior (i.e., the percentage who got a score of 1 in each category) is presented. These data are exactly equivalent to the frequency data, except in the "asked for date" category, where percentages for num-

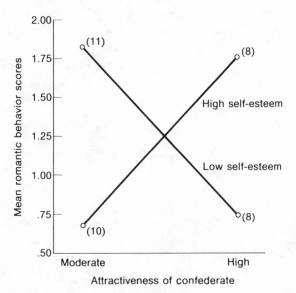

FIGURE 27.1
Romantic behavior toward another as function of self-esteem.

TABLE 27.1
Percentage of Subjects Displaying Each Type of Romantic Behavior

Romantic Behavior	Experimental condition			
	High self-esteem		Low self-esteem	
	Moderately attractive confederate (N = 10)	Highly attractive confederate (N = 8)	Moderately attractive confederate (N = 11)	Highly attractive confederate (N = 8)
Asked for date (at least one)	20	25	36	25
Asked for date (% dates of total possible)	10	19	27	13
Offered to buy coffee, etc.	0	25	18	25
Offered cigarette, gum, etc., on hand	10	25	27	0
Expressed compliments	0	25	18	0
Ignored first cue to leave	10	25	45	12
Asked for information— phone number, address, etc.	30	25	18	12

ber of subjects asking for at least one date and percentages for number of dates over total possible (two per subject) are presented separately.

The data in Table 27.1 indicate that of the 14 appropriate comparisons, 12 are in the predicted direction. Within the high-self-esteem conditions, romantic behavior is greater with the highly attractive confederate than with the moderately attractive confederate for all categories but one (asking for information). Within the low-self-esteem conditions, romantic behavior tends to be greater with the moderately attractive confederate than with the highly attractive confederate for all categories but one (offering to buy). These single comparisons are not significantly different by themselves, but the overall consistency of the results for different romantic behaviors seems to provide some additional confidence in our hypothesis.

DISCUSSION

Our results indicate that self-esteem will affect which romantic partners are chosen: Low self-esteem will lead to choice of a partner lower in attractiveness than will high self-esteem. We interpret these results as indicating *not* that high-self-esteem persons are less "realistic" or "practical," but rather that the possession of high self-esteem changes what the person believes is a realistic or practical choice. Thus our interpretation of the effect of self-esteem on romantic behavior leans heavily on the assumption that changes in perceptions of the self affect perceptions of the chances of success or failure (i.e., probable payoff) when making a romantic choice. If a highly attractive person is perceived as much more attractive than oneself, as when one has low self-esteem, then the chances of success with that person will be perceived as relatively low. If a highly attractive person is perceived as similar in attractiveness to oneself, as when one has high self-esteem, then the chances of success will be seen as relatively high.

What if, in some way, we could have made subjects believe that choosing the confederate (in all conditions) would not entail any risk of rejection or failure? First, we should find subjects choosing the highly attractive confederate more than the moderately attractive confederate—maximizing attractiveness without having to consider possible failure. However, a recent study by Walster (1965) suggests a further difference: low-self-esteem persons may have a higher need for affection than high-self-esteem persons and, moreover, may be less demanding of perfection (will rate someone as more attractive than will high-self-esteem persons). Thus their attempts to secure acceptance—given no chance of failure—should be more intense than those of high-self-esteem persons. The Walster study bears out these predictions. Female subjects who had already been asked out by a male confederate liked him more when they had low self-esteem than when they had high self-esteem. In sum then, we conclude that our results will obtain only when the chances of failure are unknown or known to be relatively high.

One further point should be considered. It is possible that, when giving the intelligence test, the experimenter manipulated acceptance of the subject, as well as perceived intellectual worthiness. If so, our low-self-esteem subjects may have felt socially rejected, as well as less intellectually capable. This would support another explanation of our results, that is, that the low-self-eseem subject liked the highly attractive girl less because she disconfirmed his expectation of himself as an undesirable person. We attempted to avoid this possibility by having the experimenter act friendly and accepting to all subjects, but only with another experiment, in which various manipulations of self-esteem are used, can this explanation be totally rejected.

References

ARONSON, E., AND CARLSMITH, J. M. Performance expectancy as a determinant of actual performance. *Journal of Abnormal Social Psychology,* 1962, **65,** 178–182.

CROWNE, D. P., AND MARLOWE, D. *The Approval Motive: Studies in Evaluative Dependence.* New York: Wiley, 1964.

HOMANS, G. C. *Social Behavior: Its Elementary Forms.* New York: Harcourt, Brace and World, 1961.

THIBAUT, J. W., AND KELLEY, H. H. *The Social Psychology of Groups.* New York: Wiley, 1959.

WALSTER, E. The effect of self-esteem on romantic liking. *Journal of Experimental Social Psychology,* 1965, **1,** 184–197.

WALSTER, E., ARONSON, V., ABRAHAMS, D., AND ROTTMAN, L. Importance of physical attractiveness in dating behavior. *Journal of Personality and Social Psychology,* 1966, **4,** 508–516.

28

Beautiful but Dangerous: Effects of Offender Attractiveness and Nature of the Crime on Juridic Judgment

Harold Sigall and Nancy Ostrove

The physical attractiveness of a criminal defendant (attractive, unattractive, no information) and the nature of the crime (attractiveness-related, attractiveness-unrelated) were varied in a factorial design. After reading one of the case accounts, subjects sentenced the defendant to a term of imprisonment. An interaction was predicted: When the crime was unrelated to attractiveness (burglary), subjects would assign more lenient sentences to the attractive defendant than to the unattractive defendant; when the offense was attractiveness-related (swindle), the attractive defendant would receive harsher treatment. The results confirmed the predictions, thereby supporting a cognitive explanation for the relationship between the physical attractiveness of defendants and the nature of the judgments made against them.

Research investigating the interpersonal consequences of physical attractiveness has demonstrated clearly that good-looking people have tremendous advantages over their unattractive counterparts in many ways. For example, a recent study by Miller (1970) provided evidence for the existence of a physical attractiveness stereotype with a rather favorable content. Dion, Berscheid, and Walster (1972) reported similar findings: Compared to unattractive people, better-looking people were viewed as more likely to possess a variety of socially desirable attributes. In addition, Dion et al.'s subjects predicted rosier futures for the beautiful stimulus person—attractive people were expected to have happier and more successful lives in store for them. Thus, at least in the eyes of others, good looks imply greater potential.

Since physical attractiveness hardly seems to provide a basis for an *equitable* distribution of rewards, one might hope that the powerful effects of this variable

---

Reprinted with permission from the authors and *The Journal of Personality and Social Psychology,* Vol. 31, No. 3, 1975. Copyright 1975 by the American Psychological Association.
This study was supported by a grant from the University of Maryland General Research Board.

would occur primarily when it is the only source of information available. Unfair or irrational consequences of differences in beauty observed in some situations would cause less uneasiness if, in other situations given other important data, respondents would tend to discount such "superficial" information. Unfortunately, for the vast majority of us who have not been blessed with a stunning appearance, the evidence does not permit such consolation. Consider, for example, a recent study by Dion (1972) in which adult subjects were presented with accounts of transgressions supposedly committed by children of varying physical attractiveness. When the transgression was severe the act was viewed less negatively when committed by a good-looking child, than when the offender was unattractive. Moreover, when the child was unattractive the offense was more likely to be seen as reflecting some enduring dispositional quality: Subjects believed that unattractive children were more likely to be involved in future transgressions. Dion's findings, which indicate that unattractive individuals are penalized when there is no apparent logical relationship between the transgression and the way they look, underscore the importance of appearance because one could reasonably suppose that information describing a severe transgression would "overwhelm the field," and that the physical attractiveness variable would not have any effect.

Can beautiful people get away with murder? Although Dion (1972) found no differences in the punishment recommended for offenders as a function of attractiveness, Monahan (1941) has suggested that beautiful women are convicted less often of crimes they are accused of, and Efran (1974) has recently demonstrated that subjects are much more generous when assigning punishment to good-looking as opposed to unattractive transgressors.

The previous findings which indicate a tendency toward leniency for an attractive offender can be accounted for in a number of ways. For example, one might explain such results with the help of a reinforcement-affect model of attraction (e.g., Byrne and Clore, 1970). Essentially, the argument here would be that beauty, having positive reinforcement value, would lead to relatively more positive affective responses toward a person who has it. Thus we like an attractive person more, and since other investigators have shown that liking for a defendant increases leniency (e.g., Landy and Aronson, 1969), we would expect good-looking (better liked) defendants to be punished less than unattractive defendants. Implicit in this reasoning is that the nature of the affective response, which influences whether kind or harsh treatment is recommended, is determined by the stimulus features associated with the target person. Therefore, when other things are equal, benefit accrues to the physically attractive. A more cognitive approach might attempt to explain the relationship between physical appearance and reactions to transgressions by assuming that the subject has a "rational" basis for his responses. It is reasonable to deal harshly with a criminal if we think he is likely to commit further violations, and as Dion's (1972) study suggests, unattractive individuals are viewed as more likely to transgress again. In addition, inasmuch as attractive individuals are viewed as possessing desirable qualities and as having relatively great potential, it makes sense to treat them leniently. Presumably they can be successful in socially acceptable ways, and rehabilitation may result in relatively high payoffs for society.

There is at least one implication that follows from the cognitive orientation which would not flow readily from the reinforcement model. Suppose that situations do exist in which, because of his high attractiveness, a defendant is viewed as more likely to transgress in the future. The cognitive approach suggests that in such instances greater punishment would be assigned to the attractive offender. We might add that in addition to being more dangerous, when the crime is attractiveness-related, a beautiful criminal may be viewed as taking advantage of a God-given gift. Such misappropriation of a blessing may incur animosity, which might contribute to severe judgments in attractiveness-related situations.

In the present investigation, the attractiveness of a defendant was varied along with the nature of the crime committed. It was reasoned that most offenses do not encourage the notion that a criminal's attractiveness increases the likelihood of similar transgressions in the future. Since attractive offenders are viewed as less prone to recidivism and has having greater potential worth, it was expected that under such circumstances an attractive defendant would receive less punishment than an unattractive defendant involved in an identical offense. When, however, the crime committed may be viewed as attractiveness-related, as in a confidence game, despite being seen as possessing more potential, the attractive defendant may be regarded as relatively more dangerous, and the effects of beauty could be expected to be cancelled out or reversed. The major hypothesis, then, called for an interaction: An attractive defendant would receive more lenient treatment than an unattractive defendant when the offense was unrelated to attractiveness; when the crime was related to attractiveness, the attractive defendant would receive relatively harsh treatment.

METHOD

Subjects and Overview

Subjects were 60 male and 60 female undergraduates. After being presented with an account of a criminal case, each subject sentenced the defendant to a term of imprisonment. One-third of the subjects were led to believe that the defendant was physically attractive, another third that she was unattractive, and the remainder received no information concerning appearance. Cross-cutting the attractiveness variable, half of the subjects were presented with a written account of an attractiveness-unrelated crime, a burglary, and the rest with an attractiveness-related crime, a swindle. Subjects were randomly assigned to condition, with the restriction that an equal number of males and females appeared in each of the six cells formed by the manipulated variables.

Procedure

Upon arrival, each subject was shown to an individual room and given a booklet which contained the stimulus materials. The top sheet informed subjects that they would read a criminal case account, that they would receive biographical information about the defendant, and that after considering the materials they would be asked to answer some questions.

The case account began on the second page. Clipped to this page was a 5 x 8 inch card which contained routine demographic information and was identical in all conditions.[1] In the attractive conditions, a photograph of a rather attractive woman was affixed to the upper right-hand corner of the card; while in the unattractive conditions, a relatively unattractive photograph was affixed. No photograph was presented in the control conditions.

Subjects then read either the account of a burglary or a swindle. The burglary account described how the defendant, Barbara Helm, had moved into a high-rise building, obtained a pass key under false pretenses, and then illegally entered the apartment of one of her neighbors. After stealing $2,200 in cash and merchandise she left town. She was apprehended when she attempted to sell some of the stolen property and subsequently was charged with breaking and entering and grand larceny. The swindle account described how Barbara Helm had ingratiated herself to a middle-aged bachelor and induced him to invest $2,200 in a nonexistent corporation. She was charged with obtaining money under false pretenses and grand larceny. In both cases, the setting for the offense and the victim were described identically. The information presented left little doubt concerning the defendant's guilt.

The main dependent measure was collected on the last page of the booklet. Subjects were asked to complete the following statement by circling a number between 1 and 15: "I sentence the defendant, Barbara Helm, to —— years of imprisonment." Subjects were asked to sentence the defendant, rather than to judge guilt versus innocence in order to provide a more sensitive dependent measure.

After sentencing had been completed, the experimenter provided a second form, which asked subjects to recall who the defendant was and to rate the seriousness of the crime. In addition, the defendant was rated on a series of 9-point bipolar adjective scales, including physically unattractive (1) to physically attractive (9),

---

[1] This information, as well as copies of the case accounts referred to below, can be obtained from the first author.

which constituted the check on the attractiveness manipulation. A postexperimental interview followed, during which subjects were debriefed.

RESULTS AND DISCUSSION

The physical attractiveness manipulation was successful: The attractive defendant received a mean rating of 7.53, while the mean for the unattractive defendant was 3.20, $F(1, 108) = 184.29$, $p < .001$. These ratings were not affected by the nature of the crime, nor was there an interaction.

The criminal cases were designed so as to meet two requirements. First, the swindle was assumed to be attractiveness-related, while the burglary was intended to be attractiveness-unrelated. No direct check on this assumption was made. However, indirect evidence is available: Since all subjects filled out the same forms, we obtained physical attractiveness ratings from control condition subjects who were not presented with a photograph. These subjects attributed greater beauty to the defendant in the swindle condition ($X = 6.65$) than in the burglary condition ($X = 5.65$), $F(1, 108) = 4.93$, $p < .05$. This finding offers some support for our contention that the swindle was viewed as attractiveness-related. Second, it was important that the two crimes be viewed as roughly comparable in seriousness. This was necessary to preclude alternative explanations in terms of differential seriousness. Subjects rated the seriousness of the crime on a 9-point scale extending from not at all serious (1) to extremely serious (9). The resulting responses indicated that the second requirement was met: In the swindle condition the mean seriousness rating was 5.02; in the burlgary condition it was 5.07 ($F < 1$).

Table 28.1 presents the mean punishment assigned to the defendant, by condition. Since a preliminary analysis demonstrated there were no differences in responses between males and females, subject sex was ignored as a variable. It can be seen that our hypothesis was supported: When the offense was attractiveness-unrelated (burglary), the unattractive defendant was more severely punished than the attractive defendant; however, when the offense was attractiveness-related (swindle), the attractive defendant was treated more harshly. The overall Attractiveness × Offense interaction was statistically significant, $F(2, 108) = 4.55$, $p < .025$, and this interaction was significant, as well, when the condition was ex-

TABLE 28.1
Mean sentence assigned, in years ($n = 20$ per cell)

Offense	Defendant condition		
	Attractive	Unattractive	Control
Swindle	5.45	4.35	4.35
Burglary	2.80	5.20	5.10

cluded, $F(1, 108) = 7.02$, $p < .01$. Simple comparisons revealed that the unattractive bur-glar received significantly more punishment than the attractive burglar, $F(1, 108) = 6.60$, $p < .025$, while the difference in sentences assigned to the attractive and unattractive swindler was not statistically significant, $F(1, 108) = 1.39$. The attractive-swindle condition was compared with the unattractive-swindle and the control-swindle conditions also, $F(1, 108) = 2.00$, *ns*. Thus, strictly speaking, we cannot say that for the swindle attractiveness was a great liability; there was a tendency in this direction but the conservative conclusion is that when the crime is attractiveness-related, the advantages otherwise held by good-looking defendants are lost.

Another feature of the data worth considering is that the sentences administered in the control condition are almost identical to those assigned in the unattractive condition. It appears that being unattractive did not produce discriminatory responses, per se. Rather, it seems that appearance had its effects through the attractive conditions: The beautiful burglar got off lightly while the beautiful swindler paid somewhat, though not significantly, more. It can be recalled that in the unattractive conditions the stimulus person was seen as relatively unattractive and not merely average looking. Therefore, the absence of unattractive-control condition differences does not seem to be the result of a weak manipulation in the unattractive conditions.

Perhaps it is possible to derive a small bit of consolation from this outcome, if we speculate that only the very attractive receive special (favorable or unfavorable) treatment, and that others are treated similarly. This is a less frightening conclusion than one which would indicate that unattractiveness brings about active discrimination.

As indicated earlier, previous findings (Efran, 1974) that attractive offenders are treated leniently can be interpreted in a number of ways. The results of the present experiment support the cognitive explanation we offered. The notion that good-looking people usually tend to be treated generously because they are seen as less dangerous and more virtuous remains tenable. The argument that physical attractiveness is a positive trait and therefore has a unidirectionally favorable effect on judgments of those who have it, would have led to accurate predications in the burglary conditions. However, this position could not account for the observed interaction. The cognitive view makes precisely that prediction.

Finally, we feel compelled to note that our laboratory situation is quite different from actual courtroom situations. Most important, perhaps, our subjects made decisions by themselves, rather than arriving at judgments after discussions with others exposed to the same information. Since the courtroom is not an appropriate laboratory, it is unlikely that actual experimental tests in the real situation would ever be conducted. However, simulations constitute legitimate avenues for investigating person perception and interpersonal judgment, and there is no obvious reason to believe that these processes would not have the effects in trial proceedings that they do elsewhere.

Whether a discussion with other jurors would affect judgment is an empirical, and researchable, question. Perhaps if even one of twelve jurors notes that some irrelevant factor may be affecting the jury's judgment, the others would see the light. Especially now when the prospect of reducing the size of juries is being entertained, it would be important to find out whether extralegal considerations are more likely to have greater influence as the number of jurors decreases.

References

BYRNE, D. AND CLORE, G. L. A reinforcement model of evaluative responses. *Personality: An International Journal*, 1970, **1,** 103–128.

DION, K. Physical attractiveness and evaluation of children's transgressions. *Journal of Personality and Social Psychology*, 1972, **24,** 207–213.

DION, K., BERSCHEID, E., AND WALSTER, E. What is beautiful is good. *Journal of Personality and Social Psychology*, 1972, **24,** 285–290.

EFRAN, M. G. The effect of physical appearance on the judgment of guilt, interpersonal attraction, and severity of recommended punishment in a simulated jury task. *Journal of Research in Personality*, 1974, **8,** 45–54.

LANDY, D., AND ARONSON, E. The influence of the character of the criminal and victim on the decisions of simulated jurors. *Journal of Experimental Social Psychology*, 1969, **5,** 141–152.

MILLER, A. G. Role of physical attractiveness in impression formation. *Psychomonic Science*, 1970, **19,** 141–243.

MONAHAN, F. *Women in crime*. New York: Washburn, 1941.

29

Social Perception and Interpersonal Behavior: On the Self-Fulfilling Nature of Social Stereotypes

Mark Snyder, Elizabeth Decker Tanke,
and Ellen Berscheid

This research concerns the self-fulfilling influences of social stereotypes on dyadic social interaction. Conceptual analysis of the cognitive and behavioral consequences of stereotyping suggests that a perceiver's actions based upon stereotype-generated attributions about a specific target individual may cause the behavior of that individual to confirm the perceiver's initially erroneous attributions. A paradigmatic investigation of the behavioral confirmation of stereotypes involving physical attractiveness (e.g., "beautiful people are good people") is presented. Male "perceivers" interacted with female "targets" whom they believed (as a result of an experimental manipulation) to be physically attractive or physically unattractive. Tape recordings of each participant's conversational behavior were analyzed by naive observer judges for evidence of behavioral confirmation. These analyses revealed that targets who were perceived (unknown to them) to be physically attractive came to behave in a friendly, likeable, and sociable manner in comparison with targets whose perceivers regarded them as unattractive. It is suggested that theories in cognitive social psychology attend to the ways in which perceivers create the information that they process in addition to the ways that they process that information.

Thoughts are but dreams
Till their effects be tried

—William Shakespeare, *The Rape of Lucrece*

Cognitive social psychology is concerned with the processes by which individuals gain knowledge about behavior and events that they encounter in social interaction, and how they use this knowledge to guide their actions. From this perspective, people are "constructive thinkers" searching for the causes of behavior, drawing inferences about people and their circumstances, and acting upon this knowledge.

Most empirical work in this domain—largely stimulated and guided by the attribution theories (e.g., Heider, 1958; Jones and Davis, 1965; Kelley, 1973)—has focused on the processing of information, the "machinery" of social cognition. Some outcomes of this research have been the specification of how individuals identify the causes of an actor's behavior, how individuals make inferences about the traits and dispositions of the actor, and how individuals make predictions about the actor's future behavior (for reviews, see Harvey, Ickes, and Kidd, 1976; Jones et al., 1972; Ross, 1977).

It is noteworthy that comparatively little theoretical and empirical attention has been directed to the other fundamental question within the cognitive social psychologist's mandate: What are the cognitive and behavioral consequences of our impressions of other people? From our vantage point, current-day attribution theorists leave the individual "lost in thought," with no machinery that links thought to action. It is to this concern that we address ourselves, both theoretically and empirically, in the context of social stereotypes.

Social stereotypes are a special case of interpersonal perception. Stereotypes are usually simple, overgeneralized, and widely accepted (e.g., Karlins, Coffman, and Walters, 1969). But stereotypes are often inaccurate. It is simply not true that all Germans are industrious or that all women are dependent and conforming. Nonetheless, many social stereotypes concern highly visible and distinctive personal characteristics, for example, sex and race. These pieces of information are usually the first to be noticed in social interaction and can gain high priority for channeling subsequent information processing and even social interaction. Social stereotypes are thus an ideal testing ground for considering the cognitive and behavioral consequences of person perception.

Numerous factors may help sustain our stereotypes and prevent disconfirmation of "erroneous" stereotype-based initial impressions of specific others. First, social stereotypes may influence information processing in ways that serve to bolster and strengthen these stereotypes.

Cognitive Bolstering of Social Stereotypes

As information processors, humans readily fall victim to the cognitive process described centuries ago by Francis Bacon:

Reprinted with permission of the authors and *The Journal of Personality and Social Psychology*, Vol. 35, No. 9, 1977. Copyright 1977 by the American Psychological Association.

Research and preparation of this manuscript were supported in part by National Science Foundation Grants SOC 75-13872, "Cognition and Behavior: When Belief Creates Reality," to Mark Snyder and GS 35157X, "Dependency and Interpersonal Attraction," to Ellen Berscheid. We thank Marilyn Steere, Craig Daniels, and Dwain Boelter, who assisted in the empirical phases of this investigation; and J. Merrill Carlsmith, Thomas Hummel, E. E. Jones, Mark Lepper, and Walter Mischel, who provided helpful advice and constructive commentary.

> The human understanding, when any proposition has been once laid down . . . forces
> everything else to add fresh support and confirmation. . . . it is the peculiar and
> perpetual error of the human understanding to be more moved and excited by affir-
> matives than negatives [pp. 23–24].

Empirical research has demonstrated several such biases in information processing. We may overestimate the frequency of occurence of confirming or paradigmatic examples of our stereotypes simply because such instances are more easily
noticed, more easily brought to mind, and more easily retrieved from memory (see
Hamilton and Gifford, 1976; Rothbart et al., unpublished). Evidence that confirms
our stereotyped intuitions about human nature may be, in a word, more cognitively "available" (Tversky and Kahneman, 1973) than nonconfirming evidence.

Moreover, we may fill in the gaps in our evidence base with information consistent with our preconceived notions of what evidence should support our beliefs.
For example, Chapman and Chapman (1967, 1969) have demonstrated that both
college students and professional clinicians perceive positive associations between
particular Rorschach responses and homosexuality in males, even though these associations are demonstrably absent in real life. These "signs" are simply those
that comprise common cultural stereotypes of gay males.

Furthermore, once a stereotype has been adopted, a wide variety of evidence
can be interpreted readily as supportive of that stereotype, including events that
could support equally well an opposite interpretation. As Merton (1948) has suggested, ingroup virtues ("We are thrifty") may become out-group vices ("They
are cheap") in our attempts to maintain negative stereotypes about disliked outgroups. (For empirical demonstrations of this bias, see Regan, Straus, and Fazio,
1974; Rosenhan, 1973; Zadny and Gerard, 1974).

Finally, selective recall and reinterpretation of information from an individual's
past history may be exploited to support a current stereotype-based inference (see
Loftus and Palmer, 1974). Thus, having decided that Jim is stingy (as are all
members of his group), it may be all too easy to remember a variety of behaviors
and incidents that are insufficient one at a time to support an attribution of stin-
giness, but that taken together do warrant and support such an inference.

Behavioral Confirmation of Social Stereotypes

The cognitive bolstering processes discussed above may provide the perceiver
with an "evidence base" that gives compelling cognitive reality to any traits that
he may have erroneously attributed to a target individual initially. This reality is,
of course, entirely cognitive: It is in the eye and mind of the beholder. But stereotype-based attributions may serve as grounds for predictions about the target's future behavior and may guide and influence the perceiver's interactions with the
target. This process itself may generate behaviors on the part of the target that
erroneously confirm the predictions and validate the attributions of the perceiver.

How others treat us is, in large measure, a reflection of our treatment of them (see Bandura, 1977; Mischel, 1968; Raush, 1965). Thus, when we use our social perceptions as guides for regulating our interactions with others, we may constrain their behavioral options (see Kelley and Stahelski, 1970).

Consider this hypothetical, but illustrative, scenario: Michael tells Jim that Chris is a cool and aloof person. Jim meets Chris and notices expressions of coolness and aloofness. Jim proceeds to overestimate the extent to which Chris's self-presentation reflects a cool and aloof disposition and underestimates the extent to which this posture was engendered by his own cool and aloof behavior toward Chris, that had in turn been generated by his own prior beliefs about Chris. Little does Jim know that Tom, who had heard that Chris was warm and friendly, found that his impressions of Chris were confirmed during their interaction. In each case, the end result of the process of "interaction guided by perceptions" has been the target person's *behavioral confirmation* of the perceiver's initial impressions of him.

This scenario makes salient key aspects of the process of behavioral confirmation in social interaction. The perceiver (either Jim or Tom) is not aware that his original perception of the target individual (Chris) is inaccurate. Nor is the perceiver aware of the causal role that his own behavior (here, the enactment of a cool or warm expressive style) plays in generating the behavioral evidence that erroneously confirms his expectations. Unbeknownst to the perceiver, the reality that he confidently perceives to exist in the social world has, in fact, been actively constructed by his own transactions with and operations upon the social world.

In our empirical research, we proposed to demonstrate that stereotypes may create their own social reality by channeling social interaction in ways that cause the stereotyped individual to behaviorally confirm the perceiver's stereotype. Moreover, we sought to demonstrate behavioral confirmation in a social interaction context designed to mirror as faithfully as possible the spontaneous generation of impressions in everyday social interaction and the subsequent channeling influences of these perceptions on dyadic interaction.

One widely held stereotype in this culture involves physical attractiveness. Considerable evidence suggests that attractive persons are assumed to possess more socially desirable personality traits and are expected to lead better lives than their unattractive counterparts (Berscheid and Walster, 1974). Attractive persons are perceived to have virtually every character trait that is socially desirable to the perceiver: "Physically attractive people, for example, were perceived to be more sexually warm and responsive, sensitive, kind, interesting, strong, poised, modest, sociable, and outgoing than persons of lesser physical attractiveness" (Berscheid and Walster, 1974, p. 169). This powerful stereotype holds for male and female perceivers and for male and female stimulus persons.

What of the validity of the physical attractiveness stereotype? Are the physically attractive actually more likable, friendly, and confident than the unattractive? Physically attractive young adults are more often and more eagerly sought out for social dates (Dermer, 1973; Krebs and Adinolphi, 1975; and Walster, et al., 1966).

Even as early as nursery school age, physical attractiveness appears to channel social interaction: The physically attractive are chosen and the unattractive are rejected in sociometric choices (Dion and Berscheid, 1974; Kleck, Richardson and Ronald, 1974).

Differential amount of interaction with the attractive and unattractive clearly helps the stereotype persevere, for it limits the chances for learning whether the two types of individuals differ in the traits associated with the stereotype. But the point we wish to focus upon here is that the stereotype may also channel interaction so that it behaviorally confirms itself. Individuals may have different styles of interaction for those whom they perceive to be physically attractive and for those whom they consider unattractive. These differences in interaction style may in turn elicit and nurture behaviors from the target person that are in accord with the stereotype. That is, the physically attractive may actually come to behave in a friendly, likable, sociable manner—not because they necessarily possess these dispositions, but because the behavior of others elicits and maintains behaviors taken to be manifestations of such traits.

Accordingly, we sought to demonstrate the behavioral confirmation of the physical attractiveness stereotype in dyadic social interaction. In order to do so, pairs of previously unacquainted individuals (designated, for our purposes, as a perceiver and a target) interacted in a getting-acquainted situation that had been constructed to allow us to control the information that one member of the dyad (the male perceiver) received about the physical attractiveness of the other individual (the female target). To measure the extent to which the actual behavior of the target matched the perceiver's stereotype, naive observer judges, who were unaware of the actual or perceived physical attractiveness of either participant, listened to and evaluated tape recordings of the interaction.

METHOD

Participants

Fifty-one male and 51 female undergraduates at the University of Minnesota participated, for extra course credit, in a study of "the processes by which people become acquainted with each other." Participants were scheduled in pairs of previously unacquainted males and females.

The Interaction Between Perceiver and Target

To insure that participants would not see each other before their interactions, they arrived at separate experimental rooms on separate corridors. The experimenter informed each participant that she was studying acquaintance processes in social relationships. Specifically, she was investigating the differences between those initial interactions that involve nonverbal communication and those, such as telephone conversations, that do not. Thus, she explained, the participant would engage in a telephone conversation with another student in introductory psychology.

Before the conversation began, each participant provided written permission for it to be tape recorded. In addition, both dyad members completed brief questionnaires concerning such information as academic major in college and high school of graduation. These questionnaires, it was explained, would provide the partners with some information about each other with which to start the conversation.

Activating the perceiver's stereotype.

The getting-acquainted interaction permitted control of the information that each male perceiver received about the physical attractiveness of his female target. When male perceivers learned about the biographical information questionnaires, they also learned that each person would receive a snapshot of the other member of the dyad, because "other people in the experiment have told us they feel more comfortable when they have a mental picture of the person they're talking to." The experimenter then used a Polaroid camera to photograph the male. No mention of any snapshots was made to female participants.

When each male perceiver received his partner's biographical information form, it arrived in a folder containing a Polaroid snapshot, ostensibly of his partner. Although the biographical information had indeed been provided by his partner, the photograph was not. It was one of eight photographs that had been prepared in advance.

Twenty females students from several local colleges assisted (in return for $5) in the preparation of stimulus materials by allowing us to take Polaroid snapshots of them. Each photographic subject wore casual dress, each was smiling, and each agreed (in writing) to allow us to use her photograph. Twenty college-age men then rated the attractiveness of each picture on a 10-point scale.[1] We then chose the four pictures that had received the highest attractiveness ratings ($M = 8.10$) and the four photos that had received the lowest ratings ($M = 2.56$). There was virtually no overlap in ratings of the two sets of pictures.

Male perceivers were assigned randomly to one of two conditions of perceived physical attractiveness of their targets. Males in the attractive target condition received folders containing their partners' biographical information form and one of the four attractive photographs. Males in the unattractive target condition received folders containing their partners' biographical information form and one of the four unattractive photographs. Female targets knew nothing of the photographs possessed by their male interaction partners, nor did they receive snapshots of their partners.

---

[1] The interrater correlations of these ratings of attractiveness ranged from .45 to .92, with an average interrater correlation of .74.

The perceiver's stereotype-based attributions.   Before initiating his get-ting-acquainted conversation, each male perceiver rated his initial impressions of his partner on an Impression Formation Questionnaire. The questionnaire was con-structed by supplementing the 27 trait adjectives used by Dion, Berscheid, and Walster (1972) in their original investigation of the physical attractiveness stereo-type with the following items: intelligence, physical attractiveness, social adept-ness, friendliness, enthusiasm, trustworthiness, and successfulness. We were thus able to assess the extent to which perceivers' initial impressions of their partners reflected general stereotypes linking physical attractiveness and personality char-acteristics.

The getting-acquainted conversation.   Each dyad then engaged in a 10-minute unstructured conversation by means of microphones and headphones con-nected through a Sony TC-570 stereophonic tape recorder that recorded each par-ticipant's voice on a separate channel of the tape.

After the conversation, male perceivers completed the Impression Formation Questionnaires to record final impressions of their partners. Female targets ex-pressed self-perceptions in terms of the items of the Impression Formation Ques-tionnaire. Each female target also indicated, on 10-point scales, how much she had enjoyed the conversation, how comfortable she had felt while talking to her partner, how accurate a picture of herself she felt that her partner had formed as a result of the conversation, how typical her partner's behavior had been of the way she usually was treated by men, her perception of her own physical attrac-tiveness, and her estimate of her partner's perception of her physical attractive-ness. All participants were then thoroughly and carefully debriefed and thanked for their contribution to the study.

Assessing Behavioral Confirmation

To assess the extent to which the actions of the target women provided behav-ioral confirmation for the stereotypes of the men perceivers, eight male and four female introductory psychology students rated the tape recordings of the getting-acquainted conversations. These observer judges were unaware of the experimen-tal hypotheses and knew nothing of the actual or perceived physical attractiveness of the individuals on the tapes. They listened, in random order, to two 4-minute segments (one each from the beginning and end) of each conversation. They heard *only* the track of the tapes containing the target women's voices and rated each woman on the 34 bipolar scales of the Impression Formation Questionnaire as well as on 14 additional 10-point scales; for example, ''How animated and enthusiastic is this person?,'' ''How intimate or personal is this person's conversation?,'' and ''How much is she enjoying herself?'' Another group of observer judges (three males and six females) performed a similar assessment of the male perceivers' behavior based upon only the track of the tapes that contained the males' voices.[2]

RESULTS

To chart the process of behavioral confirmation of social stereotypes in dyadic social interaction, we examined the effects of our manipulation of the target women's apparent physical attractiveness on (a) the male perceivers' initial impressions of them, and (b) the women's behavioral self-presentation during the interaction, as measured by the observer judges' ratings of the tape recordings.

The Perceivers' Stereotype

Did our male perceivers form initial impressions of their specific target women on the basis of general stereotypes that associate physical attractiveness and desirable personalities? To answer this question, we examined the male perceivers' initial ratings on the Impression Formation Questionnaire. Recall that these impressions were recorded *after* the perceivers had seen their partners' photographs, but *before* the getting-acquainted conversation.[3] Indeed, it appears that our male perceivers did fashion their initial impressions of their female partners on the basis of

---

[2] We assessed the reliability of our raters by means of intraclass correlations (Ebel, 1951), a technique that employs analysis-of-variance procedures to determine the proportion of the total variance in ratings due to variance in the persons being rated. The intraclass correlation is the measure of reliability most commonly used with interval data and ordinal scales that assume interval properties. Because the measure of interest was the mean rating of judges on each variable, the between-rater variance was not included in the error term in calculating the intraclass correlation. (For a discussion, see Tinsley and Weiss, 1975, p. 363.) Reliability coefficients for the coders' ratings of the females for all dependent measures ranged from .35 to .91 with a median of .755. For each dependent variable, a single score was constructed for each participant by calculating the mean of the raters' scores on that measure. Analyses of variance, including the time of the tape segment (early vs. late in the conversation) as a factor, revealed no more main effects of time or interactions between time and perceived attractiveness than would have been expected by chance. Thus, scores for the two tape segments were summed to yield a single score for each dependent variable. The same procedure was followed for ratings of male perceivers' behavior. In this case, the reliability coefficients ranged from .18 to .83 with a median of .61.

[3] These and all subsequent analyses are based upon a total of 38 observations, 19 in each of the attractive target and unattractive target conditions. Of the original 51 dyads, a total of 48 male-female pairs completed the experiment. In each of the remaining three dyads, the male participant had made reference during the conversation to the photograph. When this happened, the experimenter interrupted the conversation and immediately debriefed the participants. Of the remaining 48 dyads who completed the experimental procedures, 10 were eliminated from the analyses for the following reasons: In 4 cases the male participant expressed strong suspicion about the photograph; in 1 case, the conversation was not tape recorded because of a mechanical problem; and in 5 cases, there was a sufficiently large age difference (ranging from 6 to 18 years) between the participants that the males in these dyads reported that they had reacted very differently to their partners than they would have reacted to an age peer. This pattern of attrition was independent of assignment to the attractive target and unattractive target experimental conditions ($\chi^2 = 1.27$, *ns*).

stereotyped beliefs about physical attractiveness, multivariate $F(34, 3) = 10.19$, $p < .04$. As dictated by the physical attractiveness stereotype, men who anticipated physically attractive partners expected to interact with comparatively sociable, poised, humorous, and socially adept women; by contrast, men faced with the prospect of getting acquainted with relatively unattractive partners fashioned images of rather unsociable, awkward, serious and socially inept women, all F's $(1,36) < 5.85$, $p < .025$.

Behavioral Confirmation

Not only did our perceivers fashion their images of their discussion partners on the basis of their stereotyped intuitions about beauty and goodness of character, but these impressions initiated a chain of events that resulted in the behavioral confirmation of these initially erroneous inferences. Our analyses of the observer judges' ratings of the women's behavior were guided by our knowledge of the structure of the men's initial impressions of their target women's personality. Specifically, we expected to find evidence of behavioral confirmation only for those traits that had defined the perceivers' stereotypes. For example, male perceivers did not attribute differential amounts of sensitivity or intelligence to partners of differing apparent physical attractiveness. Accordingly, we would not expect that our observer judges would "hear" different amounts of intelligence or sensitivity in the tapes. By contrast, male perceivers did expect attractive and unattractive targets to differ in sociability. Here we would expect that observer judges would detect differences in sociability between conditions when listening to the women's contributions to the conversations, and thus we would have evidence of behavioral confirmation.

To assess the extent to which the women's behavior, as rated by the observer judges, provided behavioral confirmation for the male perceivers' stereotypes, we identified, by means of a discriminant analysis (Tatsuoka, 1971), those 21 trait items of the Impression Formation Questionnaire for which the mean initial ratings of the men in the attractive target and unattractive target conditions differed by more than 1.4 standard deviations.[4] This set of "stereotype traits" (e.g., sociable, poised, sexually warm, outgoing) defines the differing perceptions of the personality characteristics of target women in the two experimental conditions.

We then entered these 21 stereotype traits and the 14 additional dependent measures into a multivariate analysis of variance. This analysis revealed that our observer judges did indeed view women who had been assigned to the attractive target condition quite differently than women in the unattractive target condition,

---

[4] After the 21st trait dimension, the differences between the experimental conditions drop off sharply. For example, the next adjective pair down the line has a difference of 1.19 standard deviations, and the one after that has a difference of 1.02 standard deviations.

$Fm(35, 2) = 40.0003$, $p < .025$. What had initially been reality in the minds of the men had now become reality in the behavior of the women with whom they had interacted—a behavioral reality discernible even by naive observer judges, who had access *only* to tape recordings of the women's contributions to the conversations.

When a multivariate analysis of variance is performed on multiple correlated dependent measures, the null hypothesis states that the vector of means is equal across conditions. When the null hypothesis is rejected, the nature of the difference between groups must then be inferred from inspection of group differences on the individual dependent measures. In this case, the differences between the behavior of the women in the attractive target and the unattractive target conditions were in the same direction as the male perceivers' initial stereotyped impressions for fully 17 of the 21 measures of behavioral confirmation. The binomial probability that at least 17 of these adjectives would be in the predicted direction by chance alone is a scant .003. By contrast, when we examined the 13 trait pairs that our discriminant analysis had indicated did *not* define the male perceivers' stereotype, a sharply different pattern emerged. Here, we would not expect any systematic relationship between the male perceivers' stereotyped initial impressions and the female targets' actual behavior in the getting-acquainted conversations. In fact, for only 8 of these 13 measures is the differences between the behavior of the women in the attractive target condition in the same direction as the men's stereotyped initial impressions. This configuration is, of course, hardly different from the pattern expected by chance alone if there were no differences between the groups (exact binomial $p = .29$). Clearly, then, behavioral confirmation manifested itself only for those attributes that had defined the male perceivers' stereotype; that is, only in those domains where the men believed that there did exist links between physical attractiveness and personal attributes did the women come to behave differently as a consequence of the level of physical attractiveness that we had experimentally assigned to them.

Moreover, our understanding of the nature of the difference between the attractive target and the unattractive target conditions identified by our multivariate analysis of variance and our confidence in this demonstration of behavioral confirmation are bolstered by the consistent pattern of behavioral differences on the 14 additional related dependent measures. Our raters assigned to the female targets in the attractive target condition higher ratings on *every* question related to favorableness of self-presentation. Thus, for example, those who were thought by their perceivers to be physically attractive appeared to the observer judges to manifest greater confidence, greater animation, greater enjoyment of the conversation, and greater liking for their partners than those women who interacted with men who perceived them as physically unattractive.[5]

---

[5] We may eliminate several alternative interpretations of the behavioral confirmation effect. Women who had been assigned randomly to the attractive target condition were not in fact more physically

In Search of Mediators of Behavioral Confirmation

We next attempted to chart the process of behavioral confirmation. Specifically, we searched for evidence of the behavioral implications of the perceivers' stereotypes. Did the male perceivers present themselves differently to target women whom they assumed to be physically attractive or unattractive? Because we had 50 dependent measures[6] of the observer judges' ratings of the males—12 more than the number of observations (male perceivers)—a multivariate analysis of variance is inappropriate. However, in 21 cases, univariate analyses of variance did indicate differences between conditions (all p's < .05). Men who interacted with women whom they believed to be physically attractive appeared (to the observer judges) more sociable, sexually warm, interesting, independent, sexually permissive, bold, outgoing, humorous, obvious, and socially adept than their counterparts in the unattractive target condition. Moreover, these men were seen as more attractive, more confident, and more animated in their conversation than their counterparts. Further, they were considered by the observer judges to be more comfortable, to enjoy themselves more, to like their partners more, to take the initiative more often, to use their voices more effectively, to see their women partners as more attractive, and, finally, to be seen as more attractive by their partners than men in the unattractive target condition.

It appears, then, that differences in the level of sociability manifested and expressed by the male perceivers may have been a key factor in bringing out reciprocating patterns of expression in the target women. One reason that target women who had been labeled as attractive may have reciprocated these sociable overtures is that they regarded their partners' images of them as more accurate, $F(1, 28) = 6.75$, $p < .02$, and their interaction style to be more typical of the way men generally treated them, $F(1, 28) = 4.79$, $p < .04$, than did women in the unattractive target condition.[7] These individuals, perhaps, rejected their partners' treatment of them

---

attractive than those who were assigned randomly to the unattractive target condition. Ratings of the actual attractiveness of the female targets by the experimenter revealed no differences whatsoever between conditions, $t(36) = .00$. Nor, for that matter, did male perceivers differ in their own physical attractiveness as a function of experimental condition, $f(36) = .44$. In addition, actual attractiveness of male perceivers and actual attractiveness of female targets within dyads were independent of each other, $r(36) = .06$.

Of greater importance, there was no detectable difference in personality characteristics of females who had been assigned randomly to the attractive target and unattractive target conditions of the experiment. They did not differ in self-esteem as assessed by the Janis–Field–Eagly (Janis and Field, 1973) measure, $F(1, 36) < 1$. Moreover, there were no differences between experimental conditions in the female targets' self-perceptions as reported after the conversations on the Impression Formation Questionnaire ($Fm < 1$). We have thus no reason to suspect that any systematic, pre-existing differences between conditions in morphology or personality can pose plausible alternative explanations of our demonstration of behavioral confirmation.

[6] Two dependent measures were added between the time that the ratings were made of the female participants and the time that the ratings were made of the male participants. These measures were responses to the questions, "How interested is he in his partner?" and "How attractive does he think his partner is?"

[7] The degrees of freedom for these analyses are fewer than those for other analyses because they were added to the experimental procedure after four dyads had participated in each condition.

as unrepresentative and defensively adopted more cool and aloof postures to cope with their situations.

DISCUSSION

Of what consequence are our social stereotypes? Our research suggests that stereotypes can and do channel dyadic interaction so as to create their own social reality. In our demonstration, pairs of individuals got acquainted with each other in a situation that allowed us to control the information that one member of the dyad (the perceiver) received about the physical attractiveness of the other person (the target). Our perceivers, in anticipation of interaction, fashioned erroneous images of their specific partners that reflected their general stereotypes about physical attractiveness. Moreover, our perceivers had very different patterns and styles of interaction for those whom they perceived to be physically attractive and unattractive. These differences in self-presentation and interaction style, in turn, elicited and nurtured behaviors of the target that were consistent with the perceivers' initial stereotypes. Targets who were perceived (unbeknownst to them) to be physically attractive actually came to behave in a friendly, likable, and sociable manner. The perceivers' attributions about their targets based upon their stereotyped intuitions about the world had initiated a process that produced behavioral confirmation of those attributions. The initially erroneous attributions of the perceivers had become real: The stereotype had truly functioned as a self-fulfilling prophecy (Merton, 1948).[8]

We regard our investigation as a particularly compelling demonstration of behavioral confirmation in social interaction. For if there is any social–psychological process that ought to exist in "stronger" form in everyday interaction than in the psychological laboratory, it is behavioral confirmation. In the context of years of social interaction in which perceivers have reacted to their actual physical attractiveness, our 10-minute getting-acquainted conversations over a telephone must seem minimal indeed. Nonetheless, the impact was sufficient to permit outside observers who had access only to one person's side of a conversation to detect manifestations of behavioral confirmation.

Might not other important and widespread social stereotypes—particularly those concerning sex, race, social class, and ethnicity—also channel social interaction so as to create their own social reality? For example, will the common stereotype

---

[8]Our research on behavioral confirmation in social interaction is a clear "cousin" of other demonstrations that perceivers' expectations may influence other individuals' behavior. Thus, Rosenthal (1974) and his colleagues have conducted an extensive program of laboratory and field investigations of the effects of experimenters' and teachers' expectations on the behavior of subjects in psychological laboratories and students in classrooms. Experimenters and teachers led to expect particular patterns of performance from their subjects and pupils act in ways that selectively influence or shape those performances to confirm initial expectations (e.g., Rosenthal, 1974).

that women are more conforming and less independent than men (see Broverman, et al., 1972) influence interaction so that (within a procedural paradigm similar to ours) targets believed to be female will actually conform more, be more dependent, and be more successfully manipulated than interaction partners believed to be male? At least one empirical investigation has pointed to the possible self-fulfilling nature of apparent sex differences in self-presentation (Zanna and Pack, 1975).

Any self-fulfilling influences of social stereotypes may have compelling and pervasive societal consequences. Social observers have for decades commented on the ways in which stigmatized social groups and outsiders may fall "victim" to self-fulfilling cultural stereotypes (e.g., Becker, 1963; Goffman, 1963; Merton, 1948; Myrdal, 1944; Tannenbaum, 1938). Consider Scott's (1969) observations about the blind:

> When, for example, sighted people continually insist that a blind man is helpless because he is blind, their subsequent treatment of him may preclude his even exercising the kinds of skills that would enable him to be independent. It is in this sense that stereotypic beliefs are self-actualized [p. 9].

And all too often it is the "victims" who are blamed for their own plight (see Ryan, 1971) rather than the social expectations that have constrained their behavioral options.

Of what import is the behavioral confirmation process for our theoretical understanding of the nature of social perception? Although our empirical research has focused on social stereotypes that are widely accepted and broadly generalized, our notions of behavioral confirmation may apply equally well to idiosyncratic social perceptions spontaneously formed about specific individuals in the course of every day social interaction. In this sense, social psychologists have been wise to devote intense effort to understanding the processes by which impressions of others are formed. Social perceptions are important precisely because of their impact on social interaction. Yet, at the same time, research and theory in social perception (mostly displayed under the banner of attribution theory) that have focused on the manner in which individuals process information provided them to form impressions of others may underestimate the extent to which information received in actual social interaction is a product of the preceiver's own actions toward the target individual. More careful attention must clearly be paid to the way in which perceivers *create* or *construct* the information that they process in addition to the way in which they *process* that information. Events in the social world may be as much the *effects* of our perceptions of those events as they are the *causes* of those perceptions.

From this perspective, it becomes easier to appreciate the perceiver's stubborn tendency to fashion images of others largely in trait terms (e.g., Jones and Nisbett, 1972), despite the poverty of evidence for the pervasive cross-situational consistencies in social behavior that the existence of "true" traits would demand (e.g., Mischel, 1968). This tendency, dubbed by Ross (1977) as the "fundamental attribution error,'' may be a self-erasing error. For even though any target individual's behavior may lack, overall, the trait-defining properties of cross-situational consistency, the actions of the perceiver himself may produce consistency in the samples of behavior available to that perceiver. Our impressions of others may cause those others to behave in consistent traitlike fashion for us. In that sense, our trait-based impressions of others are veridical, even though the same individual may behave or be led to behave in a fashion perfectly consistent with opposite attributions by other perceivers with quite different impressions of that individual. Such may be the power of the behavioral confirmation process.

References

BACON, F. *[Novum organum]* (J. Devey, ed.). New York: P. F. Collier & Son, 1902. (Originally published, 1620.)

BANDURA, A. *Social learning theory.* Englewood Cliffs, N.J.: Prentice Hall, 1977.

BECKER, H. W. *Outsiders: Studies in the sociology of deviance.* N.Y.: Free Press, 1963.

BERSCHEID, E., AND WALSTER, E. Physical attractiveness. In L. Berkowitz (ed.) *Advances in experimental social psychology.* Vol. 7. New York: Academic Press, 1974.

BROVERMAN, I. K., VOGEL, S. R., BROVERMAN, D. M., CLARKSON, F. E., AND ROSENKRANTZ, P. S. Sex-role stereotypes: A current appraisal. *Journal of Social Issues,* 1972, **28,** 59–78.

CHAPMAN, L., AND CHAPMAN, J. The genesis of popular but erroneous psychodiagnostic observations. *Journal of Abnormal Psychology,* 1967, **72,** 193–204.

CHAPMAN, L., AND CHAPMAN, J. Illusory correlations as an obstacle to the use of valid psychodiagnostic signs. *Journal of Abnormal Psychology,* 1969, **74,** 271–280.

DERMER, M. *When Beauty fails.* Unpublished doctoral dissertation, University of Minnesota, 1973.

DION, K. K., AND BERSCHEID, E. Physical attractiveness and peer perception among children. *Sociometry,* 1974, **27**(1), 1–12.

DION, K. K., BERSCHEID, E., AND WALSTER, E. What is beautiful is good. *Journal of Personality and Social Psychology,* 1972, **24,** 285–290.

EBEL, R. L. Estimation of the reliability of ratings. *Psychometrika,* 1951, **16,** 407–424.

GOFFMAN, E. *Stigma: Notes on the management of spoiled identity.* Englewood Cliffs, N.J.: Prentice Hall, 1963.

HAMILTON, D. L., AND GIFFORD, R. K. Illusory correlation in interpersonal perception: A cognitive basis of stereotypic judgments. *Journal of Experimental Social Psychology,* 1976, **12,** 392–407.

HARVEY, J. H., ICKES, W. J., AND KIDD, R. F. *New directions in attribution research.* Hillsdale, N.J.: Erlbaum, 1976.

HEIDER, F. *The psychology of interpersonal relations.* New York: Wiley, 1958.

JANIS, I., AND FIELD, P. Sex differences and personality factors related to persuasibility. In C. Hovland and I. Janis (eds.), *Personality and persuasibility.* New Haven, Conn.: Yale University Press, 1973.

JONES, E. E., AND DAVIS, K. E. From acts to dispositions: The attribution process in person perception. In L. Berkowitz (ed.), *Advances in experimental social psychology.* Vol. 2. New York: Academic Press, 1965.

JONES et al. *Attribution: Perceiving the causes of behavior.* Morristown, N.J.: General Learning Press, 1972.

JONES, E. E., AND NISBETT, R. E. The actor and the observer: Divergent perceptions of the causes of behavior. In E. Jones, D. Kanouse, H. Kelley, S. Valins, and B. Weiner (eds.), *Attribution: Perceiving the causes of behavior.* New York: General Learning Press, 1972.

KARLINS, M., COFFMAN, T. L., AND WALTERS, G. On the fading of social stereotypes: Studies in three generations of college students. *Journal of Personality and Social Psychology*, 1969, **13**, 1–16.

KELLEY, H. H. The process of causal attribution. *American Psychologist*, 1973, **28**, 107–128.

KELLEY, H. H., AND STAHELSKI, A. J. The social interaction basis of cooperators' and competitors' beliefs about others. *Journal of Personality and Social Psychology*, 1970, **16**, 66–91.

KLECK, R. E., RICHARDSON, S. A., AND RONALD, L. Physical appearance cues and interpersonal attraction in children. *Child Development*, 1974, **45**, 305–310.

KREBS, D., AND ADINOLPHI, A. A. Physical attractiveness, social relations, and personality style. *Journal of Personality and Social Psychology*, 1975, **31**, 245–253.

LOFTUS, E., AND PALMER, J. Reconstruction of automobile destruction. *Journal of Verbal Learning and Verbal Behavior*, 1974, **13**, 585–589.

MERTON, R. K. The self-fulfilling prophecy. *Antioch Review*, 1948, **8**, 193–210.

MISCHEL, W. *Personality and assessment.* New York: Wiley, 1968.

MYRDAL, G. *An American dilemma.* New York: Harper & Row, 1944.

RAUSH, H. L. Interaction sequences. *Journal of Personality and Social Psychology*, 1965, **2**, 487–499.

REGAN, D. T., STRAUS, E., AND FAZIO, R. Liking and the attribution process. *Journal of Experimental Social Psychology*, 1974, **10**, 385–397.

ROSENHAN, D. L. On being sane in insane places. *Science*, 1973, **179**, 250–258.

ROSENTHAL, R. *On the social psychology of the self-fulfilling prophecy: Further evidence for pygmalion effects and their mediating mechanisms.* New York: M.S.S. Information Corp. Modular Publications, 1974.

ROSS, L. The intuitive psychologist and his shortcomings: Distortions in the attribution process. In L. Berkowitz (ed.), *Advances in experimental social psychology.* Vol. 10. New York: Academic Press, 1977.

ROTHBART, M., FULERO, S., JENSEN, C., HOWARD, J., AND BIRRELL, P. *From individual to group impressions: Availability heuristics in stereotype formation.* Unpublished manuscript, University of Oregon, 1976.

RYAN, W. *Blaming the victim.* New York: Vintage Books, 1971.

SCOTT, R. A. *The making of blind men.* New York: Russell Sage, 1969.

TANNENBAUM, F. *Crime and the community.* Boston: Ginn, 1938.

TATSUOKA, M. M. *Multivariate analysis.* New York: Wiley, 1971.

TINSLEY, H. E. A., AND WEISS, D. J. Interrater reliability and agreement of subjective judgments. *Journal of Counseling Psychology*, 1975, **22**, 358–376.

TVERSKY, A., AND KAHNEMAN, D. Availability: A heuristic for judging frequency and probability. *Cognitive Psychology*, 1973, **5**, 207–232.

WALSTER, E., ARONSON, V., ABRAHAMS, D., AND ROTTMAN, L. Importance of physical attractiveness in dating behavior. *Journal of Personality and Social Psychology*, 1966, **4**, 508–516.

ZADNY, J., AND GERARD, H. B. Attributed intentions and informational selectivity. *Journal of Experimental Social Psychology*, 1974, **10**, 34–52.

ZANNA, M. P., AND PACK, S. J. On the self-fulfilling nature of apparent sex differences in behavior. *Journal of Experimental Social Psychology*, 1975, **11**, 583–591.

30

Some Evidence for Heightened Sexual Attraction Under Conditions of High Anxiety

Donald G. Dutton and Arthur P. Aron

Male passersby were contacted either on a fear-arousing suspension bridge or a non–fear-arousing bridge by an attractive female interviewer who asked them to fill out questionnaires containing Thematic Apperception Test pictures. Sexual content of stories written by subjects on the fear-arousing bridge and tendency of these subjects to attempt postexperimental contact with the interviewer were both significantly greater. No significant differences between bridges were obtained on either measure for subjects contacted by a male interviewer. A third study manipulated anticipated shock to male subjects and an attractive female confederate independently. Anticipation of own shock but not anticipation of shock to confederate increased sexual imagery scores on the Thematic Apperception Test and attraction to the confederate. Some theoretical implications of these findings are discussed.

There is a substantial body of indirect evidence suggesting that sexual attractions occur with increased frequency during states of strong emotion. For example, heterosexual love has been observed to be associated both with hate (James, 1910; Suttie, 1935) and with pain (Ellis, 1936). A connection between "aggression" and sexual attraction is supported by Tinbergen's (1954) observations of intermixed courting and aggression behaviors in various animal species, and a series of experiments conducted by Barclay have indicated the existence of a similar phenomenon

Reprinted with permission of the authors and *Journal of Personality and Social Psychology,* Vol. 30, No. 4, 1974.

This research was supported by University of British Columbia Research Committee Grant 26 9840 to the first author and National Research Council Postdoctoral Fellowship 1560 to the second author.

in human behavior. In one study, Barclay and Haber (1965) arranged for students in one class to be angered by having their professor viciously berate them for having done poorly on a recent test; another class served as a control. Subsequently, both groups were tested for aggressive feelings and for sexual arousal. A manipulation check was successful, and the angered group manifested significantly more sexual arousal than did controls ($p < .01$) as measured by explicit sexual content in stories written in response to Thematic Apperception Test (TAT)-like stimuli. Similar results were obtained in two further studies (Barclay, 1969, 1970) in which fraternity and sorority members were angered by the experimenter. The 1970 study employed a female experimenter, which demonstrated that the aggression – sexual arousal link was not specific to male aggression; the 1969 study provided additional support for the hypothesis by using a physiological measure of sexual arousal (acid phosphatase content in urine samples).

Barclay has explained his findings in terms of a special aggression – sexuality link and has cited as support for his position Freud's (1938) argument that prehistoric man had to physically dominate his potential mates and also a study by Clark (1952) in which increased sexual arousal produced by viewing slides of nudes yielded increased aggression in TAT responses. Aron (1970), on the other hand, argued that an aggression – sexuality link exists, but it is only a special case of a more general relationship between emotional arousal of all kinds and sexual attraction. To demonstrate this point, he designed a study in which instead of anger, residual emotion from intense role playing was the independent variable. In this experiment, each of 40 male subjects role played with the same attractive female confederate in either a highly emotional or a minimally emotional situation. Subjects enacting highly emotional roles included significantly more sexual imagery in stories written in response to TAT-like stimuli ($p < .01$) and indicated significantly more desire to kiss the confederate ($p < .05$) than did subjects in the control condition. One possible explanation is suggested by Schachters' theory of emotion (Schachter, 1964; Schachter and Singer, 1962). He argued that environmental cues are used, in certain circumstances, to provide emotional labels for unexplained or ambiguous states of arousal. However, it is notable that much of the above-cited research indicates that a sexual attraction – strong emotion link may occur even when the emotions are unambiguous. Accordingly, taking into account both the Schachter position and findings from sexual attraction research in general, Aron (1970) hypothesized that strong emotions are relabeled as sexual attraction whenever an acceptable object is present, and emotion-producing circumstances do not require the full attention of the individual.

The present series of experiments is designed to test the notion that an attractive female is seen as more attractive by males who encounter her while they experience a strong emotion (fear) than by males not experiencing a strong emotion. Experiment 1 is an attempt to verify this proposed emotion – sexual attraction link in a natural setting. Experiments 2 and 3 are field and laboratory studies which attempt to clarify the results of Experiment 1.

EXPERIMENT 1

METHOD

Subjects

Subjects were males visiting either of two bridge sites who fit the following criteria: (a) between 18 and 35 years old and (b) unaccompanied by a female companion. Only one member of any group of potential subjects was contacted. A total of 85 subjects were contacted by either a male or a female interviewer.

Site

The experiment was conducted on two bridges over the Capilano River in North Vancouver, British Columbia, Canada. The "experimental" bridge was the Capilano Canyon Suspension Bridge, a 5-foot-wide, 450-foot-long, bridge constructed of wooden boards attached to wire cables that ran from one side to the other of the Capilano Canyon. The bridge has many arousal-inducing features such as *(a)* a tendency to tilt, sway, and wobble, creating the impression that one is about to fall over the side; *(b)* very low handrails of wire cable which contribute to this impression; and *(c)* a 230-foot drop to rocks and shallow rapids below the bridge. The "control" bridge was a solid wood bridge further upriver. Constructed of heavy cedar, this bridge was wider and firmer than the experimental bridge, was only 10 feet above a small, shallow rivulet which ran into the main river, had high handrails, and did not tilt or sway.

Procedure

As subjects crossed either the control or experimental bridge, they were approached by the interviewer.[1]

Female interviewer.   The interviewer explained that she was doing a project for her psychology class on the effects of exposure to scenic attractions on creative expression. She then asked potential subjects if they would fill out a short questionnaire. The questionnaire contained six filler items such as age, education, prior visits to bridge, etc., on the first page. On the second page, subjects were instructed to write a brief, dramatic story based upon a picture of a young woman covering her face with one hand and reaching with the other. The instructions and the

---

[1] The interviewers were not aware of the experimental hypothesis in order to prevent unintentional differential cueing of subjects in experimental and control groups.

picture (TAT Item 3GF) employed were adapted from Murray's (1943) *Thematic Apperception Test Manual.* A similar measure of sexual arousal has been employed in the Barclay studies (1969, 1970; Barclay and Haber, 1965), and in other sex-related experiments (Aron, 1970; Clark, 1952; Leiman and Epstein, 1961). The particular TAT item used in the present study was selected for its lack of obvious sexual content, since projective measures of sexual arousal based on explicit sexual stimuli tend to be highly sensitive to individual differences due to sexual defensiveness (Clark and Sensibar, 1955; Eisler, 1968; Leiman and Epstein, 1961; Lubin, 1960). If the subject agreed, the questionnaire was filled out on the bridge.

Stories were later scored for manifest sexual content according to a slightly modified version of the procedure employed by Barclay and Haber (1965). Scores ranged from 1 (no sexual content) to 5 (high sexual content) according to the most sexual reference in the story. Thus, for example, a story with any mention of sexual intercourse received 5 points; but if the most sexual reference was "girl friend," it received a score of 2; "kiss" counted 3; and "lover," 4.

On completion of the questionnaire, the interviewer thanked the subject and offered to explain the experiment in more detail when she had more time. At this point, the interviewer tore the corner off a sheet of paper, wrote down her name and phone number, and invited each subject to call, if he wanted to talk further. Experimental subjects were told that the interviewer's name was Gloria and control subjects, Donna, so that they could easily be classified when they called. On the assumption that curiosity about the experiment should be equal between control and experimental groups, it was felt that differential calling rates might reflect differential attraction to the interviewer.

Male interviewer.   The procedure with the male interviewer was identical to that above. Subjects were again supplied with two fictitious names so that if they phoned the interviewer, they could be classified into control or experimental groups.

RESULTS

Check on Arousal Manipulation

Probably the most compelling evidence for arousal on the experimental bridge is to observe people crossing the bridge. Forty percent of subjects observed crossing the bridge walked very slowly and carefully, clasping onto the handrail before taking each step. A questionnaire was administered to 30 males who fit the same criteria as the experimental subjects. Fifteen males on the experimental bridge were asked, "How fearful do you think the average person would be when he crossed this bridge?" The mean rating was 79 on a 100-point scale where 100 was equal to extremely fearful. Fifteen males on the control bridge gave a mean rating of 18 on the same scale ($t = 9.7$, $df = 28$, $p < .001$, two-tailed). In response to the question "How fearful were you while crossing the bridge?" experimental-bridge males gave a rating of 65 and control-bridge males a rating of 3 ($t = 10.6$, $p < .001$, $df = 28$, two-tailed). Hence, it can be concluded that most people are quite anxious on the experimental bridge but not on the control bridge. To prevent suspicion, no checks on the arousal of experimental subjects could be made.

Thematic Apperception Test Responses

Female interviewer.   On the experimental bridge, 23 of 33 males who were approached by the female interviewer agreed to fill in the questionnaire. On the control bridge, 22 of 33 agreed. Of the 45 questionnaires completed, 7 were unusable either because they were incomplete or written in a foreign language. The remaining 38 questionnaires (20 experimental and 18 control) had their TAT stories scored for sexual imagery by two scorers who were experienced with TAT scoring. (Although both were familiar with the experimental hypothesis, questionnaires had been coded so that they were blind as to whether any given questionnaire was written by a control or experimental subject.) The interrater reliability was $+.87$.

Subjects in the experimental group obtained a mean sexual imagery score of 2.47 and those in the control group, a score of 1.41 ($t = 3.19$, $p < .01$, $df = 36$, two-tailed). Thus, the experimental hypothesis was verified by the imagery data.

Male interviewer.   Twenty-three out of 51 subjects who were approached on the experimental bridge agreed to fill in the questionnaire. On the control bridge 22 out of 42 agreed. Five of these questionnaires were unusable, leaving 20 usable in both experimental and control groups. These were rated as above. Subjects in the experimental group obtained a mean sexual imagery score of .80 and those in the control group, .61 ($t = .36$, *ns*). Hence the pattern of result obtained by the female interviewer was not reproduced by the male interviewer.

Behavioral Data

Female interviewer.   In the experimental group, 18 of the 23 subjects who agreed to the interview accepted the interviewer's phone number. In the control group, 16 out of 22 accepted (see Table 30.1). A second measure of sexual attraction was the number of subjects who called the interviewer. In the experimental group 9 out of 18 called, in the control group 2 out of 16 called ($X^2 = 5.7$, $p < .02$). Taken in conjunction with the sexual imagery data, this finding suggests that subjects in the experimental group were more attracted to the interviewer.

TABLE 30.1
Sexual attraction under conditions of high anxiety
Behavioral responses and thematic apperception test imagery scores for each experimental group

Interviewer	No. filling in questionnaire	No. accepting phone number	No. phoning	Usable questionnaires	Sexual imagery score
Female					
Control bridge	22/23	16/22	2/16	18	1.41
Experimental bridge	23/33	18/23	9/18	20	2.47
Male					
Control bridge	22/42	6/22	1/6	20	61
Experimental bridge	23/51	7/23	2/7	20	80

Male interviewer.   In the experimental group, 7 out of 23 accepted the interviewer's phone number. In the control group, 6 out of 22 accepted. In the experimental group, 2 subjects called; in the control group, 1 subject called. Again, the pattern of results obtained by the female interviewer was not replicated by the male.

Although the results of this experiment provide prima facie support for an emotion–sexual attraction link, the experiment suffers from interpretative problems that often plague field experiments. The main problem with the study is the possibility of different subject populations on the two bridges. First, the well-advertised suspension bridge is a tourist attraction that may have attracted more out-of-town persons than did the nearby provincial park where the control bridge was located. This difference in subject populations may have affected the results in two ways. The experimental subjects may have been less able to phone the experimenter (if they were in town on a short-term tour) and less likely to hold out the possibility of further liaison with her. If this were the case, the resulting difference due to subject differences would have operated *against* the main hypothesis. Also, this difference in subject populations could not affect the sexual imagery scores unless one assumed the experimental bridge subjects to be more sexually deprived than controls. The results using the male interviewer yielded no significant differences in sexual imagery between experimental and control subjects; however, the possibility still exists that sexual deprivation could have interacted with the presence of the attractive female experimenter to produce the sexual imagery results obtained in this experiment.

Second, differences could exist between experimental and control populations with respect to personality variables. The experimental population might be more predisposed to thrill seeking and therefore more willing to chance phoning a strange female to effect a liaison. Also, present knowledge of personality theory does not allow us to rule out the combination of thrill seeking and greater sexual imagery. Accordingly, a second experiment was carried out in an attempt to rule out any differential subject population explanation for the results of Experiment 1.

EXPERIMENT 2

METHOD

Subjects

Subjects were 34 males visiting the suspension bridge who fit the same criteria as in Experiment 1.

Procedure

The chief problem of Experiment 2 was choosing a site that would allow contact with aroused and nonaroused members of the same subject population. One possibility was to use as a control group suspension-bridge visitors who had not yet crossed the bridge or who had just gotten out of their cars. Unfortunately, if a substantial percentage of this group subsequently refused to cross the bridge, the self-selecting-subject problem of Experiment 1 would not be circumvented. Alternatively, males who had just crossed the bridge could be used as a control. The problem with this strategy was that this group, having just crossed the bridge, may have felt residual anxiety or elation or both, which would confound the study. To avoid this latter problem, control subjects who had just crossed the bridge and were sitting or walking in a small park were contacted at least 10 minutes after crossing the bridge. This strategy, it was hoped, would rule out residual physiological arousal as a confounding factor. Except that a different female experimenter was used in Experiment 2 and no male interviewer condition was run, all other details of the study were identicxal to Experiment 1.

RESULTS

Check on Arousal Manipulation

As with Experiment 1, no arousal manipulation check was given to experimental subjects in order not to arouse suspicion about the real intent of the experiment. Data for a group of nonexperimental subjects of the same age and sex as experimental subjects are reported in Experiment 1.

Thematic Apperception Test Responses

In the experimental group, 25 of 34 males who were approached agreed to fill in the questionnaire. In the control group, 25 out of 35 agreed. Of the 50 questionnaires completed, 5 were unusable because they were incomplete. The remainder (23 experimental and 22 control) were scored for sexual imagery as in Experiment 1. The interrater reliability in Experiment 2 was + .79.

Subjects in the experimental group obtained a mean sexual imagery score of 2.99 and those in the control group, a score of 1.92 ($t = 3.07$, $p < .01$, $df = 36$, two-tailed). Thus the experimental hypothesis was again verified by the imagery data.

Behavioral Data

In the experimental group, 20 of the 25 subjects who agreed to the interview accepted the interviewer's phone number. In the control group, 19 out of 23 accepted. In the experimental group, 13 out of 20 called, while in the control group, 7 out of 23 phoned ($X^2 = 5.89$, $p < .02$). Thus the behavioral result of Experiment 1 was also replicated.

Experiment 2 enables the rejection of the notion of differential subject populations as an explanation for the control–experimental-bridge differences for female interviewers in Experiment 1. However, some additional problems in the interpretation of the apparent anxiety–sexual attraction link require the superior control afforded by a laboratory setting.

First, although the female experimenter was blind to the experimental hypothesis and her behavior toward the subjects was closely monitored by the experimenter, the possibility of differential behavior toward the subjects occurring was not excluded. Distance of the interviewer from the subjects was controlled in both Experiments 1 and 2, but more stable nonverbal forms of communication (such as eye contact) could not be controlled without cueing the female interviewer to the experimental hypothesis.

Second, even if the interviewer did not behave differentially in experimental and control conditions, she may have appeared differently in the two conditions. For example, the gestalt created by the experimental situation may have made the interviewer appear more helpless or frightened, virtually a "lady in distress." Such would not be the case in the control situation.

If this different gestalt led to differences in sexual attraction, the apparent emotion–sexual arousal link might prove artifactual. Accordingly, a laboratory experiment was run in which tighter control over these factors could be obtained. This experiment involved a 2 × 2 factorial design, where (*a*) the male subject expected either a painful or nonpainful shock (subject's emotion was manipulated) and (*b*) the female confederate also expected either a painful or nonpainful shock (the lady-in-distress gestalt was manipulated).

EXPERIMENT 3

METHOD

Subjects

Eighty male freshmen at the University of British Columbia took part in this experiment. All subjects were volunteers.

Much of the initial phase of the procedure was patterned after that used in Schachter's (1959) anxiety and affiliation research. Subjects entered an experimental room containing an array of electrical equipment. The experimenter welcomed the subject and asked him if he had seen another person who looked like he was searching for the experimental room. The experimenter excused himself "to look for the other subject," leaving the subject some Xeroxed copies "of previous studies in the area we are investigating" to read. The articles discussed the effects of electric shock on learning and pain in general.

The experimenter reentered the room with the "other subject," who was an attractive female confederate.[2] The confederate took off her coat and sat on a chair three feet to the side and slightly in front of the subject. The experimenter explained that the study involved the effects of electric shock on learning and delivered a short discourse on the value and importance of the research. At the end of this discourse, the experimenter asked if either subject wanted out of the experiment. As expected, no subject requested to leave.

The experimenter then mentioned that two levels of shock would be used in the experiment, describing one as quite painful and the other level as a "mere tingle, in fact some subjects describe it as enjoyable," and concluded by pointing out that the allocation of subjects to shock condition had to be "completely random so that personality variables won't affect the outcome." At this point, the experimenter asked both subjects ro flip a coin to determine which shock level they would receive.[3] Hence, the subject reported "heads/tails," the confederate reported "heads/tails," and the experimenter said, "Today heads receives the high shock level." The

---

[2] The female confederate knew that the study involved sexual attraction but did not know the experimental hypothesis. Her every action in the experimental room was carefully rehearsed to avoid any possibility of differential behavior among experimental conditions. Spacing of the confederate's chair from the subject's was carefully controlled, and the confederate was instructed to avoid any eye contact with the subject after their initial introduction. Hence, eye contact was restricted to the confederate's entering the room and returning to her chair after removing her coat. Both the confederate's and the subject's chairs faced the same direction (toward the experimenter), so that eye contact was easily avoided. In addition, the confederate's chair was somewhat closer to the experimenter than was the subject's chair, so that the subject could see the confederate while the experimenter delivered the instructions.

[3] Toward the end of the experiment, the confederate was told to report either the same result as the subject or a different result (of the coin flip) to facilitate obtaining equal *ns* for experimental conditions as quickly as possible.

experimenter then described the way in which the shock series would take place, the method of hooking subjects into electrodes, etc.

The experimenter then asked if the subjects had any questions, answered any that arose, and then said:

> It will take me a few minutes to set up this equipment. While I'm doing it, I would like to get some information on your present feelings and reactions, since these often influence performance on the learning task. I'd like you to fill out a questionnaire to furnish us with this information. We have two separate cubicles down the hall where you can do this—you will be undisturbed and private, and I can get this equipment set up.

The confederate then got up, walked in front of the subject to her coat, which was hanging on the wall, rummaged around for a pencil, and returned to her chair. The experimenter then led the subject and the confederate to the cubicles, where they proceeded to fill out the questionnaires.

RESULTS

A three-part questionnaire constituted the dependent measure of this study. Part 1 (feelings about the experiment included a check on the anxiety manipulation, Part 2 (feelings toward your co-subject) included two attraction questions found to be most sensitive in experimental situations of this sort (Aron, 1970), and Part 3 included the TAT picture used in Experiments 1 and 2, which was again scored for sexual imagery.

Anxiety

Anxiety was measured by the question "How do you feel about being shocked?" (cf., Schachter, 1959) to which subjects could respond on a 5-point scale where scores greater than 3 indicated dislike. (The greater the score, the greater the anxiety.) Table 30.2 presents the results on this measure. In conditions where the subject anticipated receiving a strong shock, subjects reported significantly more anxi-

TABLE 30.2
Reported anxiety in experimental conditions

Subject expects:	Female con-federate to get strong shock	Female con-federate to get weak shock	No female confederate
Strong shock	3.17	3.05	3.80
Weak shock	2.42	2.28	

Note:n per cell = 20.

TABLE 30.3
Attraction ratings by experimental condition

Subject expects:	Female confederate to get strong shock	Female confederate to get weak shock
Strong shock	3.7	3.4
Weak shock	2.9	2.7

Note: Strongest attraction rating is 5.

ety than in conditions where the subject anticipated receiving a weak shock ($t = 4.03$, $p < .001$, $df = 39$, one-tailed). In conditions where the subject anticipated receiving a strong shock with the female co-subject present, subjects reported significantly less anxiety than in a control condition ($n = 20$), where two male subjects were run ($t = 2.17$, $p < .025$, $df = 19$, one-tailed). No significant differences in the subject's anxiety occurred as a function of the confederate receiving a strong versus a weak shock (see Table 30.2).

Attraction to Confederate

Two questions assessed attraction to the confederate in this study: (*a*) How much would you like to ask her out for a date? and (*b*) how much would you like to kiss her? (An alternative set of questions was provided for those subjects who ostensibly had a male copartner. The experimenter instructed subjects in this condition to overlook these.) Attraction ratings were established by taking the mean rating made by subjects on these two questions. Table 30.3 shows the results, by condition, of those ratings. A 2×2 analysis of variance revealed a significant main effect for subjects anticipating strong shock to themselves on attraction ratings ($F = 22.8$, $p < .001$). Subjects' expectations of strong versus weak shock to the female confederate produced no significant increase in attraction ($F = 2.61$, *ns*). (There was no significant interaction.) Hence, the lady-in-distress effect on attraction did not seem to appear in this study.

Thematic Apperception Test Responses

Sexual imagery scores on the TAT questionnaire were obtained as in Experiments 1 and 2 and are shown in Table 30.4. In the present study, sexual imagery was higher when the subject expected strong shock but only when the female confederate also expected strong shock ($F = 4.73$, $p < .05$). When the female confederate expected weak shock, differences in sexual imagery scores as a function of strength of shock anticipated by the subject failed to achieve significance ($F = 4.22$, $P = .07$).

TABLE 30.4
Sexual imagery scores by experimental condition

Subject expects:	Female confederate to get strong shock	Female confederate to get weak shock
Strong shock	2.27	2.19
Weak shock	1.52	1.69

Note: Strongest imagery score is 5.

GENERAL DISCUSSION

The results of these studies would seem to provide a basis of support for an emotion–sexual attraction link. The Barclay studies (Barclay, 1969, 1970; Barclay and Haber, 1965) have already demonstrated such a link for aggression and sexual arousal, and the present findings seem to suggest that the link may hold for fear as well. Indeed, the present outcome would seem to be particularly satisfying in light of the very strong differences obtained from the relatively small subject populations, and because these results were obtained, in Experiments 1 and 2, outside of the laboratory in a setting in which real-world sexual attractions might be expected to occur.

The strong result of Experiment 3 supports the notion that strong emotion per se increases the subject's sexual attraction to the female confederate. Brehm, Gatz, Goethals, McCrimmon, and Ward (1967) obtained results consistent with Experiment 3 in a similar study. They also had male subjects threatened by impending electric shock in two experimental groups and not threatened in the control group. They obtained an F significant at the 5 percent level for differences between the threat and no-threat groups.

The theoretical implications of these results are twofold. In the first place, they provide additional support in favor of the theoretical positions from which the original hypothesis was derived: the Schachter and Singer (1962) tradition of cognitive labeling of emotions and the Aron (1970) conceptual framework for sexual attraction processes. In the second place, these data seem to be inconsistent with (or at least unpredictable by) standard theories of interpersonal attraction. Both the reinforcement (Byrne, 1969) and the cognitive consistency (Festinger, 1957; Heider, 1958) points of view would seem to predict that a negative emotional state associated with the object would *decrease* her attractiveness; and neither theory would seem to be easily capable of explaining the arousal of a greater sexual emotion in the experimental condition of the present experiments.

Although the present data support the cognitive relabeling approach in general, they are consistent with more than one interpretation of the mechanics of the process. The attribution notions of Nisbett and Valins (1972), self-perception theory (Bem, 1972), and role theory (Sarbin and Allen, 1968) can all provide possible explanations for the anxiety–sexuality link. A further possible explanation is that heightened emotion, instead of being relabeled as sexual, serves merely to disinhibit the expression of preexistent sexual feelings. It is known that inhibition and sexual defensiveness influence sexual content in TAT stories (Clark, 1952), and this alternative cannot be ruled out by the present data. Yet another alternative suggested by Barclay (personal communication, 1971), is that the aggression–sexuality and anxiety–sexuality links may be independent phenomena and not necessarily subcases of a general emotion–sexuality link.

Some evidence for the mechanics of the anxiety–sexual arousal link in the current research may be obtained from the fear ratings made by subjects in Experiment 3. When subjects anticipated receiving a strong shock and the female confederate was present during the anxiety manipulation, subjects reported significantly less fear than when no potential sexual object was present ($t = 2.17$, $df = 19$, $p < .025$). Since the questionnaires were filled out in private in both groups, it is unlikely that subjects' reporting merely reflects appropriate behavior in the presence of the opposite or same sex. One possible explanation for this result is that, having relabeled anxiety as sexual arousal, the subject is less likely to feel anxious. A more conclusive explanation of the mechanics of the anxiety–sexual arousal link must await the conclusion of present laboratory studies designed specifically to investigate this problem. However, regardless of the interpretation of the mechanics of this link, the present research presents the clearest demonstration to date of its existence.

References

ARON, A. Relationship variables in human heterosexual attraction. Unpublished doctoral dissertation, University of Toronto, 1970.

BARCLAY, A. M. The effect of hostility on physiological and fantasy responses. *Journal of Personality,* 1969, **37,** 651–667.

BARCLAY, A. M. The effect of female aggressiveness on aggressive and sexual fantasies. *Journal of Projective Techniques and Personality Assessment,* 1970, **34,** 19–26.

BARCLAY, A. M., AND HABER, R. N. The relation of aggressive to sexual motivation. *Journal of Personality,* 1965, **33,** 462–475.

BEM, D. Self-perception theory. In L. Berkowitz (ed.), *Advances in experimental social psychology.* Vol. 6. New York: Academic Press, 1972.

BREHM, J. W., GATZ, M., GOETHALS, G., McCRIMMON, J., AND WARD, L. Psychological arousal and interpersonal attraction. Unpublished manuscript, Duke University, 1967.

BYRNE, D. Attitudes and attraction. In I. L. Berkowitz (ed.), *Advances in experimental social psychology.* Vol. 4. New York: Academic Press, 1969.

CLARK, R. A. The projective measurement of experimentally induced levels of sexual motivation. *Journal of Experimental Psychology,* 1952, **44,** 391–399.

CLARK, R. A., AND SENSIBAR, M. R. The relationship between symbolic and manifest projections of sexuality with some incidental correlates. *Journal of Abnormal Social Psychology,* 1955, **50,** 327–334.

EISLER, R. M. Thematic expression of sexual conflict under varying stimulus conditions. *Journal of Consulting and Clinical Psychology,* 1968, **32,** 216–220.

ELLIS, H. *Studies in the Psychology of Sex.* New York: Random House, 1936.

FESTINGER, L. *A theory of cognitive dissonance.* Evanston, Ill.: Row, Peterson, 1957.

FREUD, S. *Basic writings.* New York: Modern Library, 1938.

HEIDER, F. *The psychology of interpersonal relations.* New York: Wiley, 1958.

JAMES, W. *The principles of psychology.* Vol. 2. New York: Holt, 1910.

KELLY, H. Attribution theory in social psychology. In D. Levine (ed.), *Nebraska Symposium on Motivation: 1967,* Lincoln: University of Nebraska Press, 1967.

LEIMAN, A. H., AND EPSTEIN, S. Thematic sexual responses as related to sexual drive and guilt. *Journal of Abnormal and Social Psychology,* 1961, **63,** 169–175.

LUBIN, B. Some effects of set and stimulus properties on T.A.T. stories. *Journal of Projective Techniques,* 1960, **24,** 11–16.

MURRAY, H. A. *Thematic Apperception Test manual.* Cambridge, Mass.: Harvard University Press, 1943.

NISBETT, R., AND VALINS, S. *Perceiving the causes of one's own behavior.* New York: General Learning Press, 1972.

SARBIN, T. R., AND ALLEN, V. L. *Role theory: Handbook of social psychology.* Reading, Mass.: Addison-Wesley, 1968.

SCHACHTER, S. *The psychology of affiliation.* Stanford, Calif.: Stanford University Press, 1959.

SCHACHTER, S. The interaction of cognitive and physiological determinants of emotional state. In L. Berkowitz (ed.), *Advances in experimental social psychology.* Vol. 1. New York: Academic Press, 1964.

SHACHTER, S., AND SINGER, J. E. Cognitive, social and physiological components of the emotional state. *Psychological Review,* 1962, **69,** 379–399.

SUTTIE, I. D. *The origins of love and hate.* London: Kegan Paul, 1935.

TINBERGEN, N. The origin and evolution of courtship and threat display. In J. S. Huxley, A. C. Hardy, and E. B. Ford (eds.), *Evolution as a process.* London: Allen & Unwin, 1954.

VIII

INTERPERSONAL COMMUNICATION AND SENSITIVITY

31

When Familiarity Breeds Respect: The Effects of an Experimental Depolarization Program on Police and Student Attitudes Toward Each Other

Michael Jay Diamond and W. Charles Lobitz

During student riots at Stanford University in the spring of 1970, 164 students (of whom 95 were control subjects with no contact) and 37 local policemen were brought together to facilitate nonviolent interactions and promote understanding between students and police. Three forms of contact were utilized: students riding in police squad cars, police having dinner and "rap sessions" with students, and encounter groups. Self-report questionnaires assessed the attitudes of members of each group toward the other both before and after the contact. Significant attitudinal depolarization toward the other group occurred as a result of the three types of contact. These findings are discussed in terms of the reduction of autistic hostility between groups as well as an increase in self-disclosure. Methodological problems inherent in such social action projects are considered and suggestions made for future projects of this kind.

It is no secret that we are living in a time of much unrest in our urban communities. While the causes of such disorder are undoubtedly numerous and are certainly a focus for a great deal of speculation and discussion, one factor is well agreed upon; that is, there has been a breakdown in trust and meaningful com-

Reprinted with permission of the authors and *Journal of Social Issues*, Vol. 29, No. 4, 1973.

The authors wish to thank Greg Morris and Robert Schatz of the Mountain View Police Department, Philip G. Zimbardo of Stanford University, and Stanford students Rick Bowers, Claudia Cohen, Diane Fisher, Hunt Kooiker, Gretchen Lobitz, Herb Murphy, Neil Morse, Hilde Ann Olds, and Elsa Rosenberg for their efforts during the Depolarization Project. Additional thanks go to Jerry Brennan and Carol Kanayama for their help in the data tabulation and to Michael H. Bond, Jerrold L. Shapiro, Samuel I. Shapiro, and Roland Tharp for their cogent suggestions concerning preparation of this manuscript, and to the Social Science Research Institute, University of Hawaii, for typing assistance. A version of this paper was presented at the meeting of the Western Psychological Association, San Francisco, April 1971.

munication between police and various segments of the community. Moreover, after extensive study of urban unrest, the U.S. Riot Commission (Kerner et al., 1968) reported that abrasive contacts between police and citizens were a major cause of the disorders of recent years.

Fortunately the problem has not gone entirely unnoticed either by concerned members of the police force, community leaders, or social scientists. Beginning with the development of the National Center on Police and Community Relations at Michigan State University in 1955, there has been growing nationwide interest in developing and coordinating programs designed to improve police-community relations (Brandstatter and Radelet, 1968). Typically these programs have developed under the guise of so-called "Police Human Relations Training" projects (Siegel, Federman, and Schultz, 1963). In such programs, police are trained in developing such interpersonal skills as those involved in family crisis intervention (Bard, 1969), in understanding and relating to ethnic minority communities (Bell et al., 1969), and in handling psychiatric and other mental health problems (Danish and Brodksy, 1970). However, in spite of an increasing amount of literature on police-community relations, there is still insufficient empirical evidence concerning the efficacy of such programs. Hopefully the recent (1970) formation of the SPSSI Committee on Police-Community Relations will promote more systematic criterion-oriented evaluation of these programs. There is reason to believe that the assessment issue is already being more carefully considered (Reiser, 1970).

Although programs aimed at improving the relations between police and selected ethnic and professional (i.e, mental health workers) minority groups are underway in many cities, relatively little attention has been paid to enhancing relations between police and the recently increasing minority group of late adolescent and college-age young people. Increasing hostility between police and these young people—often labeled hippies, yippies, radicals, and/or activists—has been well documented. Dissent and violent protest are not especially unusual occurrences on college campuses; in fact, recent events suggest that confrontations between the police and students have already moved outside the campus confines.

In spite of such explosive examples and despite warnings by numerous social observers (Henig and Furst, 1970), there has been little apparent effort on the part of police administrators to foster an understanding of protesting youth. An exception to this would be the "Community Awareness" program in Minneapolis (Flint, 1971). Furthermore, United States law enforcement agencies have failed to differentiate clearly between civil and criminal violations (Rokeach, 1968). This failure to attend to the root of the problem, coupled with the fact that police seem to be quite unattracted to the ideas and the types of people that inhabit campus settings (Johansson, 1970) makes continued lack of understanding and future violence all the more probable between socially concerned young people and the police. Similarly, there is little evidence to suggest that students are any more well informed as to what policemen do, think, and feel than policemen are as to students. Hence there is reason to believe that the attitudes of each group toward the other group are characterized not only by ignorance but by polarization, extremeness, and excessive stereotyping (Flint, 1971).

That such ignorance and polarization have apparently survived over the years can undoubtedly in part be accounted for by Newcomb's (1947) notion of "autistic hostility." According to this hypothesis, when persons for whatever reasons develop hostility toward each other, they tend to restrict or avoid mutual communication; therefore the hostility cannot be corrected through open communicative contact. This becomes particularly salient when one considers that negative stereotypes toward one another are in part reinforced in the course of the law enforcement situations wherein these young citizens and police officers typically interact (Flint, 1971). According to such notions, the only way in which this chain can be broken is by instituting a program designed for open communication.

THE DEPOLARIZATION PROJECT

During the spring of 1970, massive student rioting on campuses throughout the nation broke out as a protest against President Nixon's announced invasion of Cambodia by U.S. troops. Stanford University experienced the worst disorder in its history. Police from all over the San Francisco Bay Area were called to protect University property. During the peak week of the crisis, students and police clashed nightly; students threw rocks at the police, and the police clubbed and arrested students. By the following week, physical contact between students and police had ceased. However, strong antipolice sentiment pervaded the campus. Furthermore, local police departments reported intense antistudent feelings among their officers. These tensions threatened to erupt into renewed violence. In response, the "Police Student Depolarization Project" was initiated.

In order to reduce the chances for subsequent violent confrontation, the project's primary aim was to reduce campus tensions and promote understanding between students and police. In addition the authors were actively concerned with assessing the influence of the project on the polarized attitudes of both police and students. In this way it was hoped a necessary beginning to bridging the police–student gap could be made.

Assuming the existence of "autistic hostility" between the groups, the project was structured to promote nonviolent, communicative interactions between involved students and police. With the cooperation of several involved local police departments, three types of mutual communicative-oriented, nonviolent contact situations were arranged between volunteer students and police. Both police and student project participants were willing to submit to evaluative procedures; in this sense they became experimental subjects.

It was expected that such contact, fostering mutual communication, would also increase understanding (i.e., tendency to see the other as other sees himself) as measured by self-report questionnaires. Our hypotheses can be formally stated as follows:

> 1a. Each of the three types of contact employed will increase student understanding of police role, activities, and attitudes.
> 1b. The encounter session will increase police understanding of students' activities, concerns, and attitudes.

Since it was not possible to assess police understanding in either of the other contact modes, no hypotheses were formulated concerning police changes as a result of these contacts.

It should, however, be recognized that not all attitudes and behaviors will become depolarized, since they are sometimes realistically based upon what is likely to occur in law enforcement situations. Moreover, in certain cases increased understanding could result in greater attitudinal and behavioral polarization rather than depolarization. Newman and Steinberg (1970) have wisely warned social interventionists of this possibility. However, since violent police–student clashes on campus had done much to polarize the two groups, the nonviolent contacts employed in the project were expected to generally have a depolarizing effect. This effect could be demonstrated by a shift from a negative to a positive reaction (attitude) toward the other group. Thus the second set of hypotheses were:

> 2a. Student reactions (attitudes) toward police will shift in a positively valued direction as a result of each of three types of contact employed.
> 2b. Police reactions (attitudes) toward students will shift in a positively valued direction as a result of the encounter session.

In addition, depolarization could be demonstrated by a shift in behavior towards the other group. Unfortunately, it was impossible to assess actual behavior, so behavioral intentions were solicited from the students. It was not possible to obtain police intentions over time. The third set of hypotheses were as follows:

> 3a. Students' self-reported covert and overt behaviors will become less hostile toward the police as a result of the contacts employed.
> 3b. Students will indicate an increased willingness to encounter police in a positive way as a result of such contact.
> 3c. Students' self-reported fear responses toward the police will decrease as a result of such contact.

It should be noted that while there are many methodological problems inherent in the nature of such a social action project, these problems were not overlooked and will be considered when discussing the findings. Furthermore, attempts were made to overcome some of these problems by providing a student no-treatment control group (a control group of police subjects was unattainable at the time), by requesting behavioral intentions rather than just relying on self-reported attitudes

(Wicker, 1969), and by requesting honesty on all questionnaire responses (Bowers, 1967).

METHOD

Subjects

A total of 164 students and 37 policemen volunteered to participate in this project. The students were male and female undergraduate and graduate students from Stanford University. The police were obtained from police departments in Los Altos, Menlo Park, Mountain View, Palo Alto, and San Jose, and from the Santa Clara County Sheriffs Office. Of the 164 students, 95 were "controls" who did not participate in any of the contact sessions. Thus, there were 69 "experimental" subjects. Twenty-seven students and 12 policemen participated in the encounter session, 28 students and 13 policemen were engaged in the squad-car riding portion of the project, and 14 students and 12 policemen participated in the dinner "rap session" program.

Types of Contact

The three different types of contact structured to facilitate nonviolent communicative interactions between students and police were: (a) student squad-car riding, (b) police–student dinners and rap sessions, and (c) an encounter session for police and students together.

Student squad-car riding.   Students who rode in squad cars were able to observe the policeman in his "typical" role. The student could ask about the policeman's professional responsibilities, as well as his personal beliefs. The students were paired and accompanied an individual policeman for a period ranging from four to eight hours on the policeman's regular beat.

Dinners and rap sessions.   Students who had a policeman over for dinner or for a rap session were able to discuss informally such common concerns and intrerests as Vietnam, drugs, and civil disobedience. These sessions were held in students' homes, apartments, and dormitories. Typically, one or two policemen would join with three or four students for a session lasting about two hours.

Encounter sessions.   The encounter sessions consisted of small group discussions between members of both groups over the course of a three-hour period. The sessions were held at a local high school auditorium. Participants were broken down into triads, which later combined into sextets; finally the entire group reassembled. Police were dressed in street clothes. Groups were encouraged by a professional facilitator to express their views verbally about themselves and the social situation. A few nonverbal sensory awareness encounter techniques were used in several of the groups.

Procedure

Twenty-four hours prior to their contact, students who either went riding or had dinner rap sessions with police completed a survey on their attitudes toward policemen. This questionnaire was designed to tap students' cognitive and affective reactions to police, as well as their knowledge of policemen's attitudes and behavior. For example, to assess students' cognitive and affective reactions, students were asked to check off on a 5-point scale (strongly agree to strongly disagree) whether "police are anti-intellectual," "callous and insensitive," and whether "police deserve to be called pigs." Similarly, they were asked to check whether "police consider themselves professionals" and whether "police use excessive brutality." Twenty-four hours following their contact these experimental subjects again completed the survey. Twenty-seven control volunteer subjects who had no contact with police also completed the survey twice, the second time 48 hours after the first. In addition, both experimental and control subjects completed a self-report questionnaire of their behavioral intentions towards policemen prior to their contact and again after a 5-week period. On this questionnaire, students responded on a 5-point scale (never to always) to items like the following: "When talking with friends, I refer to police as 'pigs' "; When I drive past a police car and another car parked beside the road, I assume that the policeman is giving the other driver a ticket"; and, "When I am stopped by a policeman for no apparent reason, I respond with anxiety, e.g., increased heartrate, sweating, tightening in my stomach."

Students who participated in the encounter session completed an attitudinal survey just prior to and immediately following the session. There were no control subjects. The format was similar to the attitude survey discussed above. Additional items included: "Most police are biased against males with long hair"; "Most police consider themselves professionals, comparable to doctors, teachers"; and "Most police are vehemently opposed to the use of drugs (including marijuana) by anyone except when prescribed by a doctor." In addition, police who participated in the encounter sessions completed pre-post attitudinal surveys of a similar format. For example, questions included: "Most students take drugs to escape reality"; "Most students do not understand why we need rules and laws"; and, "Most students are impressionable and easily swayed." Open-ended responses to the session as a whole were also solicited from all participants. These included the following questions: "What do you feel can be done to improve police-student community relations?''; ''What do you think was the value of tonight's meeting?''; ''Would you like more of these?''; and ''Have you changed because of tonight's meeting, and if so, how?''

RESULTS

On the 53 questionnaire items answered by both experimental and control student subjects, only two items yielded significant ($p < .05$) precontact differences. Correlated t-test comparisons were used on these two items. On the remaining 51 items, t-test comparisons were performed on the pre–post test change scores. On those 45 items where no control group was available, t-tests were performed comparing the precontact with the postcontact score. Since no attempts were made to compare the separate modes of contact as to their effectiveness, the pre–post questionnaire items from the different types of contact are considered together and examined with the context of the three separate hypotheses sets. Results of major comparisons are presented in Table 31.1.

TABLE 31.1
Comparison of Pooled Mean Self-Report Scores Before and After Contact

Pooled items	Experimental group		Control group[a]		t
	Before	After	Before	After	
Student understanding of police role, activities, and attitudes	2.64	2.60	2.76	2.78	1.31*
Police understanding of students' activities, concerns, and attitudes	3.57	3.71			1.97**
Student reactions toward police	2.84	3.21	3.09	3.07	8.15***
Police reactions toward students and student related issues	3.27	3.49			5.09***
Student overt and covert hostile behaviors toward police	3.22	3.23	3.28	3.17	1.04
Student willingness to encounter police positively	3.86	3.61	3.86	3.87	2.75***
Student fear responses toward police (reverse scored)	2.00	2.21	2.57	2.54	1.77**

Note: The scores are based on the subjects responses to a 5-point rating scale ranging from 1 (strongly agree or always, about 100 percent of time event occurs) to 2 (agree) to 3 (ambivalent or undecided) to 4 (disagree) to 5 (strongly disagree or never, about 0 percent of time event occurs).
[a] When there are no figures recorded in this column there was no appropriate control group available.
*$p < .10$
**$p < .05$
***$p < .01$

Hypothesis 1

It was hypothesized that the three types of contact employed would increase student understanding of police role, activities, and attitudes (Hypothesis 1a), and similarly that the encounter session would increase police understanding of students' activities, concerns, and attitudes (Hypothesis 1b).[1]

Eight of the 32 items involved yielded significant ($p < .05$) increases in student understanding of police roles, activities, and attitudes following contact. The binomial probability of this proportion of significant items was itself significant ($p < .05$), suggesting that increased understanding did occur. For example, there was a significant increase ($p < .05$) in their tendency to agree with the statement that "Police are very familiar with criminal laws and civil rights." A t-test on the difference scores of the pooled items between the experimental and control group yielded an overall change in student understanding following contact, which approached significance ($t = 1.31$, $df = 72$, $p < .10$, see Table 31.1).

The results of the police responses prior to and following the encounter group indicated that significant changes ($p < .05$) occurred on only 2 of the 8 items, a nonsignificant proportion. However, certain items suggested that some increased understanding of students resulted. For example there was a significant reduction ($p < .05$) in the tendency for police to agree with the statement that "Most students take drugs to escape reality." As Table 31.1 indicates, a t-test on these items pooled together shows there to be a significant overall change in understanding following the encounter session ($t = 1.97$, $df = 11$, $p < .05$).

Hypothesis 2

It was predicted that both student (Hypothesis 2a) and police (Hypothesis 2b) reactions (attitudes) toward one another would shift in a positively valued direction as a result of contact, a hypothesis for which test results provide support.

These results suggest that student reactions toward police became much more positive as a result of contact. Significant ($p < .05$) pre–post test differences pertained on the majority (i.e., 12 of 22) of questionnaire items ($p < .001$). For example, students were more likely to disagree with statements asserting that police are impersonal, rigid, callous, and insensitive, as well as biased against minority races and long-haired males (all p's $< .05$). Similarly, in contrast to control subjects, students receiving contact were more likely to disagree with items stating that "Police use excessive brutality" ($p < .001$) and are "hostile" ($p < .05$). To

---

[1] In order to determine how the "other sees himself," representative policemen and Stanford University students discussed what they felt to be the typical activities, roles, concerns, and attitudes of the members of their own group. Understanding, as measured by the questionnaire items, was scored in accordance with this information.

assess the impact of contact in altering overall reactions toward police, a t-test on the difference scores of pooled items between the experimental and control groups was performed. Again (Table 31.1) there was a significant effect due to contact ($t = 8.15$, $df = 162$, $p < .001$).

The results of the analyses indicated that police reactions toward students and student related issues became significantly ($p < .05$) more positive as a result of the encounter session on 5 of 17 items ($p < .05$). Following contact, police were significantly more likely ($p < .05$) to disagree with statements asserting students to be impressionable and easily swayed, soft and inexperienced, as well as "too promiscuous." Moreover, police were more likely to differentiate "rioters" from "dangerous criminals" ($p < .05$). An overall change in attitudes toward students was demonstrated (see Table 31.1) by a t-test performed on the pooled items ($t = 5.09$, $df = 11$, $p < .001$).

Hypothesis 3

The third prediction was that contact would alter students' intended behavior toward police in terms of being reported as less hostile (Hypothesis 3a), reflecting an increased willingness for positive interaction (Hypothesis 3b), and reducing self-reported fear responses toward police (Hypothesis 3c). Pre–post test changes between experimental and control subjects were examined by t-test to test this hypothesis.

A significant difference ($p < .05$) between groups in their tendency to assume that policemen stopping cars along the road are "giving the other driver a ticket" gives partial support to Hypothesis 3a. However, although the trends are mainly in the predicted direction, only 2 of the remaining 7 items showed statistically significant ($p < .05$) differences, a nonsignificant ratio.

Hypothesis 3b is more strongly supported in that 4 of 9 items showed increases ($p < .10$) in willingness for positive interaction ($p < .01$). For example, students who received contact were more likely to report that they "say hello, smile, or otherwise acknowledge" policemen ($p < .05$). A t-test on the difference scores of the pooled items between the groups (Table 31.1) showed an overall significant difference between contact and no-contact subjects in their self-reported willingness to encounter police positively ($t = 2.75$, $df = 72$, $p < .01$).

Finally, there is some evidence that self-reported fear responses toward police decreased following contact. For example, following contact, experimental subjects reported a decreased tendency to respond with anxiety when "stopped by a policeman for no apparent reason" ($p < .05$). A t-test on the difference scores of the pooled items between groups (see Table 31.1) was statistically significant ($t = 1.77$, $df = 72$, $p < .05$).

DISCUSSION

The results indicate that, in general, the project achieved its hypothesized objectives of increasing understanding (i.e., the tendency to see the other as the other sees himself) between police and students, depolarizing police and student attitudes towards each other (changing attitudes in a positive direction), and changing students' intended behavior towards police in a positive direction. Overall, depolarization trends were indicated on 70 out of 82 items that appeared to involve the depolarization–polarization dimension (32 of which were statistically significant), whereas only 6 items suggested increased polarization and 6 no change ($\chi^2 = 96.81$, $df = 2$, $p < .005$). Furthermore, none of the items that indicated polarization approached statistical significance.

Although no formal attempt was made to compare the three modes of contact, the encounter sessions seemed to be the most efficient in depolarizing police and student attitudes in a short time span. Undoubtedly, the more structured nature of the encounter session brought the issues into focus more quickly. An unexpected consequence of the encounter session was the liberalization of police attitudes on issues important to students (e.g., drugs, the Indochina War, and student demonstrations). The most significant change ($p < .05$) on this dimension was that, after the group session, police reported that student rioters who refuse to obey the law are less dangerous than people who are committing crimes against other people (e.g., robbers).

While the encounter session seemed to be most efficient in overcoming "autistic hostility" between the groups, the dinners, rap sessions, and squad-car riding programs were also rather effective. Both students and police demonstrated their commitment to understanding and influencing each other nonviolently. While this study did not assess what components of the contacts were most responsible for the effects, it is likely that a willingness to confront each other nonviolently was most important. It is doubtful that such changes would have occurred among nonvolunteer police and student subjects. The generalizability of the findings to nonvolunteer individuals is still unknown.

Besides commitment to nonviolent interaction, the success of the project would seem to be a function of the fact that there is much that police and students from a large university have in common. This was noted by several members from both groups. Members of both groups are known to be victims of prejudice and stereotyping (consider the often-used terms "pigs" and "hippies"). Members of both groups tend to live in "inverted" societies characterized by clans within the same group and a general ostracism by those outside the group. Members of both groups are made constantly aware of the inequities in the legal system as well as society at large and are similarly frustrated in the availability of responses to such problems, so that police are impelled to enforce the laws rather than change them, while students are often powerless to effect such change. And finally, members of each group share a somewhat biased and distorted view of members of the other group, a view that resists disconfirmation in part due to a lack of appropriate non-threatening contact.

The provision of such contact in the project seemed to enable members of each group to establish a common ground with members of the other by sharing feelings about these similarities. Moreover, the contacts allowed the participants to come to know each other on a person-to-person basis and to develop trust in each other. By talking and listening to one another as human beings rather than just as representatives of their respective groups, the participants were able to share personal viewpoints and experiences. Such beneficial interpersonal effects of self-disclosure have been excellently discussed by Jourard (1964). That such effects occurred as a result of the contacts employed in the project is suggested by an examination of the open-ended police and student responses to the encounter session. When asked if they would like more of such sessions, 12 of 12 police said "yes" and 26 of 27 students said "yes." Typical police reactions to the question concerning the value of such a session included such statements as "the recognition of both sides of the humanity of each other," "knowing students as people," and "establishing honesty." When asked if they changed as a result of the encounter, several policemen mentioned they did as they realized that "longhairs are not necessarily up to no good." Similarly, typical student statements to the first question on the session's value included, "getting to meet the kinds of people we don't usually try to meet," "finding out we are all human," "helped me to become aware of what it is like to be a cop," and "by relating some of my personal experiences, I think I opened some doors to understanding for him." In response to the question concerning change as a result of the session, student responses ranged from "Yes, I found out some of these guys have a sense of humor and can be civil" to "Yes, I now have more compassion for cops."

Although the results indicated that the project was successful in depolarizing the attitudes of police and students, the study suffered from several methodological flaws, including the lack of an adequately matched student and police control group, the reliance on multiple t-tests for data analysis, a set of rather vague experimental hypotheses, and the reliance on self-report rather than behavioral assessment devices. Unfortunately, such problems were an unavoidable consequence of the emergency nature of his social action project. Future research in this area should attempt to rectify these problems while addressing itself to answering several questions raised by this study, including: (a) which contact mode is superior in achieving depolarization; (b) what are the effects of such contacts when employed over a longer time span (e.g., weekly encounter sessions and squad-car rides over a 3-month period); and (c) how could such contacts be institutionalized in the absence of a crisis situation.

The present study has only addressed itself to the general question of "How can the gap between police and students be bridged?" As indicated by the data, some general answers have been provided. Moreover, the success of the program becomes even more important when one realizes that it was conducted in the midst of a serious campus crisis. Because of the emergency nature of the crisis, the depolarization project was more of a social action program than a research study. In stressing the importance of using the knowledge of the social sciences to reverse the course of social decline, Walker (1970) has suggested that constructive social change depends upon larger scale collaborations between social scientists and politicians. The depolarization project has demonstrated that effective social action can occur even without the cooperation of politicians. A handful of concerned social scientists, police administrators, and students have brought two polarized groups closer together. It is hoped that this kind of grass roots action will be undertaken elsewhere with the goal being the "primary prevention" of such crises situations.

Walker (1970) has also warned of a danger for psychologists who engage in social action: In their zeal for effecting social change, they may forget their role as psychologists. The easiest way for this to happen is if psychologists neglect to use their skills as scientists. On the other hand, if they attempt to rigorously evaluate the effects of their action, they can maintain the dual role of scientist and activist. This study has made a crude but sincere attempt to evaluate the effects of a socially relevant project. Hopefully, others will improve on the model. Some of the means have been suggested above. Whatever the outcomes, psychology as a science can exist comfortably with psychology as a constructive social force if psychologists are willing to apply their research skills in attempting to solve society's social problems.

References

BARD, M. Family intervention police teams as a community mental health resource. *Journal of Criminal Law, Criminology and Police Science,* 1969, **60,** 247–250.

BELL, R. L., CLEVELAND, S. E., HANSON, P. G., AND O'CONNELL, W. E. Small group dialogue and discussion: An approach to police community relationships. *Journal of Criminal Law, Criminology and Police Science, 1969,* **60,** 242–246.

BOWERS, K. S. The effects of demands for honesty on reports of visual and auditory hallucinations. *International Journal of Clinical and Experimental Hypnosis,* 1967, **15,** 31–36.

BRANDSTATTER, A. F., AND RADELET, L. A. *Police and community relations: A sourcebook.* Beverly Hills: Glencoe, 1968.

DANISH, S. F., AND BRODSKY, S. L. Training of policemen in emotional control and awareness. *American Psychologist,* 1970, **25,** 368–369.

FLINT, R. T. Initiating community awareness among policemen: Community awareness training in the Minneapolis Police Academy. Paper presented at the meeting of the Midwestern Psychological Association, Detroit, May 1971.

HENIG, P., AND FURST, R. Cops: Same role, new tactics. In A. Niederhoffer and A. S. Blumberg (eds.), *The ambivalent force: Perspectives on the police.* Waltham, Mass.: Ginn, 1970.

JOHANSSON, C. B. Policemen and recruits: Vocationally risky, mechanical, and military. Paper presented at meeting of the American Psychological Association, Miami Beach, September, 1970.

JOURARD, S. M. *The transparent self.* Princeton, N.J.: Van Nostrand, 1964.

KERNER, O., et al. *Report of the National Advisory Commission on Civil Disorders.* New York: Bantam, 1968.

NEWCOMB, T. M. Autistic hostility and social reality. *Human Relations,* 1947, **1,** 69–86.

NEWMAN, L. E., AND STEINBERG, J. L. Consultation with police on human relations training. *American Journal of Psychiatry,* 1970, **126,** 1421–1429.

REISER, M. Psychological research in an urban police department. Paper presented at the meeting of the American Psychological Association, Miami Beach, September 1970.

ROKEACH, M. Police and community: As viewed by a psychologist. In A. F. Brandstatter and L. A. Radelet (eds.), *Police and community relations: A sourcebook.* Beverly Hills: Glencoe, 1968.

SIEGEL, A. I., FEDERMAN, P. J., AND SCHULTZ, D. G. *Professional police–human relations training.* Springfield, Ill.: Charles C. Thomas, 1963.

WALKER, E. L. Relevant psychology is a snark. *American Psychologist,* 1970, **25,** 1081–1086.

WICKER, A. W. Attitudes versus actions: The relationship of verbal and overt behavioral responses to attitude objects. *Journal of Social Issues,-* 1969, **25**(4), 41–78.

32

The Reduction of Prejudice
Through Laboratory Training

Irwin Rubin

An experiment was conducted to test the hypothesis that increases in self-acceptance, resulting from sensitivity training, have the theoretically predictable but indirect effect of reducing an individual's level of ethnic prejudice. The role of an individual's level of psychological anomy,[1] hypothesized to condition the influences of sensitivity training, was also examined. The results suggest that sensitivity training may well be a powerful technique in the reduction of ethnic prejudice, particularly among those who are low in psychological anomy.

INTRODUCTION

Robert Kahn has stated (1963, p. 14), "The theory of T Groups implies that reduction in prejudice should be one of the results of a general increase in sensitivity to the needs of others and insight into one's own motives and behavior as it affects others. No research is available, however, to test this prediction."

Prior research (Bunker, 1963, 1965; Gordon, 1950) has shown that one of the effects of sensitivity training is an increased level of self-acceptance among the participants. In addition, it has been demonstrated that the way a person feels about himself is positively related to the way he feels about others (e.g., Stock, 1949; Sheerer, 1949). These two factors, when combined, suggest the following question:

---

[1] For the definition of this term, see p. 427.

Reprinted with permission from the author and *The Journal of Applied Behavioral Science,* Vol. 3, No. 1, pp. 29–49. Copyright 1967, NTL Institute.

This research was supported by a Ford Foundation Dissertation Fellowship in Business Administration. The author is grateful to Professors Edgar Schein and William McKelvey and to David Meredith, all of M.I.T., for their many helpful comments on various drafts of this paper.

Does raising a person's level of self-acceptance have the theoretically predictable but indirect effect of raising his level of acceptance-of-others?

The crux of this experiment is not that sensitivity training per se can be demonstrated to increase acceptance-of-others. The salient point to be tested is that demonstrated changes in a theoretically related variable (self-acceptance) produce this effect.

A second area of interest concerns the factors that might condition the kinds of learning an individual experiences as a result of sensitivity training. Certain personality types may be more susceptible than others to the influences of sensitivity training (Miles, 1960; Steele, 1965). The personality variable chosen for investigation in this study was psychological anomy. The rationale for this choice will be discussed in detail later.

HYPOTHESES

1. As a result of sensitivity training, an individual's level of self-acceptance will increase.
    a. An individual's focus during the T-Group sessions (as determined by trainer ratings), leaning toward more personal areas, will be associated with increased self-acceptance.
2. As a result of sensitivity training, an individual's level of acceptance-of-others will increase.
3. Those low in anomy will increase more in self-acceptance and acceptance-of-others than those high in anomy.
    a. An individual's level of anomy will be unaffected by sensitivity training.
4. Those who increase in self-acceptance will increase more in acceptance-of-others than those who do not change or decrease in self-acceptance.
5. Changes in self-acceptance *will lead* to changes in acceptance-of-others.

Sensitivity Training

The major independent variable in this study is what has come to be known as sensitivity training or laboratory training.[2] In a broad sense, it can be defined as

> . . . an educational strategy which is based primarily on the experiences generated in various social encounters by the learners themselves and aims to influence attitudes and develop competencies toward learning about human interactions (Schein and Bennis, 1965, p. 4).

Many phenomena occur with the T Group, and it is not within the scope of this study to examine the differential impact of each of these upon the variables of

---

[2] For a complete discussion of all that is involved in a sensitivity-training experience, see Schein and Bennis (1965).

"self-acceptance" and "acceptance-of-others." An attempt, however, was made to control for the effect of two specific aspects of all that occurred within the T Group. The trainers involved were asked to provide *for each individual*—at the end of the laboratory—the following information: (1) To what extent did the person explicitly discuss the topic of race relations (on a scale from "not at all" to "very much," i.e., 50 percent of the time)? (2) What was the nature of the individual's focus during the T Group (on a 7-point scale from Group Process = 1 to Personal Development = 7)?

Self-Acceptance

The term "self-acceptance," as it is used in this paper, involves a willingness to confront ego-alien as well as ego-syntonic aspects of the self and to accept rather than deny their existence. Implicitly, it connotes some sense of rationality or "realistic acceptance" as opposed to, for example, a person's claim, "I am superman. I accept myself as superman. Therefore all of you are underlings!"

The Dorris, Levinson, Hanfmann Sentence Completion Test (S.C.T.) (Dorris, R. J., Levinson, D., and Hanfmann, E., 1954) was used to measure the effect of sensitivity training upon an individual's level of self-acceptance. The S.C.T. includes 50 sentence stems. Half the stems use first-person pronouns and half, a third-person pronoun or proper name.[3] The first- and third-person items are matched in content,[4] e.g.:

When he gets angry he

When I get angry I

The measure of self-acceptance used in this study was derived in the following manner: Individual stem completions were coded[5] for ego-threatening content.[6] The

---

[3] First- and third-person items randomly distributed rather than appearing sequentially.

[4] The person is instructed to complete each of the stems as quickly as he can, using more than one word. After finishing all the items, he is asked to go back, reread his responses, and place a (+) sign next to those sentences that he feels refer to some personal experience or that reflect the way he might feel or act under the specified circumstances. If a sentence has no personal relevance, a (−) sign is used. In introducing the self-reference technique, the authors assumed that the denial of self-reference may be indicative of the subject's lack of awareness of the personal tendency expressed in the completion.

[5] Each pair of items was copied on separate pieces of paper. The respondent's identification number was placed on the *reverse side*. This procedure made it impossible for the coders to know whether the response was "pre" or "post." It also eliminated the halo effect that might have been created by reading an individual's total record.

[6] The correlation coefficient between two independently coded samples was 0.89. (See Johnson, 1949, p. 97, for the formula used to compute this coefficient.) The author gratefully acknowledges the assistance provided by his colleague, Tim Hall, in this phase of the study.

term, "ego-threatening," was defined as follows: "Any item which states or strongly implies any attitude, feelings, or action, which if accepted by[7]——as *applying to oneself,* would involve confronting at least a mild degree of psychological pain." For example, expression of fears, socially unacceptable responses, admission of inferiority or incompetence, extreme hostility or aggression, and so on were coded as threatening.

The assumption was then made that the more willing a person is to admit the personal relevance of ego-threatening material, the greater his level of self-acceptance. Therefore, the number of ego-threatening responses next to which the respondent placed a (+), divided by the total number of ego-threatening responses (#ET), yields the measure of self-acceptance, (ETA)[8] used in this study.

It is important to note that, by this definition, self-acceptance (ETA) can increase because the numerator increases or the denominator decreases. To clarify this point, it is hypothesized that the absolute number of statements coded as being ego-threatening will *not* change as a result of sensitivity training. The rationale here is that sensitivity training will not rid a person of his basic conflicts and anxieties nor does it attempt to help him make light of his times of crises. Instead, in some ideal sense, sensitivity training may help a person to find in himself the natural tools that enable him to effectively cope with these things. This will result, for example, from positive, nonevaluative feedback, the opportunity to test ideas and beliefs (increased "reality testing" about oneself), and a high level of trust and openness resulting in greater authenticity. An environment is created within which there should be a reduction of an individual's need to use projective defense mechanisms which act to distort his perception of himself and others.

Acceptance-of-Others

Harding and Schuman (1961) conceptualize prejudice as the departure from or failure to adhere to three ideal norms of behavior: the norm of rationality, the norm of justice, and the norm of human-heartedness. In this experiment it was decided to focus upon the norm of human-heartedness (HH)[9] which enjoins a person's emotional *acceptance-of-others* in terms of their common humanity, no matter how different they may seem from oneself. The major dependent variable in this study,

---

[7] For the females, the phrase, "the majority of women associated with the nursing profession," was inserted because virtually all the females in the experimental population fell into that category. For the males, who were more heterogeneous, the phrase, "the average male in our culture," was inserted. Two forms of the scale—male and female—were used for this research.

[8] Throughout the remainder of this paper, the following symbols will be used:

  1. ET means ego-threatening.
  2. #ET means absolute number of sentences scored as ego-threatening.
  3. ETA means self-acceptance as defined above.

[9] Throughout the remainder of this paper, the symbol HH is used to represent an index of a person's level of acceptance-of-others.

in other words, is not prejudice per se but only the affective component of the individual's attitude.

The scale is made up of 15 items[10] of the following type:

> The white school board in a community builds two new schools and fixes the school lines so that almost all the Negro children go to one new school and all the white children to the other new school. Now do you suppose most of the Negroes in the community would react to this?
>
> ——a. While there are some exceptions, many Negroes are mainly concerned with getting money for food, rent, and other things, and so do not have too much interest in the matter of school one way or the other.
>
> ——b. Every community is different, and it is almost impossible for someone not living in it to know enough about the situation to judge.
>
> ——c. The average Negro parent would not like what the school board has done about drawing school lines.
>
> ——d. The average Negro parent would simply be pleased to have a new school for his children, especially if it were equal to the white school in every way.

The measure of human-heartedness used in this study was derived in this manner: The respondent was asked to rank each of the four choices following an item from *1* ("most likely reaction") to *4* ("least likely reaction"). Each respondent's series of ranks was then compared with a theoretically ideal set of ranks[11] and the absolute difference between ranks was computed. The sum of these differences across the 15 experimental items yielded the respondent's human-heartedness score (HH). This score could range from 0-120 (i.e., 15 items times a maximum difference of 8 points for any item).

Psychological Anomy

The personality variable chosen for investigation in this research was psychological anomy,[12] defined as a sense of normlessness, "the feeling that the world and oneself are adrift, wandering, lacking in clear rules and stable moorings . . . a feeling of moral emptiness" (McClosky and Schaar, 1965, p. 14). This definition is analogous to Seeman's second major usage of the alienation concept—*meaninglessness* wherein "the individual is unclear as to what he ought to believe—when the individual's minimal standards for clarity in decision making are not met" (Coser and Rosenberg, 1964, p. 530).

---

[10] In addition, four control items are included to check on the extent to which response set is operating.

[11] Howard Schuman and the writer *independently* ranked all items as to how the "most human-hearted person" would assign his ranks. We agreed on 100 percent of the first and second ranks and 88 percent of the third and fourth ranks, yielding an overall percent agreement of 94 percent.

[12] The scale used to measure this variable is a nine-item Guttman scale developed by McClosky and Schaar (1965). The items are of the following form:

a. People were better off in the old days when everyone knew just how he was expected to act.

b. It seems to me that other people find it easier to decide what is right than I do.

McClosky and Schaar (1965) present evidence to suggest that anomic responses are powerfully governed by cognitive and personality factors independent of or in combination with social influences. They conclude that anomy "results from impediments to interaction, communication, and learning, and is a sign of impaired socialization." In other words, given that anomic feelings result from a lack of learning, "whatever interferes with one's ability to learn a community's norms, or weakens one's socialization into its central patterns of belief, must be considered among the determinants of anomy" (p. 20).

In a real sense, the T group represents for its members a new community or society with a set of norms unlike those to which the members have become accustomed. The individual participant, if he is to benefit from sensitivity training, must be able to see and understand the norms of this new culture. Only then. will he be able to decide rationally[13] whether they are personally relevant and functional and if so, to truly internalize these new learnings.

The highly anomic person might experience difficulty in understanding and internalizing the dominant norms of the T Group. Furthermore, due to the relatively short duration (two weeks) of the experiment and the here-and-now focus of the T Group, no change was expected in a person's level of anomy.

THE STUDY

Subjects

The laboratory population studied in this research were the participants in the Osgood Hill[14] 1965 summer program in sensitivity training. The program was two weeks in length (June 25–July 7), and the participants "lived in" in the sense that they slept on the premises and ate virtually all their meals together.

There were 50 participants—30 females and 20 males. They ranged in age from 23 to 59, with a mean age of 33 years. The majority had at least a B.S. degree and a few had advanced degrees. The majority came from the New England area, but several came from Miami, Cleveland, and Chicago. There were eight Negroes in the population, and the trainers made certain that each of the five T Groups[15] that were formed had at least one Negro and an even proportion of males and females.

Occupationally, the males were a relatively heterogeneous group that included several businessmen, teachers, policemen, clerics, graduate students, government

---

[13] Bennis, W. G., Schein, E. H., Berlew, D. E., and Steele, F. I. (1964) discuss this point in terms of a possible meta-goal of sensitivity training—"expanded consciousness and sense of choice."

[14] Osgood Hill is in Andover, Massachusetts. It is owned and operated by Boston University. The author wishes to acknowledge the cooperation and assistance provided by the entire staff group of Osgood Hill in the successful completion of this study.

[15] Two of the trainers were females—one of whom was a Negro—and the remaining four were males. (One group had two trainers.)

employees, a male nurse, and a dentist. The females were much more homogeneous, the majority of them being associated with the nursing professions (students, teachers, practicing nurses, and nursing supervisors).

Experimental Design and Procedure

One of the problems facing the researcher interested in evaluating the effects of sensitivity training is that of finding a relevant control group. The participants in a laboratory are, in one sense, a self-selected group—a circumstance which negates the relevance, for control purposes, of just any group of warm bodies.

Thus the experimental design utilized in this study was one in which the subjects served as their own controls. Herbert Hyman (Hyman, H., Wright, C. R., and Hopkins, T. K., 1962, p. 42) utilized this approach in his evaluation of the effects of citizenship camps, as did Carl Rogers in his attempts to evaluate the effects of psychotherapy. As Hyman points out:

> With such a procedure, matching of experimental subjects and controls presents no difficulty, for the same persons constitute both groups. By determining how much instability there is in the group's attitudes, opinions, or other characteristics *during a normal period of time* we could then estimate how much of the change manifested during the experimental period exceeds the normal change resulting from other factors.

Within this design, the total available experimental group ($N = 50$) was randomly split into two groups of equal size. The smaller group ($N = 14$) was tested (0_{1C}) via mail questionnaires two weeks prior to their arrival at Osgood Hill. The entire group was then tested (for controls: 0_{2C} and for experimentals: 0_{1E}) upon their arrival, but before the first T-Group session. The final "after" measures (for controls: 0_{3C} and for experimentals: 0_{2E}) were obtained the morning of the next-to-last day of the laboratory.[16] This timing was necessary in order to provide a feedback session to all participants prior to their departure at the end of the laboratory.

This design can be depicted in the following manner:[17]

June 11	*June 25*	*July 5*
$0_{1C} \dfrac{\text{controls}}{\text{2 weeks}}$	$0_{2C} \dfrac{\text{controls}}{\text{T Group}}$	0_{3C}
($N = 11$)		
	$0_{1E} \dfrac{\text{experimentals}}{\text{T Group}}$	0_{2E}
	($N = 30$)	

---

[16] All administrations, other than 0_{1C}, were conducted by the author on a group basis.

[17] Of the available control group of 14, two persons never arrived and one returned an unusable questionnaire, leaving a final control group of 11. Of the available experimental group of 36, one missed the pretest and five returned unusable questionnaires, leaving 30 for the final experimental group.

RESULTS

Control Groups

Table 32.1 presents the test-retest scores for the control group (0_{1C}, 0_{2C}) and the initial test scores for the experimental group (0_{1E}). A series of t tests were performed that compared scores for 0_{1E} versus 0_{1C} and 0_{2C} in order to determine empirically the degree of similarity between experimentals and controls. None of the resulting t's reached statistical significance, with p's being greater than 0.50. On the basis of these results, it is assumed that the members of the control group represent a population comparable with the experimentals on the major variables.

It can also be seen from Table 32.1 that among the members of the control group $\overline{Ap}$ increased slightly, $\overline{\#ET}$ increased slightly, and $\overline{ETA}$ and $\overline{HH}$ both decreased slightly. In using a t test for dependent samples (Blalock, 1960), it was observed that none of the resulting t's reached the 0.60 level of significance. On the basis of these results, it is assumed that the controls do not change significantly from 0_{1C} to 0_{2C} on any of the major variables. It is assumed, therefore, that any changes found among experimentals cannot be attributable to the main effects of instrument instability and/or practice.

TABLE 32.1
Before-after scores for control group (0_{1C}, 0_{2C}) and before scores for experimental group (0_{1E})

0_{1C} (N = 11)	0_{2C}*	0_{1E}* (N = 30)
(a) $\overline{Ap}$ = 5.5	$\overline{Ap}$ = 5.8	$\overline{Ap}$ = 6.5
(b) $\overline{\#ET}$ = 11.0	$\overline{\#ET}$ = 12.0	$\overline{\#ET}$ = 13.5
(c) $\overline{ETA}$ = 66.0	$\overline{ETA}$ = 65.0	$\overline{ETA}$ = 55.0
(d) $\overline{HH}$ = 46.5	$\overline{HH}$ = 47.5	$\overline{HH}$ = 46.2

(a) $\overline{Ap}$ represents mean level of anomy. Scores ranged from 1–10, with a low score representing a low level of anomy.
(b) $\overline{\#ET}$ represents the mean absolute number of statements scored as being ego-threatening, with the range from 5 to 23.
(c) $\overline{ETA}$ represents mean level of self-acceptance, i.e., the number of ego-threatening statements accepted divided by absolute number of ego-threatening statements. Scores ranged from 0 to 100 percent, with a low score indicating a low level of self-acceptance.
(d) $\overline{HH}$ represents mean level of human-heartedness. Scores ranged from a low of 18 to a high of 80. The lower the score, the closer the respondent's set of ranks was to the theoretically perfect set of ranks and, therefore, the higher his level of human-heartedness.
*0_{2C} + 0_{1E} were gathered at the same point in time, just prior to the first T-Group session.

TABLE 32.2
Scores for experimental group
(N = 30) before (0_{1E}) and after (0_{2E})
a 2-week T Group

0_{1E}	0_{2E}
$\overline{Ap}$ = 6.5	$\overline{Ap}$ = 5.9
$\overline{\#ET}$ = 13.5	$\overline{\#ET}$ = 13.2
$\overline{ETA}$ = 55.0	$\overline{ETA}$ = 67.0
$\overline{HH}$ = 47.2	$\overline{HH}$ = 42.0

Experimental Group

It was hypothesized that Ap and #ET would not change as a result of sensitivity training. Examination of Table 32.2 reveals that $\overline{\#ET}$ decreased slightly over this two-week period. Using a t test for dependent samples, it was found that for Δ Ap (change in Ap), t = 0.84 with an associated p < 0.40 two-tail (N = 30); and for Δ #ET (change in #ET), t = 0.70 with an associated p < .045 two-tail (N = 30). We are unable to reject the null hypothesis of no difference and can therefore assume that sensitivity training had no appreciable effect upon an individual's level of anomy (Ap) or upon the absolute number of ego-threatening statements generated by an individual on our sentence completion test.

The next major hypothesis concerns Δ ETA[18] (change in self-acceptance). The prediction here was that self-acceptance would increase as a result of sensitivity training. Examination of Table 32.2 reveals that Ap and #ET a mean of 55.0 percent to a mean of 67.0 percent. The differences between these means (t test for dependent samples) is significant at the 0.01 level one-tail (N = 30, t = 2.58, p < 0.01). It is therefore concluded that as a result of sensitivity training, an individual exhibits a greater willingness to accept the personal relevance of ego-threatening material; i.e., his ETA increases.

With respect to Δ HH[19] (change in human-heartedness), it was predicted that an individual's level of human-heartedness would increase. Operationally, this means that his "after" HH score should be lower than his "before" HH score. Table 32.2 reveals that HH decreased from 47.2 to 42.0. The difference between these means (t test for dependent samples) is significant at the 0.01 level one-tail (N = 30, t = 2.54, p < 0.01). In other words, the rankings an individual assigned after the laboratory corresponded more closely with expert rankings than those he assigned before the laboratory—he was found to be more human-hearted.[20]

---

[18] Δ ETA refers to change in self-acceptance score—ETA score after the laboratory, minus ETA score before the laboratory.

[19] Δ HH refers to change in human-heartedness score—HH score after the laboratory minus HH score before the laboratory.

[20] The critical test here is whether Δ ETA and Δ HH among the experimentals differ from Δ ETA and Δ HH among the controls. A Mann-Whitney U-Test (Siegel, 1956, pp. 116–127) was therefore

Conditioning Influence of Anomy

We turn now to an examination of the conditioning influence of anomy with respect to the observed changes in ETA and HH. It was predicted that those E's low in anomy (Ap) would change more on ETA and HH than those high in anomy (Ap). The skewed nature of the distribution of Ap scores (the majority of respondents scored either 1 or 8, 9, 10, with virtually no scores in the middle) suggested that the most relevant test of these hypotheses would be to split the group at the median Ap score and to compare the magnitude and direction of ETA and HH differences among groups.

Utilizing the Mann-Whitney U-Test, it is observed that those below the median in Ap increased significantly more on ETA than those above the median in Ap ($N_1 = 19$, $N_2 = 19$,[21] $Z = 1.77$, $p < .04$ one-tail). A similar trend was found with respect to HH scores ($N_1 = 19$, $N_2 = 19$, $wz = 1.56$, $p < .06$ one-tail). In absolute terms, those low in Ap increased 17 percent on the average. With respect to HH, those low in Ap decreased six points on the average, while the high Ap's decreased only two points. In summary, strong support is provided for the hypothesized conditioning influence of Ap on changes in self-acceptance (ETA), and marginal support is provided with respect to changes in human-heartedness (HH).

Central Hypotheses

In the light of the results of these preliminary analyses, we are now in a position to examine the central hypotheses of this study:

> 1. Those who increase in self-acceptance will increase more in human-heartedness than those who either do not change or decrease in self-acceptance.[22]
>
> 2. Changes in self-acceptance will lead to change in human-heartedness.

With respect to the first, of the 38 members of the total experimental group, 23 increased on ETA, six did not change, and nine decreased in ETA. The sample was therefore split into $+\Delta$ ETA (positive changers in self-acceptance, $N = 23$) and 0Δ ETA (zero or negative changers in self-acceptance, $N = 15$). On the aver-

---

performed on the difference between the changes. This analysis yielded a $Z = 1.76$ for the Δ ETA's ($N_1 = 11$, $N_2 = 30$, $p < .05$ one-tail) and a $Z = 1.76$ for the Δ HH's ($N_1 = 11$, $N_2 = 30$, $p < .04$ one-tail). In other words, *not only* do the experimentals change while the controls do not, but the *experimentals also change significantly more* than the controls.

[21] For the purposes of this and the following analyses, the eight of 11 control group members who returned usable responses after the laboratory (0_{3C}) were added to the 30 experimentals. These eight persons changed as much (percentage-wise) in ETA and HH after the laboratory as did the experimentals. In addition, like the experimentals, they did not change in Ap or #ET. This raises our available population from $N = 30$ to $N = 38$.

[22] The *initial* correlation between ETA versus HH was $R = -0.32$ ($N = 41$, $p < .05$ one-tail). The minus sign is explained by the fact that a high level of HH is represented by a low score.

age, the $+\Delta$ ETA group decreased five points in HH, a result which is statistically significant at the $p<.01$ level one-tail ($N=23$, $t=2.80$). The 0Δ ETA group also decreased in HH an average of three points, but this change does not reach significance ($N=15$, $t=1.03$, $p<0.20$ one-tail). However, the difference between these changes is *not* significant (Mann-Whitney U-Test, $N_1=15$, $N_2=23$, $=1.0$, $p<.16$ one-tail). The hypothesis in its present form cannot be unequivocally supported.

In order to shed some light on the reasons for this result, individual change scores on ETA were examined more closely. There appeared to be a sharp discontinuity in the distribution of scores. Several persons increased a moderate amount in ETA (8 to 14 percent), but then the next highest change was 21 percent. There were 13 persons who increased 21 percent or more in self-acceptance. When we examined this group of high $+\Delta$ ETA's versus the remainder of the sample, the following results emerged: The high $+\Delta$ ETA group decreased an average of 8.0 points on HH ($N=13$, $t=3.0$, $p<.01$ one-tail), while the remainder of the sample decreased an average of 2.0 points on HH ($N=25$, $t=1.3$, $p<.12$ one-tail). A Mann-Whitney U-Test on the difference between these differences yielded a $Z=1.76$ ($N_1=13$, $N_2=25$, $p<.04$ one-tail). In other words, those who increase a great deal in self-acceptance (Δ ETA >21 percent) will increase significantly more in human-heartedness than those who decrease in self-acceptance or increase only a moderate amount.

One way to test the hypothesis that changes in self-acceptance lead to changes in human-heartedness is to utilize the method of partial correlation.[23] The three-variable[24] model to be tested can be depicted in the following manner:

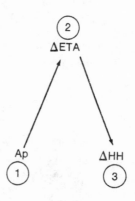

Within the framework of this research, we should like to know the direction of the causal arrow in the relationship between Δ ETA moved and Δ HH. In order

---

[23] The utilization of partial correlations to infer causality rests upon several assumptions. In addition, all other possible models must be eliminated. A complete discussion of these assumptions and the methods for eliminating irrelevant models can be found in Simon (1954) and Blalock (1960).

[24] Anomy (AP) was chosen as the third variable because, as discussed earlier, it was unaffected by the training experience but was related both to changes in self-acceptance and changes in human-heartedness. Other ways exist to prove causality but, for these, different experimental designs are required.

to infer that Δ ETA is causing Δ HH, the following mathematical condition must be satisfied[25] (Simon, 1954; Blalock, 1960):

> The correlation of Ap versus Δ HH with the effect of Δ ETA removed should be less than the zero order correlation of Ap versus Δ HH; i.e., $R_{13.2} < R_{13}$.

Table 30.3 presents the data from which the required zero order correlations are computed. The dichotomous nature of the Ap scores suggested that a tetrachoric correlation method would be most appropriate. Under appropriate conditions (Guilford, 1956), this method "gives a coefficient that is numerically equivalent to a Pearson r and may be regarded as an approximation to it." In every case, the high versus low split was based upon those above and below the median.[26]

Substitution of the zero order correlations into the partial correlation formula (Blalock, 1960) yields an $R_{13.2} = +0.09$ and the mathematical condition stated above is therefore satisfied.[27] It is important to note that this analysis does not enable one to rule out a direct effect of sensitivity training on HH. Nor does it eliminate the possibility that sensitivity training influences another variable which may be termed "feeling-orientation" which, in turn, influences ETA and HH. All it suggests is that some change in HH does result from a change in ETA.[28]

TABLE 32.3
Contingency tables necessary to compute tetrachoric correlations between Ap, Δ HH, and Δ ETA

A.

	Low Ap	High Ap
High Δ HH	13	10
Low Δ HH	6	9

^{R}Ap, Δ HH $= -0.255$ (R_{13})

B.

	Low Ap	High AP
High Δ ETA	13	6
Low Δ ETA	6	13

^{R}Ap, Δ ETA $= -0.550$ (R^{12})

C.

	High ETA	Low ETA
High Δ HH	15	8
Low Δ HH	4	11

$^R\Delta$ ETA, Δ HH $= 0.575$ (R_{23})

---

[25] Numerical subscripts are used for simplicity.
  1 = Ap
  2 = Δ ETA
  3 = Δ HH
[26] Median Ap = 5.0.
Median ETA = $+8$; i.e., 8 percent increase in self-acceptance.
Median HH = 2.0; i.e., 2-point decrease in HH score.
[27] A more conservative approach here is to split the total sample at the median Δ ETA score and compute the tetrachoric correlation between Ap versus ΔHH within each subsample. The split was made, and the results are almost identical with those obtained when the partial correlation formula was used.
[28] A Kruskall-Wallis one-way analysis of variance (Siegel, 1956, pp. 184–193) among the five T

Trainer Ratings

Trainers were asked, at the end of the laboratory, to characterize the nature of each individual's participation during the T-Group session on a scale from 1 (Group Process Orientation) to 7 (Personal Development). In addition, the trainers rated, for each individual, the "Salience of the Topic of Race Relations" (i.e., percent of time spent discussing the Topic).

It was hypothesized that changes in self-acceptance (Δ ETA) would be associated with an "individual orientation" leaning toward Personal Development. Again, this hypothesis is supported only when we compare the high $+ \Delta$ ETA group with the remainder of the sample. The average trainer rating for the high $+ \Delta$ ETA's was 5.2 (i.e., leaning toward Personal Development), as compared with 3.8 (i.e., leaning toward Group Process) for the remainder of the sample. This difference is significant ($N^{29} = 30$, $t = 2.16$, $p < .02$ one-tail).

No directional hypotheses were made concerning the effect of "Salience of the Topic" on an individual's change in human-heartedness. The 20 persons for whom these ratings were available were split into two groups—high (20 to 50 percent of time) versus low (0 to 20 percent) salience, and changes in HH within the two groups were examined. The low-salience group decreases an average of eight points in HH, while the high-salience group decreases an average of only one point in HH ($N_1 = 10$, $N_2 = 10$, $Z = 1.65$ $p < .10$ two-tail, Mann-Whitney U-Test). In other words, there appears to be somewhat of a negative relationship between the amount of time spent discussing the topic of race relations and the change in human-heartedness.[30]

DISCUSSION

Generalizability of Results

One question which comes up immediately is the extent to which the findings of this study are generalizable. It was pointed out earlier that the members of the experimental population all shared a certain level of "motivation to attend a laboratory." It is not yet known what personality variables, for example, differentiate those who are "motivated to attend" from those who are not. Even if knowledge of these parameters did exist, it would then have to demonstrate that they have

---

Groups on all major variables was performed, and none of the resulting HH's reached the 0.50 level of significance two-tail. From this result, it can be assumed that there was no significant trainer effect, nor can the observed changes be attributed to some other factor unique to any one of the T Groups.

[29]The sample is reduced here because one set of trainer-rating forms was never returned to the researcher.

[30]The correlation between "Salience of Topic" and initial HH score was zero, as was the correlation between "Individual Orientation" and the initial ETA score.

relevance in terms of differential learnings resulting from training. This broad issue is beyond the scope of this study. However, several related sub-issues are manageable.

Concerning the distribution of initial self-acceptance scores, a reasonably normal distribution of scores with a mean value close to 50 percent was observed. Unfortunately, no norms exist to indicate what the expected average score might be. Two comparison samples, however, are available: The average ETA score among the college sophomore group studied by Dorris, et al. (1954) was 53 percent, and among a pretest group of 30 Sloan Fellows at M.I.T. (with a simplified index of self-acceptance being used), the mean score was 50 percent. In addition, the results of the present study suggest that even some of those who were initially very low in self-acceptance could be "reached" by sensitivity training.

Concerning human-heartedness scores, Schuman and Harding (1963) found in their main standardization sample that the average HH score (with a simplified measure being used) leaned toward the "unhuman-hearted" end of the scale. The distribution of initial scores observed in this study was skewed in the other direction—toward the human-hearted end of the scale. The atypical[31] educational level of the Osgood Hill sample, with the majority having at least a bachelor's degree, helps to explain this difference. It may be that a certain level of education is a necessary prerequisite to learning via sensitivity training. This proposition is as yet untested empirically.

What Kind of Sensitivity Training

Another question of importance deals with the impact of different emphases in sensitivity training.[32] The results of this study highlight the importance of a "personal development" as opposed to a "group process" orientation. The greatest increasers in self-acceptance and, consequently, in human-heartedness were those whose predominant focus during the T-Group sessions was in more personal areas.

From a pragmatic viewpoint, if one wishes to use sensitivity training as a means to reduce prejudice, then, within the Schein and Bennis (1965) framework, the individual should be viewed as the client, and learning about self and others should be stressed at the levels of awareness and changed attitudes.[33] Furthermore, given the specific goal of prejudice reduction and a personal focus, a shorter laboratory

---

[31] The terms "typical" and "atypical," used in this section, have as their frame of reference "a random sample of adults drawn from the general population."

[32] Schein and Bennis (1965) present a three-dimensional schema for classifying the goals of a laboratory in these terms: What is the learning about? Who is the ultimate client? What is the level of learning?

[33] For an excellent description of this form of sensitivity training, see Irving R. Weschler, Fred Massarik, and Robert Tannenbaum, The self in process: A sensitivity training emphasis, in *Issues in human relations training*, No. 5 in NTL's Selected Readings Series, Washington, D.C.: National Training Laboratories, 1962, pp. 33–46.

might be feasible. Much research is needed to determine the optimal mix of group process versus personal development orientation, the relative impacts of various kinds of supplementary cognitive inputs, and the effect of laboratory duration on the amount of change observed.

One of the most interesting findings in this study involved the strong conditioning influence of anomy with respect to changes in self-acceptance. The success of sensitivity training as an educational strategy rests upon an individual's ability to see and understand the dominant norms of self-exposure, openness, and feedback which develop with the T Group. What remains to be demonstrated by future research is the role of anomy as a conditioning variable for learning criteria other than increased self-acceptance.

The roles played by discussion of the topic of race relations and the presence of Negroes in the T Group are still unclear. Pure discussion does not help those who are doing the talking. This situation does not mean that the observed changes in human-heartedness could have occurred without any such discussion. The nontalkers[34] may have benefited immensely from listening to the more vocal members of the group. On the other hand, the talkers may have been "intellectualizing"—a technique commonly employed in T Groups to keep the discussion on a less threatening level. This negative effect of participation has been observed by other researchers,[35] and further research is necessary to better understand the dynamics of the relationship between participation (amount and content) and change.

Concerning the effect of racially mixed groups, it may be that for a majority of the white participants the T-Group experience was the first opportunity they ever had to meaningfully interact with a Negro. During the T-Group discussions, many insights may have occurred that served to highlight a feeling of "oneness" of common humanity. For example, "He [a Negro] has feelings and emotions just the same as I!" Research is needed to examine in greater detail the specific patterns of interaction (e.g., Negro to white) and discussion content within a mixed T Group and their effects on the attitudes people have toward one another, as well as the effects of an all-white group.

Change in Self-Acceptance Versus Change in Human-Heartedness

One of the central hypotheses in this study was that those who increase in self-acceptance will increase more in human-heartedness than those who decrease or do not change in self-acceptance. The data suggest that this hypothesis, in its original form, was too broad. It appears instead that some minimum increase in self-acceptance (20 percent in this study) is necessary in order for any significant change

---

[34] "Nontalker" does not mean "silent member," but refers instead only to the substance or content of an individual's discussion. The most vocal members, in terms of total participation, may never have mentioned the topic of race relations.

[35] Personal communication from David Kolb of M.I.T. concerning some research he is conducting on individual change within T Groups, 1965.

in human-heartedness to be immediately observable.[36] Perhaps, where sensitivity training really "took" (in the sense of great increase in self-acceptance), those involved may have been better able to immediately make the mental transfer from self-acceptance to human-heartedness. The others may have needed some period of incubation in order for this transfer to occur.

Support for this interpretation is provided by Katz (Katz, D., Sarnoff, I., and McClintock, C. M., 1956, 1957) who found that as a result of a self-insight manipulation no changes in prejudice were observed immediately after the experimental induction, but that highly significant shifts occurred several weeks afterwards. In other words, a "sleeper effect" appeared to be operating. The written case study utilized by Katz, et al. (1956) to increase self-insight is certainly less intensive than a two-week sensitivity training laboratory and may well be less powerful. It is possible, therefore, that changes in human-heartedness will persist after the laboratory and, in fact, may become more marked among the group who experienced only moderate increases in self-acceptance.[37] This hypothesis could not be tested because it was necessary to provide a full feedback session[38] for the laboratory participants prior to their departure.

Finally, the reader had undoubtedly noticed that by changing a few words, e.g., "T Group" to "therapy group" and "trainer" to "therapist," this study could have been concerned with the effect of client-centered psychotherapy upon prejudiced attitudes. Both the T Group and the therapy group provide the elements of psychological safety, support, and opportunities for reality testing assumed necessary to effect an increase in an individual's level of self-acceptance and consequently, by our model, to decrease one's level of ethnic prejudice. To the extent that future research and practical experience substantiate the conclusions drawn from the present study, a step has been taken toward solving a problem posed by Adorno (Adorno, T. W., Frenkel-Brunswick, E., Levinson, D. J., and Sanford, R. N., 1950, p. 976) some 17 years ago.

> Although it cannot be claimed that psychological insight (self-insight) is any guarantee of insight into society, there is ample evidence that people who have the greatest difficulty in facing themselves are the least able to see the way the world is made. Resistance to self-insights and resistance to social facts are contrived, most essentially, of the same stuff. It is here that psychology may play its most important role. Techniques for overcoming resistance, developed mainly in the field of individual psychotherapy, can be improved and adapted for use with groups and even for use on a mass scale.

---

[36] The risk of maximizing change variations by examining a small subgroup of the total population is reduced considerably by the findings concerning individual focus during the T-Group sessions. The great changers in self-acceptance were also those whose focus during the T Group was in more personal areas.

[37] The Bunker studies (1963, 1965) discussed earlier suggest that many of the learnings derived from sensitivity training *do* remain with an individual over a long period of time.

[38] The reason for this was only partially based upon ethical considerations. Of equal importance was the fact that the data which were fed back to the participants became topics for discussion in the few remaining T-Group sessions and therefore, hopefully, enhanced the learning value of their training experience.

References

ADORNO, R. W., FRENKEL-BRUNSWICK, E., LEVINSON, D. J., AND SANFORD, R. N. *The authoritarian personality.* New York: Harper & Row, 1950.

BENNIS, W. G., SCHEIN, E. H., BERLEW, D. E., AND STEELE, F. I. *Interpersonal dynamics.* Chicago: Dorsey, 1964.

BLALOCK, H. M. *Social statistics.* New York: McGraw-Hill, 1960.

BUNKER, D. The effect of laboratory education upon individual behavior. *Proc. of the 16th Annual Meeting,* Industrial Relat. Res. Ass., December 1963. Pp. 1–13.

BUNKER, D. Individual applications of laboratory training. *Journal of Applied Behavioral Science,* 1965, **1**(2), 131–148.

COSER, L. A., AND ROSENBERG, B. (eds.) *Sociological theory—A book of readings.* New York: Macmillan, 1964.

DORRIS, R. J., LEVINSON, D., AND HANFMANN, E. Authoritarian personality studied by a new variation of the sentence completion technique. *J. abnorm. soc. Psychol.,* 1954, **49**, 99–108.

GORDON, T. What is gained by group participation? *Educ. Leadership,* January 1950, 220–226.

GUILFORD, J. P. *Fundamental statistics in psychology and education.* New York: McGraw-Hill, 1956.

HARDING, J., AND SCHUMAN, H. An approach to the definition and measurement of prejudice. Unpublished manuscript, Harvard Univer., January 1961.

HYMAN, H., WRIGHT, C. R., AND HOPKINS, T. K. *Application of methods of evaluation.* Los Angeles: Univer. of California Press, 1962.

JOHNSON, P. C. *Statistical methods in research.* New York: Prentice-Hall, 1949.

KAHN, R. Aspiration and fulfillment: Themes for studies of group relations. Unpublished manuscript, Univer. of Michigan, 1963.

KATZ, D., SARNOFF, I., AND McCLINTOCK, C. M. Ego defense and attitude change. *Human Relat.,* 1956, **9**, 27–45.

KATZ, D., SARNOFF, I., AND McCLINTOCK, C. M. The measurement of ego defense as related to attitude change. *J. Pers.,* 1957, **25**, 465–474.

McCLOSKY, H., AND SCHAAR, J. H. Psychological dimensions of anomy. *Amer. soc. Rev.,* 1965, **30**(1), 14–40.

MILES, M. B. Human relations training: Processes and outcomes. *J. counsel. Psychol.,* 1960, **7**(4), 301–306.

SCHEIN, E. H., AND BENNIS, W. G. *Personal and organizational change through group methods: The laboratory approach.* New York: Wiley, 1965.

SCHUMAN, H., AND HARDING, J. Sympathetic identification with the underdog. *Pub. Opin. Quart.,* Summer 1963, 230–241.

SHEERER, E. T. The relationship between acceptance of self and acceptance of others. *J. consult. Psychol.,* 1949, **13**, 169–175.

SIEGEL, S. *Nonparametric statistics for the behavioral sciences.* New York: McGraw-Hill, 1956.

SIMON, H. A. Spurious correlation: A causal interpretation. *J. Amer. Stat. Ass.,* 1954, **49**, 467–479.

STEELE, F. I. The relationships of personality to changes in interpersonal values effected by laboratory training. Unpublished doctoral dissertation, Massachusetts Institute of Technology, 1965.

STOCK, D. An investigation into the interrelations between the self-concept and feelings directed toward other persons and groups, *J. consult. Psychol.,* 1949, 13.

33

Changes in Locus of Control as a Function of Encounter Group Experiences: A Study and Replication

Michael Jay Diamond and Jerrold Lee Shapiro

The present study was designed to investigate the effects of encounter group experience on locus of control. Two phases are reported. In the first phase, 39 graduate students were assigned to three professionally led experimental encounter groups and one non-treatment control group. In the second phase (replication), 44 equivalent subjects were assigned to four groups. The three experimental encounter groups were led by supervised graduate students. The Rotter Internal-External Locus of Control scale was administered to each subject before and after the encounter experience. Significant increases in internal locus of control occurred as a result of encounter group experience. These experiences were discussed as a potent means for inducing cognitive behavioral change. Implications of the modification of generalized expectancies were noted, and future studies were considered.

Recently, evidence has been accumulating to demonstrate differential effects of encounter group training. Several studies have indicated that professionally led groups effect a wide range of member behaviors. For example, encounter group training has been shown to increase sensitivity to verbal (Bunker, 1965) and nonverbal behavior (Delaney and Heimann, 1966), interpersonal competence (Argyris, 1962) and job effectiveness (Miles, 1965), attitude change (Diamond and Lobitz, 1973; Rubin, 1967) and cognitive structuring (Harrison, 1966), and hypnotic suscepti-

Reprinted with permission from the author and *The Journal of Abnormal Psychology,* Vol. 82, No. 3, 1973. Copyright 1973 by the American Psychological Association.

This study was supported in part by University of Hawaii Research Council Grant 21-G-71-028-F330-0-068. The authors wish to express their gratitude to Tom Glass, Bob Hunt, Linda Tillich, Roger Katsutani, Minnie Komagome, Tom Loomis, Paul Marano, Judy Nakashima, Natalie Vanderburg, and Janet Wright, who served as the group leaders, to Harold Ayabe, who consulted on statistics, to Bonnie R. Strickland for her constructive comments, and to the dedicated research assistants of the Laboratory of Cognitive Behavior Control, Departments of Psychology and Educational Psychology, University of Hawaii. Order of authorship was determined by a coin toss.

bility (Shapiro and Diamond, 1972). In addition, numerous studies (Culbert, Clark, and Bobele, 1968; Reddy, 1972; Rueveni, Swift, and Bell, 1969) have demonstrated increased self-actualization as measured by Shostrom's (1966) Personal Orientation Inventory. Consistent with these findings, training has enhanced self-ideal self congruence (Peters, 1966; Shapiro, 1970), self-confidence (Bunker and Knowles, 1967), and self-acceptance (Bunker, 1961).

One of the better predictor constructs for a wide range of competence and independence behaviors is locus of control (Joe, 1971; Lefcourt, 1966; Rotter, 1966). Locus of control represents a generalized expectancy concerning the contingencies of reinforcement following behavior. An "internal" individual perceives life events as being largely under his personal control. Conversely, an "external" person perceives life events as being outside his personal control and understanding and more likely a result of fate, luck, or chance.

Of relevance to therapeutic change and personal growth is evidence demonstrating that internals, in contrast to externals, are more achievement oriented (Rotter, 1966), less anxious (Feather, 1967), less dogmatic (Joe, 1971), more trusting and less suspicious of others (Hamsher, Geller, and Rotter, 1968), less apt to use sensitizing modes of defenses, more self-confident and insightful (Joe, 1971), and less maladjusted (Hersch and Scheibe, 1967). In addition, internals have been shown to be more willing to remedy personal problems, while evidencing a greater tendency to seek information and adopt behavior patterns which facilitate personal control over their environment (Lefcourt, 1966; Rotter, 1966). Since encounter groups are also geared toward increased openness to experience, trust, interpersonal satisfaction, insight into one's feelings, and risk taking (Rogers, 1970), it seems likely that successful encounter group experiences will affect locus of control. Specifically, a successful encounter group would be more likely to create an environment in which an internal orientation can be learned. In particular, members of a group are encouraged to take responsibility for their verbal and nonverbal behavior, to try on novel behavior, and to attempt to resolve personal conflicts by focusing primarily on their own feelings and behaviors.

At least two studies suggest that therapeutic experiences produce changes in locus of control. Gillis and Jessor (1970) combining group and individual psychotherapy reported that successful outcome is related to increases on the internal dimension. Unfortunately, methodological inadequacies (e.g., undefined independent variable, reliance on internal analyses, and use of inadequate control group) detract from their conclusions. Similarly, Smith (1970) reported that six-week "crisis intervention therapy" yielded increased internal scores in comparison with an unmatched control group. Contrary results have been reported by Aronson (1970). In this study, no changes in locus of control resulted from exposure to leaderless (audiotape) encounter groups.

The present investigation consists of two studies which examined the effects of encounter group training on locus of control as measured by Rotter's (1966) Internal–External Locus of Control scale. The first study, employing highly experienced professional group leaders of different theoretical orientations, was designed to demonstrate the effects of encounter group training on locus of control. It was predicted that members of these groups would score as more internal in comparison with a matched-control group. Study 2 was designed as a replication and extension of the first study. In this study, the group leaders were relatively inexperienced, closely supervised graduate students in counseling psychology. It was again predicted that group members would become more internal in comparison with controls.

STUDY 1

METHOD

Subjects

Thirty-one volunteer University of Hawaii graduate students in one fall semester class in counseling psychology were matched for age, sex, and counseling experience and randomly assigned to three experimental groups (A, B, C). Nine volunteer students, matched for age, sex, and counseling experience, from an equivalent class, but not involved in any group experience during this time, were assigned to the control condition (D). Since there were no pretest differences between experimental and control subjects and since experimental and control subjects had equivalent graduate training, it was presumed that there were no systematic differences between these groups. Due to illness, one group member had to leave school before the groups had completed and had to be eliminated from the study. Final numbers for the group were: for Group A, $n = 9$; for Group B, $n = 11$; for Group C, $n = 10$; and for Group D (control), $n = 9$.

Leaders

Three expert group leaders typifying different approaches to encounter groups were each chosen to lead one group. In addition, each group had a graduate student co-leader. Each leader was instructed to lead his (her) group in his (her) typical fashion. None was aware of the hypothesis under investigation nor of students' locus of control score. Each leader encouraged a "here-and-now" focus, spontaneous expression of and dealing with feelings, openness to novel experiences, and responsibility for one's behavior.

Procedure

Prior to any group experiences, all subjects were administered the Internal–External scale in conjunction with additional personality tests.[1] All subjects were truthfully informed that the test battery was part of a departmental program evaluation and were not told that the encounter groups themselves were the subject of assessment.

Each experimental group consisted of eight two-hour weekly sessions and one 10-hour marathon session in an 11-week period. No member of the control group received any group experience during this period.

All groups met in a 12 × 16 foot room that contained only a rug and pillows. In addition, each session was videotaped to facilitate the training experiences for the graduate student leaders. The videotapes were preserved for only a two-week period.

After the conclusion of the groups, all subjects were again administered the test battery. Debriefing occurred one week after all subjects had completed the tests. There was no evidence that any subject became aware of the true purpose of the Internal–External scale prior to debriefing.

STUDY 2

METHOD

The same method and procedure were used as in Study 1.

Subjects

Thirty-three students from the same held in the Spring semester were similarly matched and assigned to three experimental groups. There were 11 control subjects. All controls met the criteria described in Study 1. Thus, $n = 11$ for Groups E, F, G, and H (control).

---

[1]This study represents one part of a continuous investigation of encounter group effects. A battery of personality paper and pencil tests has been routinely administered to all students involved in the graduate program in counseling and was included herein. The Internal–External scale was specifically included for the first time to measure a belief dimension left untapped by other questionnaires. A summary report of the continuous findings regarding assessed personality change is available (see Shapiro and Diamond, 1973).

Leaders

Six counseling psychology graduate students, all of whom had co-led a group with an experienced professional and had either co-led or observed a group in Study 1, served as group leaders. Two students co-led each group and were supervised by an experienced professional. These leaders encouraged the same in-group behaviors as in Study 1.

RESULTS AND DISCUSSION

Analyses of variance indicated no pretest differences in Internal–External score between experimental and control groups for either Study 1 or 2 ($F < 1$). Similarly, there were no pretest differences between Groups A–H ($F < 1$). There were no significant posttest differences between groups ($F = 2.16$). However, repeated-measures analyses of variance indicated several withingroup differences. There was a significant change toward greater internal locus of control for the three experimental groups (A, B, C) in Study 1 ($F = 4.05$, $p < .05$), while no such change pertained for the control group (D) ($F = .08$). Similarly, the three experimental groups employed in Study 2 (E, F, G) yielded results in the predicted direction ($F = 2.99$, $.05 < p < .10$). Once again, no significant changes occurred in the control group (H) for Study 2 ($F = .81$).

Pretest and posttest mean Internal–External scores are presented in Table 33.1. An inspection of this table indicates that five of the six experimental groups decreased their mean Internal–External score (i.e., become more internal). Neither of the control groups demonstrated this decrease. Taken together, the six experimental groups demonstrate a significant change toward increased internal locus of control ($F = 6.72$, $p < .01$), while the two control groups yield no such change ($F = .93$).

These findings indicate that subjects exposed to professionally led encounter groups developed a more internal orientation. Subjects not exposed to training did not demonstrate these changes. The results of Study 2 suggest that these changes pertain to some extent even with supervised, graduate student leaders.

Aronson's (1970) recent data further suggested that leadership style is a critical variable in encounter group outcome. These findings combined with the present data imply that internal generalized expectancies are more likely to be developed when an experienced leader is conducting the group. Recent findings (Shapiro, Marano, and Diamond, 1973) demonstrate several differences between experienced and inexperienced leader groups. Members exposed to experienced leaders showed greater trust and positive self-concept change. These changes were examined with reference to leader modeling behavior. In the present study, experienced leaders appeared more self-confident, less anxious, less suspicious, and less defensive as well as more highly skilled at encouraging novel behavioral explo-

TABLE 33.1

Comparison of means and standard deviations for internal–external scores and F values

Group	Pretest		Posttest		F
	$\overline{X}$	SD	$\overline{X}$	SD	
Study 1					
Experimental (A, B, C) (n = 30)	10.90	4.00	9.78	4.05	4.05**
A (n = 9)	9.56	3.34	8.00	2.16	
B (n = 11)	12.09	4.88	11.27	3.67	
C (n = 10)	10.90	3.42	9.80	5.47	
Control (D) (n = 9)	12.00	3.61	12.00	3.58	.08
Study 2					
Experimental (E, F, G)	11.36	3.94	10.38	3.60	2.99*
E (n = 11)	11.00	4.28	11.18	3.78	
F (n = 11)	11.45	3.98	9.80	3.90	
G (n = 11)	11.63	3.52	10.18	3.04	
Control (H) (n = 12)	11.18	3.80	12.09	3.85	.81
Studies 1 and 2					
Experimental (A, B, C, E, F, G)	11.16	3.96	10.14	3.82	6.72***
Control (D, H)	11.51	3.74	12.05	3.60	.93

*$.05 < p < .10$.
**$p < .05$.
***$p < .01$.

ration. In short, the experienced leaders themselves were better models for an internal locus of control. An audiotaped, leaderless group (cf. Aronson, 1970) provides no such model.

In order to ascertain the mediating mechanisms accounting for the change in generalized expectancies, future studies will investigate the effects of leader modeling on members' behaviors and expectancies. Similarly, the specific leader behaviors which facilitate such cognitive behavioral change also warrant further investigation.

One implication of the present study is that particular environmental manipulations (e.g., encounter groups, psychotherapy, educational programs, etc.) can be employed to modify generalized expectancies and thus allow individuals to perceive themselves to have greater control over their lives. This perception of increased control is positively related to a wide variety of competence and adjustment behaviors. Furthermore, enhanced feelings of self-control may serve to counteract those feelings of alienation and powerlessness so common in complex, modern societies.

References

ARGYRIS, C. *Interpersonal competence and organizational effectiveness.* Chicago: Irwin-Dorsey, 1962.

ARONSON, S. R. A comparison of cognitive vs. focused-activities techniques in sensitivity group training. Unpublished doctoral dissertation, University of Connecticut, 1970.

BUNKER, D. R. Individual applications of laboratory training. *Journal of Applied Behavioral Science,* 1965, **1,** 131–148.

BUNKER, D. R., AND KNOWLES, E. S. Comparison of behavioral changes resulting from human relations training laboratories of different lengths. *Journal of Applied Behavioral Science,* 1967, **3,** 505–523.

BUNKER, G. L. The effects of group perceived esteem on self and ideal self-concepts in an emergent group. Unpublished master's thesis, Brigham Young University, 1961.

CULBERT, S. A., CLARK, J. V., AND BOBELE, H. K. Measures of change toward self-actualization in two sensitivity training groups. *Journal of Counseling Psychology,* 1968, **15,** 53–57.

DELANEY, D. M., AND HEIMANN, R. Z. Effectiveness of sensitivity training on the perception of nonverbal communications. *Journal of Counseling Psychology,* 1966, **13,** 436–440.

DIAMOND, M. J., AND LOBITZ, W. C. When familiarity breeds respect: The effects of an experimental "depolarization" program on police and student attitudes toward each other. *Journal of Social Issues,* 1973, **29,** in press.

FEATHER, N. T. Some personality correlates of external control. *Australian Journal of Psychology,* 1967, **19,** 253–260.

GILLIS J. S., AND JESSOR, R. Effects of psychotherapy on belief in internal control: An exploratory study. *Psychotherapy: Theory, Research, and Practice,* 1970, **7,** 135–137.

HAMSHER, J. H., GELLER, J. D., AND ROTTER, J. B. Interpersonal trust, internal-external control, and the Warren Commission Report. *Journal of Personality and Social Psychology,* 1968, **9,** 210–215.

HARRISON, R. Cognitive change in a sensitivity training laboratory. *Journal of Consulting Psychology,* 1966, **30,** 517–520.

HERSCH, P. D., AND SCHEIBE, K. E. On the reliability and validity of internal-external control as a personality dimension. *Journal of Consulting Psychology,* 1967, **31,** 609–614.

JOE, V. C. Review of the internal-external control construct as a personality variable. *Psychological Reports,* 1971, **28,** 619–640 (Monograph Supplement 3-V28).

LEFCOURT, H. M. Internal versus external control of reinforcement: A review. *Psychological Bulletin,* 1966, **65,** 206–220.

MILES, M. B. Changes during and following laboratory training: A clinical-experimental study. *Journal of Applied Behavioral Science,* 1965, **1,** 215–242.

PETERS, D. R. Identification and personal change in laboratory training groups. Unpublished doctoral dissertation, Massachusetts Institute of Technology, 1966.

REDDY, W. B. On affection, group composition, and self actualization in sensitivity training. *Journal of Consulting and Clinical Psychology,* 1972, **38,** 211–214.

ROGERS, C. R. *Carl Rogers on encounter groups.* New York: Harper & Row, 1970.

ROTTER, J. B. Generalized expectancies for internal versus external control of reinforcement. *Psychological Monographs,* 1966, **80,** (1, Whole No. 609).

RUBIN, I. The reduction of prejudice through laboratory training. *Journal of Applied Behavioral Science,* 1967, **5,** 233–238.

RUEVENI, J., SWIFT, M., AND BELL, A. A. Sensitivity training: Its impact on mental health workers. *Journal of Applied Behavioral Science,* 1969, **4,** 600–602.

SHAPIRO, J. L. An investigation into the effectiveness of sensitivity training procedures. Unpublished doctoral dissertation, University of Waterloo, 1970.

SHAPIRO, J. L., AND DIAMOND, M. J. Increases in hypnotizability as a function of encounter group training: Some confirming evidence. *Journal of Abnormal Psychology,* 1972, **79,** 112–115.

SHAPIRO, J. L., AND DIAMOND, M. J. Toward the long-term scientific study of encounter group phenomena: II. Empirical findings. Paper presented at the 53rd Annual Convention of the Western Psychological Association, Anaheim, California, April 1973.

SHAPIRO, J. L. MARANO, P., AND DIAMOND, M. J. An investigation of encounter group outcome and its relationship to leadership experience. Paper presented at the 19th Annual Meeting of the Southeastern Psychological Association, New Orleans, April 1973.

SHOSTROM, E. L. *Personal Orientation Inventory Manual,* San Diego, Calif.: Educational and Industrial Testing Service, 1966.

SMITH, R. E. Changes in locus of control as a function of life crisis resolution. *Journal of Abnormal Psychology,* 1970, **75,** 328–332.

Name Index

Subject Index